Sustainable Agriculture Reviews

Volume 54

Series Editor

Eric Lichtfouse
Aix-Marseille University, CNRS, IRD, INRAE, Coll France, CEREGE
Aix-en-Provence, France

Other Publications by Dr. Eric Lichtfouse

Books
Scientific Writing for Impact Factor Journals
https://www.novapublishers.com/catalog/product_info.php?products_id=42242

Environmental Chemistry
http://www.springer.com/978-3-540-22860-8

Sustainable Agriculture
Volume 1: http://www.springer.com/978-90-481-2665-1
Volume 2: http://www.springer.com/978-94-007-0393-3

Book series
Environmental Chemistry for a Sustainable World
http://www.springer.com/series/11480

Sustainable Agriculture Reviews
http://www.springer.com/series/8380

Journal
Environmental Chemistry Letters
http://www.springer.com/10311

Sustainable agriculture is a rapidly growing field aiming at producing food and energy in a sustainable way for humans and their children. Sustainable agriculture is a discipline that addresses current issues such as climate change, increasing food and fuel prices, poor-nation starvation, rich-nation obesity, water pollution, soil erosion, fertility loss, pest control, and biodiversity depletion.

Novel, environmentally-friendly solutions are proposed based on integrated knowledge from sciences as diverse as agronomy, soil science, molecular biology, chemistry, toxicology, ecology, economy, and social sciences. Indeed, sustainable agriculture decipher mechanisms of processes that occur from the molecular level to the farming system to the global level at time scales ranging from seconds to centuries. For that, scientists use the system approach that involves studying components and interactions of a whole system to address scientific, economic and social issues. In that respect, sustainable agriculture is not a classical, narrow science. Instead of solving problems using the classical painkiller approach that treats only negative impacts, sustainable agriculture treats problem sources.

Because most actual society issues are now intertwined, global, and fast-developing, sustainable agriculture will bring solutions to build a safer world. This book series gathers review articles that analyze current agricultural issues and knowledge, then propose alternative solutions. It will therefore help all scientists, decision-makers, professors, farmers and politicians who wish to build a safe agriculture, energy and food system for future generations.

More information about this series at http://www.springer.com/series/8380

Vinod Kumar Yata • Ashok Kumar Mohanty
Eric Lichtfouse
Editors

Sustainable Agriculture Reviews 54

Animal Biotechnology for Livestock Production 1

Springer

Editors
Vinod Kumar Yata
Animal Biotechnology Centre
National Dairy Research Institute
Karnal, Haryana, India

Ashok Kumar Mohanty
Animal Biotechnology Centre
National Dairy Research Institute
Karnal, Haryana, India

Eric Lichtfouse
Aix-Marseille University
CNRS, IRD, INRAE
Coll France, CEREGE
Aix-en-Provence, France

ISSN 2210-4410 ISSN 2210-4429 (electronic)
Sustainable Agriculture Reviews
ISBN 978-3-030-76528-6 ISBN 978-3-030-76529-3 (eBook)
https://doi.org/10.1007/978-3-030-76529-3

This Springer imprint is published by the registered company Springer Nature Switzerland AG
The registered company address is: Gewerbestrasse 11, 6330 Cham, Switzerland

Preface

Many developed countries are actually facing a ban of animal food products, promoted mainly by urban and vegetarian activists who have never experienced living in farms and who do not acknowledge that their modern way of life is largely the result of hard work of their elders with the help of farm animals. However, since the start of society, animal production has been an essential agricultural sector worldwide providing food, labor, aesthetics and social values, and even today many farmers would not survive without animals. This book entitled 'Animal Biotechnology for Livestock Production 1' is our first volume providing advanced knowledge on biotechnological methods to improve the livestock production, with focus on animal reproduction, health, diagnosis and nutrition. Chapter 1 presents on artificial insemination in cattle, with focus on physiology aspects of the estrous cycle, estrus synchronization program, ovulation synchronization program for timed artificial insemination, strategies for improving fertility and use of sexed semen in artificial insemination. Chapter 2 reviews biotechnological applications for production of dromedary camels, with details on camel herd reproduction, reproduction control and artificial insemination. Sperm dilution, thawing, conservation, and insemination techniques are also discussed. Recent biotechnological applications for livestock production are summarized in Chap. 3, with emphasis on somatic cell nuclear transfer, artificial insemination, embryo transfer, embryonic stem cell technology and marker assisted selection.

Cattle production in France. Copyright 2021 Eric Lichtfouse

Chapter 4 reviews applications of stem cells in livestock, with emphasis on mesenchymal stem cells. Immunomodulatory, antimicrobial activity, migration and reparative functions of stem cells are detailed. Chapter 5 presents techniques for profiling proteins and metabolites associated with feed efficiency in dairy cattle. Recent findings on key metabolites and proteins of metabolic pathways are also disclosed. Chapter 6 focuses on processing, packaging, and safety of dairy products. Applications of biotechnologies in food diagnosis are also explained. Chapter 7 reviews 'on-farm point-of-care' diagnostic technologies in animals. This chapter covers various point-of-care and on-farm diagnostic technologies for monitoring animal health and disease with focus on molecular, electrochemical-biosensors diagnostics. Chapter 8 presents biotechnological applications in the poultry industry. This chapter covers the concepts and developments of biotechnologies for poultry production, breeding, feed and nutrition. This chapter also discusses applications in poultry vaccines, biologics, disease diagnosis and food processing.

We express our thanks to all authors who have contributed high quality chapters. Our special thanks are due to the Indian Council of Agricultural Research (ICAR), the Government of India and the Director of the ICAR National Dairy Research Institute (NDRI), Karnal, India for providing the institutional support. We would like to acknowledge Dr. Sudarshan Kumar, Scientist, ICAR-NDRI, Karnal, India for his help in choosing contributors and reviewers. We would like extend our thanks to the staff of Springer Nature, for their generous assistance, constant support, and patience in initializing and publication of this book. We acknowledge our thanks to

Department of Biotechnology, Government of India for providing financial support from "DBT-RA Program in Biotechnology & Life Sciences".

Karnal, India Vinod Kumar Yata

Karnal, India Ashok Kumar Mohanty

Aix-en-Provence, France Eric Lichtfouse

Contents

1 **Artificial Insemination Program in Cattle** 1
Fábio Morotti, Elis Lorenzetti, and Marcelo Marcondes Seneda

2 **Reproduction Management and Artificial Insemination in Dromedary Camel** 55
Djallel Eddine Gherissi and Ramzi Lamraoui

3 **Biotechnological Advancements in Livestock Production** 107
Bhaskar Sharma, Dixita Chettri, and Anil Kumar Verma

4 **Applications of Stem cells Technology in Livestock Production** 131
Vinay Bhaskar, Satish Kumar, and Dhruba Malakar

5 **Metabolomics and Proteomics Signatures in Feed-Efficient Beef and Dairy Cattle** 153
Ahmed A. Elolimy, Mohamed Zeineldin, Mohamed Abdelmegeid, Alzahraa M. Abdelatty, Abdulrahman S. Alharthi, Mohammed H. Bakr, Mona M. M. Y. Elghandour, Abdelfattah Z. M. Salem, and Juan J. Loor

6 **Biotechnological Applications in Dairy Products and Safety** 167
Ayushi Kapoor, Monica Yadav, Aparna Verma, and Kiran Ambatipudi

7 On-Farm Point-of-Care Diagnostic Technologies for Monitoring Health, Welfare, and Performance in Livestock Production Systems 209
Mohamed Zeineldin, Ahmed A. Elolimy, P. Ravi Kanth Reddy, Mohamed Abdelmegeid, Miguel Mellado, Mona M. M. Y. Elghandour, and Abdelfattah Z. M. Salem

8 Biotechnological Applications in Poultry Farming 233
S. M. Lutful Kabir and S. K. Shaheenur Islam

Index 273

About the Editors

Vinod Kumar Yata is an interdisciplinary researcher working at National Dairy Research Institute, Karnal, India. Previously, he worked as an Assistant professor at the Department of Biotechnology, Dr. B R Ambedkar National Institute of Technology Jalandhar, Punjab, India. He received his Ph.D. in Biotechnology from Indian Institute of Technology Guwahati. He specializes in interdisciplinary research which includes Nanotechnology, Microfluidics, Animal Biotechnology, Cancer biology and Bioinformatics. He has developed a microfluidic device for the separation of live and motile spermatozoa form cattle semen samples. He opened up a new avenue to prodrug enzyme therapy by introducing the nanocarriers for the delivery of non-mammalian prodrug activating enzymes. He elucidated the structural features and binding interactions of several bio molecules by in silico methods. He has published three books as an editor and one book as an author with Springer Nature publisher. He has published several research papers in peer reviewed international journals and presented papers in several international conferences.

Ashok Kumar Mohanty is a principal scientist at Animal Biotechnology Centre, ICAR-National Dairy Research Institute, Karnal, India. His group is involved in various basic and applied research related to animal production systems. His research group has made pioneering contributions in the field of Animal Biotechnology, with emphasis on Gene cloning, Expression and Functional characterization of Animal Proteins, Proteomics in Animal Production, Cell and Molecular Biology and Structural Biology of Proteins. His group has developed a Buffalo Mammary Epithelial cell line for the first time which can be used a model system to understand lactation biology in animal as well as human. His team has also developed a pregnancy diagnostic kit for the early detection of pregnancy cattle and buffalo. His group is also extensively involved in developing low cost technology for semen sexing in cattle. He has organized a number of national and international workshops and international conferences. He is a recipient of several awards which includes, DBT Overseas Associateship by Ministry of Science & Technology, Govt. of India., Jawaharlal Nehru Award (gold medal) by Indian Council of Agricultural Research (ICAR), New Delhi for outstanding postgraduate research in the field of Animal Biotechnology and Young Scientist Award sponsored by International union of Crystallography (IUCr) and Dept. of Science and Technology (DST), Govt. of India for attending IUCr congress at Geneva, Switzerland. He is the Fellow of National Academy of Dairy Sciences, India and executive member of proteomics society of India and associate fellow of National Academy of Veterinary Science, India. He has supervised more than 50 graduate, PhD students and post docs. He published more than 200 peer reviewed research and review papers. He has also authored 8 book chapters in the areas of animal and food biotechnology published by in national and international publishers.

Eric Lichtfouse is a professor of environmental sciences and scientific writing at Aix Marseille University and Xi'an Jiaotong University. He has invented carbon-13 dating, a molecular-level method allowing to study the dynamics of organic compounds in temporal pools of complex environmental media. He has discovered temporal pools of individual substances in complex media such as soils. He is Chief Editor and founder of the journal Environmental Chemistry Letters, and the book series Sustainable Agriculture Reviews and Environmental Chemistry for a Sustainable World. He is the author of the book Scientific Writing for Impact Factor Journals, which includes an innovative writing tool: the micro-article. He has awards in analytical chemistry and scientific editing.

Contributors

Alzahraa M. Abdelatty Department of Nutrition and Clinical Nutrition, Faculty of Veterinary Medicine, Cairo University, Giza, Egypt

Mohamed Abdelmegeid Department of Animal Medicine, College of Veterinary Medicine, Kafr-Elsheikh University, Kafrelsheikh, Egypt

Abdulrahman S. Alharthi Department of Animal Production, College of Food and Agriculture Sciences, King Saud University, Riyadh, Saudi Arabia

Kiran Ambatipudi Department of Biosciences and Bioengineering, Indian Institute of Technology Roorkee, Roorkee, Uttarakhand, India

Mohammed H. Bakr Animal Production Department, Faculty of Agriculture, Cairo University, Giza, Egypt

Vinay Bhaskar Animal Biotechnology Centre, National Dairy Research Institute, Karnal, Haryana, India

Dixita Chettri Department of Microbiology, Sikkim University, Gangtok, Sikkim, India

Mona M. M. Y. Elghandour Facultad de Medicina Veterinaria y Zootecnia, Universidad Autonoma del Estado de Mexico, Toluca, Mexico

Ahmed A. Elolimy Department of Animal Production, National Research Centre, Giza, Egypt

Djallel Eddine Gherissi Institute of Agronomic and Veterinary Sciences, University of Souk-Ahras, Souk Ahras, Algeria

Laboratory of Animal Productions, Biotechnologies and Health, University of Souk-Ahras, Souk Ahras, Algeria

Nesrein M. Hashem Animal and Fish Production Department, Faculty of Agriculture, Alexandria University, Alexandria, Egypt

Ayushi Kapoor Department of Biosciences and Bioengineering, Indian Institute of Technology Roorkee, Roorkee, Uttarakhand, India

Satish Kumar Animal Biotechnology Centre, National Dairy Research Institute, Karnal, Haryana, India

Ramzi Lamraoui Institute of Agronomic and Veterinary Sciences, University of Souk-Ahras, Souk Ahras, Algeria

Department of Biology of Living Organisms, Faculty of Natural and Life Sciences, University of Batna, Batna, Algeria

Juan J. Loor Department of Animal Sciences, University of Illinois, Urbana, IL, USA

Elis Lorenzetti Laboratory of Animal Virology (LABVIRAL), State University of Londrina-UEL, Londrina, PR, Brazil

Post Graduate Program in Animal Health and Production, Universidade Pitágoras Unopar, Arapongas, PR, Brazil

S. M. Lutful Kabir Department of Microbiology and Hygiene, Bangladesh Agricultural University, Mymensingh, Bangladesh

Dhruba Malakar Animal Biotechnology Centre, National Dairy Research Institute, Karnal, Haryana, India

Miguel Mellado Department of Animal Nutrition, Autonomous Agrarian University Antonio Narro, Saltillo, Mexico

Fábio Morotti Laboratory of Animal Reproduction and Biotechnology, State University of Londrina-UEL, Londrina, PR, Brazil

P. Ravi Kanth Reddy Veterinary Dispensary, Taticherla, Andhra Pradesh, India

Abdelfattah Z. M. Salem Facultad de Medicina Veterinaria y Zootecnia, Universidad Autonoma del Estado de Mexico, Toluca, Mexico

Marcelo Marcondes Seneda Laboratory of Animal Reproduction and Biotechnology, State University of Londrina-UEL, Londrina, PR, Brazil

S. K. Shaheenur Islam Department of Microbiology and Hygiene, Bangladesh Agricultural University, Mymensingh, Bangladesh

Department of Livestock Services, Krishi khamar Sarak, Dhaka, Bangladesh

Bhaskar Sharma School of Life and Environmental Sciences, Faculty of Science, Engineering, and Built Environment, Deakin University, Geelong, VIC, Australia

Department of Biotechnology, TERI School of Advanced Studies, Delhi, India

Anil Kumar Verma Department of Microbiology, Sikkim University, Gangtok, Sikkim, India

Aparna Verma Department of Biosciences and Bioengineering, Indian Institute of Technology Roorkee, Roorkee, Uttarakhand, India

Monica Yadav Department of Biosciences and Bioengineering, Indian Institute of Technology Roorkee, Roorkee, Uttarakhand, India

Mohamed Zeineldin Department of Animal Medicine, College of Veterinary Medicine, Benha University, Benha, Egypt

Chapter 1
Artificial Insemination Program in Cattle

Fábio Morotti, Elis Lorenzetti, and Marcelo Marcondes Seneda

Abstract Artificial insemination is a key technique to increase reproductive efficiency and the genetic improvement in cattle. The first report of artificial insemination has been dated to more than 220 years ago; after that, many studies have been published, and this biotechnology is widely used in domestic animals. Despite the advances of embryo transfer programs *in vivo* and *in vitro* in cattle, artificial insemination still represents the main form of assisted reproduction in many herds, being one of the lowest cost strategies to improve genetic merit in the farms. Artificial insemination has many advantages compared to natural breeding, such as health benefits, e.g. control of infectious diseases; genetic improvement, e.g. use of genetically improved bulls; reproductive management improvements; and more stringent control of the zootechnical and economic features of the herd, e.g. standardized management and animals.

In addition, we highlight the main advances related to the use of artificial insemination in cattle, such as the development of other reproductive biotechniques, such as timed artificial insemination and resynchronization of estrus or ovulation. In addition, the sexed semen has evolved satisfactorily with promising results both in insemination and in embryo production. An important advance in artificial insemination is related to the hormonal control of the estrous cycle that allows the elaboration of highly efficient protocols for the synchronization of estrus and ovulation. The hormonal protocols have contributed to the dissemination of the artificial insemination by allowing the insemination of many females in a short period of time. Based on many studies evaluating timed artificial insemination protocols and

F. Morotti (✉) · M. M. Seneda
Laboratory of Animal Reproduction and Biotechnology, State University of Londrina-UEL, Londrina, PR, Brazil
e-mail: fabiomorotti@uel.br

E. Lorenzetti
Laboratory of Animal Virology (LABVIRAL), State University of Londrina-UEL, Londrina, PR, Brazil

Post Graduate Program in Animal Health and Production, Universidade Pitágoras Unopar, Arapongas, PR, Brazil

V. K. Yata et al. (eds.), *Sustainable Agriculture Reviews 54*, Sustainable Agriculture Reviews 54, https://doi.org/10.1007/978-3-030-76529-3_1

early and accurate pregnancy diagnosis techniques, it was also possible to establish estrus and ovulation resynchronization strategies, which have greatly contributed to economic profitability. Finally, here we review many strategies associated with manipulation of the dominant follicle, estrus behavioral and antral follicle count have been elaborated and they represent an important progress to improve the use of timed artificial insemination in cattle. Therefore, this chapter aims to present the main advances of artificial insemination and timed artificial insemination, highlighting the technical parameters, advantages, and influencing factors, and to discuss practical and current strategies for the improvement of the herd via artificial insemination programs.

Keywords Cattle · Estrous cycle · Synchronization · Artificial insemination · Timed artificial insemination · Resynchronization · Sexed semen · Pregnancy · Antral follicle · Fertility

Abbreviations

TAI	Timed Artificial Insemination
TCR	Temporary calf removal
AFC	Antral follicle count
AI	Artificial Insemination
BCS	Body condition score
CL	Corpus luteum
EB	Estradiol benzoate
EC	Estradiol cypionate
eCG	equine chorionic gonadotropin
EV	Estradiol valerate
FSH	Follicle stimulating hormone
GnRH	Gonadotropin-releasing hormone
hCG	human chorionic gonadotropin
IVF	*In vitro* fertilization
LH	Luteinizing hormone
P4	Progesterone
PGF2α	Prostaglandins
RSS	Reverse-sorted semen
RTS	Reproductive tract scoring system
US	Ultrasound

1.1 Introduction

For definition artificial insemination (AI) is a procedure characterized by manually semen depositionin the female reproductive tract. This practice is considered the oldest reproductive biotechnology and is one of the more efficient strategies for genetically improving livestock and increasing their reproductive efficiency. Therefore, AI is still considered the most widely used biotechnology in the world due to its genetic, health and economic benefits for beef and dairy cattle.

The productive and reproductive performance of the herd are considered the main factors responsible for affecting the livestock economic scenario. However, reproductive traits have low heritability and can be highly affected by environmental, nutritional, and health factors. Thus, adequate management of all livestock sectors is requiredto increase productive and reproductive efficiency and achieve economic autonomy. Therefore, to use reproductive strategies with maximum efficiency, it is necessary to control influencing factors as well as ensure that the overall management is suitable for livestock production.

Conventional AI programs that using estrus detection and insemination procedure have been used for a long time and still represents the main form of reproduction in many herds, mainly for dairy cattle (Smith et al. 2018). However, low service rates due to estrus detection failures represents one of the main obstacles to improving reproductive efficiency for both beef and dairy cattle. This problem is most evident in *Bos indicus* herds that have a short duration of estrus and high percentage of estrus behavior that occurs at night in relation *Bos taurus* females (Pinheiro et al. 1998; Galina et al. 1996). In addition to the failures in the estrus detection and the particularities of estrus behavior, a high proportion of cows managed in tropical pastures face nutritional problems that intensify the postpartum anestrus, hindering the return of cyclicality and compromising the efficient use of conventional AI (Baruselli et al. 2004a, 2018).

On the other hand, timed artificial insemination (TAI) have emerged as an alternative to conventional AI programs, in addition to representing the most effective strategy for employment and dissemination of AI and its benefits, mainly for large-scale programs, because the TAI aims for efficient control of estrus and ovulation of all females destined for breeding. After the onset of programs for the synchronization of ovulation, the estrus detection was dispensed, and TAI became widely used because it allows ovulation control and insemination of large numbers of females in a predetermined amount of time. Many advantages have been added to the AI technique associated with synchronization of ovulation, such as elimination of estrus detection, improvement of livestock management, planning of services and financial resources, and improvement of pregnancy rates depending on categories and reproductive characteristics of the herd (Colazo and Mapletoft 2014; Lamb et al. 2010; Baruselli et al. 2004a). TAI protocols synchronize follicular recruitment and growth, regression of the corpus luteum (CL), and ovulation of the dominant follicle, improving reproductive performance because all females are inseminated regardless of their estrus behavior (Lamb and Mercadante 2016). Therefore, TAI programs have been used as a reproductive tool that better enables the expansion of AI in dairy and beef cattle.

After the first protocol of ovulation synchronization, a wide variety of protocols have been developed focusing on adjustmentsin female category to minimize management and cost and to obtain more reproductive efficiency. Among various hormonal combinations for TAI in cattle, the association of progesterone/progestin (P4) or gonadotropins with estrogen and prostaglandins (PGF2α) is noteworthy (Baruselli and Teixeira 2018; Baruselli et al. 2004a). Other reproductive strategies such as estrus or ovulation resynchronization have been successfully used to reduce the number of females that fail pregnancy after the first AI procedure. In addition, resynchronization has been used to increase pregnancy rates in the short time of the breeding season and reduces the use of bulls for natural breeding (Marques et al. 2015; Campos et al. 2013). This chapter aims to present the main advances of AI and TAI, highlighting mainly the technical parameters, advantages, and influencing factors and to discuss practical and current strategies for improving reproductive performance in cattle.

1.2 Artificial Insemination Program: Benefits and Advantages

Genetic improvement is considered one of the main benefits of using AI (Fig. 1.1). In cattle, the genetic improvement is achieved by the choice of proven bulls with desirable characteristics for enhancement in milk or meat production. In addition, AI also provides genetic gain in a short amount of time and low cost compared to natural breeding because semen storage is an economically viable alternative

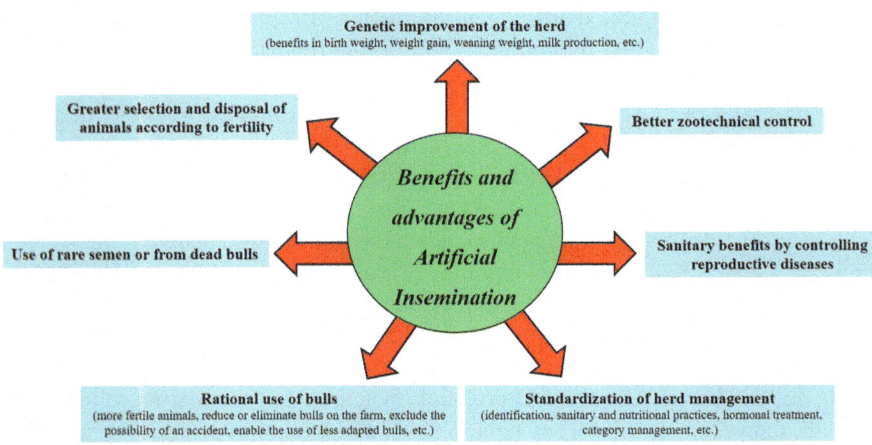

Fig. 1.1 Main benefits and advantages associated with the use of artificial insemination programs in cattle

compared to purchase and keeping a bull with high zootechnynical potential in the herd (Baruselli et al. 2018).

Another benefit associated with the use of AI programs is the intensification and standardization of reproductive management of the herd for better zootechnical control of animals. Particularly, of this is important in tropical countries, because the majority of beef and dairy production comes from herds kept almost exclusively in grazing conditions. This strategy represents a low-cost production system; however, some management practices need some adjustments to achieve success after using reproductive biotechnologies. Thus, optimized management practices and better zootechnical control allows for more efficient production in addition to the selection of the most fertile animals and for discarding females with infertility (Marques et al. 2018).

Many infectious agents such as bacteria, viruses, protozoa, and fungi affect reproductive performance in cattle and can potentially be transmitted through natural breeding (Anderson 2007; Yoo 2010). Reproductive diseases such asBrucellosis, Leptospirosis, Campylobacteriosis, Trichomoniasis, Bovine viral diarrhea and Infectious bovine rhinotracheitis (Infectious pustular vulvovaginitis) and Trichomoniasis can provide severe consequences to reproductive performance and economic gain due to the increased occurrence of early and late embryonic loss and fetal death (Alfieri et al. 2019). In this context, another benefit of AI is herd sanity because of the elimination of natural breeding and the use of bulls that can carry contagious infectious diseases (Eaglesome and Garcia 1997). Thus, along with genetic improvement, AI aims to use semen commercialized by qualified companies, which select sires based on sanitary program analysis to control reproductive diseases.

The implementation of AI programs also allows for the reduction or elimination of bulls on farm (Marques et al. 2018), releasing pastures for other animal categories. Furthermore, replacing the natural breeding program to AI program can to avoiding the possibility of bull attacks (accidents) on workers involved in dairy activities, which is common. In addition, the insemination facilitates the selection of bulls with low expected progeny differences in birth weight, reducing the occurrence of dystocia due to the birth of smaller calves (Mee 2008).

In countries with a tropical climate, with high temperatures, the use of some bull breeds represents a great challenge for the breeder's good performance in the natural breeding. The same situation can be observed in countries with cold climates and that intend to make use of bulls of Zebu breeds, which are often more adapted and regions with hot climates (Menegassi and Barcellos 2015). In this contex, the use of AI also allows for the use of bulls less adapted to some climatic conditions, i.e., it is common to observe a reduction in libido of some breeds of bulls when placed in high temperature environments. Therefore, AI enables the use of *Bos taurus* bulls for crossbreeding, which are less adapted to high temperatures and require more management. Furthermore, AI also favors heterosis by crossing *Bos taurus* and *Bos indicus* cattle (Buckley et al. 2014), which represents an interesting strategy due to the improvement in zootechnical performance, quality of progeny carcasses (Gama et al. 2013), parasitic resistance, adaptability, and good acceptance by consumers.

Finally, despite AI having many advantages for the herd, there are some limitations, such as the need for specialized labor, failure in applying biotechnology, and nutrition and sanitarian factors (Demetrio et al. 2007; Orihuela 2000). Understanding these limitations is necessary for implementation of strategies to enable the use of AI programs.

1.3 Physiology Aspects of the Estrous Cycle in Cattle

The estrous cycle reflects a cyclical pattern of ovarian activity that makes the female receptive to the male, favoring mating and further gestation (Forde et al. 2011). Domestic cattle (*Bos indicus* and *Bos taurus*) are considered annual poliestric animals and exhibit estrus behavior on average every 21 days, varying from 18 to 24 days depending on the number of waves of follicular growth (Figueiredo et al. 1997; Roche 1996; Ginther et al. 1989b). However, this cyclical behavior is only possible from puberty, when the female reached appropriate weight and age for each breed.

1.3.1 Puberty

The pubertal process is characterized by an activation of physiological events in the hypothalamic-adenohypophyseal-ovarian axis which culminate in reproductive maturity. In this process, the females undergo by a sequence of events that begins with the expression of estrus, followed by the occurrence of ovulation and the beginning of a luteal phase (Seneda et al. 2019; Atkins et al. 2013; Rawlings et al. 2003). The onset of puberty is variable according to age, weight, and subspecies of heifer, for example taurine animals have greater sexual precocity than zebuine animals. In *Bos taurus*, puberty can occur at approximately 6 to 12 months of age or when the heifer reaches 40 to 50% of the adult weight (Moran et al. 1989). However, in *Bos indicus*, puberty is achieved at approximately 15 to 18 months of age or close to 60 to 70% of the adult weight (Nogueira 2004). In addition, factors such as season of the year, growth rate, nutritional intake, social cues and treatment with exogenous progestins can change the age to the puberty.

In prepubertal females, the low concentration of estradiol produced in the gonads due a negative feedback mechanism prevents the hypothalamic preovulatory center from secreting adequate pulses of Gonadotropin-releasing hormone (GnRH; Amstalden et al. 2011). The absence of GnRH also inhibits the release of Follicle stimulating hormone (FSH) and Luteinizing hormone (LH) from the anterior pituitary, thus causing an absence of follicular waves and ovulation (Senger 1997). On the other hand, when the hypothalamus starts the GnRH neuron function (influenced mainly by nutritional control), there is a decrease in the negative feedback of estradiol, which leads to an increase in the frequency of release of the GnRH pulses

(Amstalden et al. 2011; Kinder et al. 1995). In this context, the follicle growth occurs due to increased levels of intrafollicular estradiol and release of a GnRH discharge through of the center of the hypothalamus (Jainudeen and Hafez 2013). Subsequently, a GnRH discharge results in an increase in the tonic release of LH pulses that is the main endocrine factor that regulates the onset of puberty in heifers. Increased frequency of LH release increases the development of ovarian follicles that produce estradiol sufficient to induce behavioral estrus and a preovulatory peak of gonadotrophins (Foster et al. 2006; Kinder et al. 1995).

Commonly the first behavioral estrus is not preceded by significant concentrations of P4. Thus, the first luteal structure that forms disappears and puberty is reached only after the occurrence of the first behavioral estrus. From that, there is the ovulation and the development of a CL with a typical lifespan. Finally, in puberty, all components of the hypothalamic-adenohypophyseal-ovarian axis will be properly normalized so that reproductive activity occurs cyclically and regularly, and sexual maturity is achieved. The prepubertal heifers present a specific pattern of endocrine and ovarian follicular dynamics. The first ovulation usually occurs at 56 weeks of age, being usual non regular cycles at the beginning of ovarian activity. The short duration of ovulatory cycle was associated with low P4 concentrations and small CL (Evans et al. 1994).

In this context, it is important to emphasize that the occurrence of puberty is considered one starting point to the productive life of a female (Brumatti et al. 2011). Then, early attainment of puberty predicts anticipation of this process (Ferraz et al. 2018). However, studies have shown that the optimal age at first birth is 24 months (Haworth et al. 2008). Below this age range, it is improbable that heifers are of enough body size to express their full genetic potential, either for the production of milk throughout life or even to deliver and maintain a healthy calf. In addition, breeding costs are increased for heifers calving later than those heifers calving early (Boulton et al. 2017). Heifers calving later have lower chances of completing the first lactation (Brickell and Wathes 2011) and relatively low fertility with longer delivery intervals (Wathes et al. 2014). In this way, producers that breed late femalesmay be wasting resources with less likely to give birth in an ideal age or to survive beyond a first lactation in order to recover the costs of breeding (Boulton et al. 2017). In this context, studies have been developed with the purpose of generating alternatives to promote puberty as early as possible in heifers, without disrupting future productive and reproductive performance of these females (Seneda et al. 2019). Genetic selection by animals with greater sexual precocity and the use of hormonal protocols to induce puberty represent good alternatives for the selection of earlier and more productive matrices in the livestock (Gonzalez et al. 2020).

1.3.2 *Neuroendocrine Control of the Estrous Cycle*

The estrous cycle, a series of reproductive events and ovarian functions in cattle, such as recruitment, follicular growth, ovulation, luteinizing and luteolysis, are regulated by a perfect hormonal interaction among hormones from the hypothalamus (GnRH), anterior pituitary (FSH and LH), ovary (P4, estradiol, inhibins and oxytocin), and uterus (prostaglandin F2α – PGF2α). In addition, non-hormonal factors, such as insulin-like growth factors, also have an important function in ovarian follicular activity. The hormones act through a positive and negative feedback system for controlling estrous cycle, allowing reproductive events by specific control of ovary and uterus activity (Forde et al. 2011; Crowe 2008; Roche 1996).

GnRH is one of the main hormones involved in the estrous cycle control by stimulating the production and secretion of gonadotropins (Fig. 1.2). This hormone acting on the anterior pituitary or adenohypophysis to stimulate FSH and LH production and release. GnRH is produced in the nucleus of hypothalamus neurons and transported by axons to the median eminence, which is where it is stored (Yin and Gore 2010). Subsequently, GnRH is released and transported to the hypophysis by the portal hypothalamic-hypophyseal system, where GnRH binds to cell surface receptors and signals the release of FSH and LH (Vizcarra et al. 1999; Moenter et al. 1992; Keri et al. 1985). FSH is stored in secretory granules located in the cytoplasm for shorter periods, whereas LH is stored for longer periods during the estrous cycle (Farnworth 1995).

The GnRH functions depend of stage of the estrous cycle. At the beginning of the cycle, GnRH stimulates FSH secretion to follicular recruitment and initial growth of small follicles. At the end of the cycle GnRH stimulates LH secretion to induce final

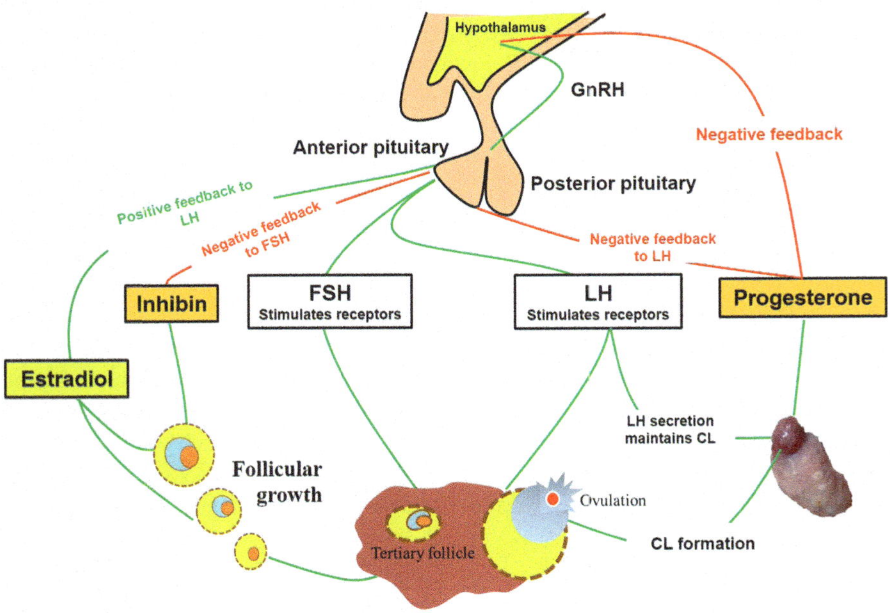

Fig. 1.2 Schematic representation of the hormonal interaction of the hypothalamic-pituitary-gonadal axis to stimulate and/or inhibit ovarian follicular activity in cattle. Gonadotropin releasing hormone (GnRH), Luteinizing hormone (LH), Follicle-stimulating hormone (FSH), and corpus luteum (CL)

growth and ovulation of the dominant follicle (Crowe and Mullen 2013). The CL regression during the follicular phase of the estrous cycle results in basal P4 concentrations. So, there is a concomitant increase in estradiol concentration due to the development of a dominant follicle, which induces increasing release of GnRH and allows estrus behavior during which females are sexually receptive to mounts (Frandson et al. 2009). However, ovulation of the dominant follicle only occurs with basal P4 plasma concentration and LH pulse frequencies occurring every 40–70 min for 2–3 days (Roche 1996). The dominant follicle ovulates approximately 10 to 14 h after the end of the estrus behavior. After breakage of the preovulatory follicle, metaestrous is initiated and by 3 to 4 days is characterized by the formation of CL from the hemorrhagic corpus (Forde et al. 2011).

Immediately after ovulation, CL formation occurs from luteinization of granulosa and internal theca cells. This ovarian structure is responsible for P4 production, which is required for the diestrous phase or maintaining a pregnancy (Schams and Berisha 2004; Butcher et al. 1992). During the diestrous phase, P4 concentrations remain elevated, and recruited follicles continue to develop due to FSH being released from the pituitary. However, growing follicles are inhibited due to high levels of P4 during this phase, which does not allow appropriate frequencies and/or amplitudes of LH pulses due to negative feedback with this hormone. Therefore, the dominant follicle does not ovulate and undergoes atresia, gradually decreasing its diameter, during diestrous (Crowe 2008; Manikkam and Rajamahendran 1997; Savio et al. 1993; Taylor and Rajamahendran 1991). On the other hand, during the proestrous phase, P4 concentrations decrease due to CL regression in response to PGF2α uterine secretion; then, this hormonal condition (low P4 and high estradiol) allows for LH pulses and consequently ovulation (Fig. 1.3) (Crowe and Mullen 2013; Forde et al. 2011; Goff 2004).

1.3.3 Estrous Cycle Phases and Estrus Behavior

The estrous cycle in cattle includes a normal length of 18 to 24 days with two distinct phases: (1) the luteal or progesteronic phase and (2) follicular or estrogenic phase. The luteal phase lasts 14 to 18 days and is the period that follows ovulation, being characterized by formation and CL presence (also known as metestrus and diestrus phases). The follicular phase extends from CL regression (luteolysis) until ovulation and lasts 4 to 6 days. This phase is also known as proestrous and estrus, being a required event for ovulation of dominant follicle. In this phase, the dominant follicle undergoes final maturation, and the ovulatory follicle releases the oocyte to fertilization at the uterine tube (Forde et al. 2011; Adams et al. 2008; Rawlings et al. 2003; Sunderland et al. 1994).

The signs of estrous behavior are manifested during the follicular phase and characterized by primary and secondary signs. Unrest, increased physical activity (vocalization and hyperactivity), swelling and reddening of the vulva, friendly contact with herd mates, smelling and licking the genitalia, clear mucus secretion, and

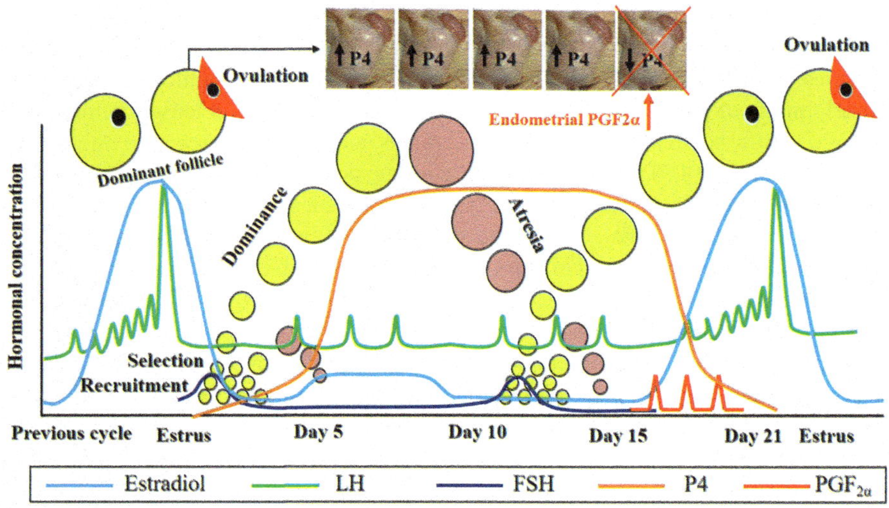

Fig. 1.3 Schematic description of the growth pattern of ovarian follicles during follicular dynamics [recruitment, selection, dominance, and atresia (luteal phase) or ovulation (follicular phase)] and secretion pattern of estradiol, Luteinizing hormone (LH), Follicle-stimulating hormone (FSH), progesterone (P4), and prostaglandin F2α (PGF2α) during the estrous cycle in cattle

attempts to mount constitute secondary signs. However, being receptive to mounting is the most reliable signal of estrus in cattle, which is considered the primary sign (Løvendahl and Chagunda 2010; Galina et al. 1982). Frequently, secondary signs precede (proestrous) estrus and may continue after females accept the mount. Therefore, the identification of secondary signals also assists in the process of identifying the estrus itself.

The estrogen is the main hormone responsible for estrus behavior, which is produced by the dominant follicle and signals the hypothalamus to induce the secondary signs and behavior of receptivity, an event necessary for successful mating. However, an increase in the estradiol concentration and a reduction of P4 concentration is required for manifestation of estrus behavior. This hormonal condition occurs at the end of the luteal phase, being induced by luteolysis between 15 and 17 days of the estrous cycle (Allrich 1994; Vailes et al. 1992). Thereafter, the P4 plasma level is reduced below 1 ng/mL, and this event is responsible for estrus return in nonpregnant females.

Estrus behavior duration and intensity can be affected by several factors, such as, age, production level, environmental factors, nutritional factors, health status, size of the sexually active group, presence of a bull, climatic factors (strong rain and wind), management of animals in pastures and stalls, and subspecies (*Bos taurus* versus *Bos indicus*) (Crowe 2008; Galina et al. 1996). In beef cattle, an average duration of 8 to 10 h has been reported (Crowe 2008; Bó et al. 2003). In dairy cattle, the average estrus duration is 8 to16 h in cows and 12 to 14 h in heifers, but this duration tends to decrease as milk production increases (Forde et al. 2011; Crowe

2008; Allrich 1994). This negative relation between milk production and duration of estrus has been reported in high-producing dairy cows (above 40 kg of milk per day) because there is a decreasing circulating estradiol level due to increased metabolism of this hormone (Wiltbank et al. 2006).

Differences in the characteristics of estrus between *Bos taurus* and *Bos indicus* females has been well reported (Pinheiro et al. 1998; Galina et al. 1996). In general, zebu cattle present a short duration of estrus and high rates of estrus expression at night, which makes it difficult to identify the estrus for conventional insemination program. Under Brazilian management conditions, the use of a radiotelemetry device in heifers has demonstrated a shorter average estrus duration for *Bos indicus* (12.9 h in Nelore) compared to that of *Bos taurus* (16.3 h in Angus), but no difference was found between *Bos indicus* x *Bos taurus* crossbred (12.4 h in Nelore x Angus) and *Bos indicus* (Mizuta 2003). In addition, it was reported that 53.8% of the estrus expression in *Bosindicus* (Nelore cows) begins at night (between 6 p.m. and 6 a.m.) with 30% of its beginning and finishing occurring at night (Pinheiro et al. 1998).

1.3.4 Events of the Ovarian Follicular Dynamics

Follicular dynamics is one of the most important events of ovarian physiology, being widely studied in *Bos taurus* and *Bos indicus* females (Cuervo-Arango et al. 2011; Mackey et al. 2000; Figueiredo et al. 1997; Ginther et al. 1989a). The follicular dynamics involve growth, development, and maturation of ovarian follicles that are essential processes for reproductive efficiency in farm animals. These dynamics events are totally dependent on the perfect functioning of the hypothalamus-pituitary-gonadal axis that acts through positive and negative feedback mechanisms to stimulate development of ovarian follicles by gonadotropins secretion (FSH and LH).

During the estrous cycle, 2 or 3 waves of follicular growth may be present, but only the last wave is considered an ovulatory wave (Figueiredo et al. 1997; Ginther et al. 1989b). Every wave consists of the recruitment of a group of small follicles (4 to 6 mm in diameter) by FSH stimulation. These follicles grow and future dominant follicle is selected, which suffers a more efficient action of FSH in relation to the other follicles of the wave. So, there is follicular divergence and a dominant follicle emerges, which undergoes atresia or ovulation depending on the circulating levels of P4 that may or may not allow the estrogen trigger on LH pulsatility (Atkins et al. 2013; Forde et al. 2011). Follicular waves are initially established during the prepubertal period; however, the dominant follicle undergoes atresia and does not ovulate. Therefore, only after puberty the ovarian follicular dynamics become a cyclic event characterized by estrus and ovulation (Atkins et al. 2013; Bergfeld et al. 1994).

In general, ovarian follicular development through a non-gonadotropin-dependent phase and a gonadotropin-dependent phase (Webb et al. 2004). During fetal life, for example, follicle development occurs for 3–4 months and the two phases are present. During the gonadotropin-dependent stage, follicular development occurs in waves (2–3 waves of growth per estrous cycle), being that each follicular wave comprises emergence, selection, dominance, and atresia or ovulation (Fig. 1.3) (Ginther et al. 1989a). Frequently, dairy cows have two waves and heifers, or beef cattle have three waves (Forde et al. 2011). In *Bosindicus* (Nelore), there is a predominance of two waves for cows (83.3%) and three waves for heifers (64.7%) (Figueiredo et al. 1997).

The emergence of the wave is the first event gonadotropin dependent, which is characterized by recruitment of a group of follicles (1 to 50 follicles ≥5 mm) that starts growth in response to increased FSH concentration (Sunderland et al. 1994). Follicular growth and cell proliferation in this period occur primarily by FSH stimulation, being characterized by the presence of FSH receptors located in the granulosa cells of follicles at day 3 of the follicle wave (Ginther et al. 2002; Richards et al. 1998). Following follicular development, the next important event is follicular selection, a mechanism that reducesthe number of recruited follicles to the ovulatory quota of each species, usually one follicle in bovine species (Sunderland et al. 1994). In this phase, there is a transient increase in FSH concentration, which induces an increase in the aromatase enzyme activity of granulosa cells responsible for conversion of androgen to estrogen (Hillier 1994). This process increases estrogen and inhibin concentrations of healthy follicles, assisting the selection process and gonadotropin secretion (Ginther et al. 2002).

Once the follicular selection process has been established, the dominant follicle diameter increases, and divergence is started when the largest follicle reaches an average diameter of 5.7 to 6.1 mm in *Bosindicus* (Sartorelli et al. 2005) and 8.0 to 8.5 mm in *Bos taurus* (Ginther et al. 1999). In sequence, the follicular dominance is started by growth of the dominant follicle of approximately 1 to 2 mm per day (Figueiredo et al. 1997; Knopf et al. 1989). Then, its diameter is enlarged due to an increase in the follicular fluid. This fluid contains high estrogen concentration, in addition to high levels of inhibin that is associated with negative feedback in the pituitary, reducing FSH levels to basal concentrations (Sunderland et al. 1994). However, due to the presence of LH receptors in the granulosa and theca cells, the dominant follicle becomes sensitive to LH, and even at a low FSH concentration, it continues to grow and accumulate estrogen in the follicular fluid (Xu et al. 1995).

With the growth of the dominant follicle, now being more stimulated by the action of LH than FSH, the destination of the dominant follicle is dependent on frequency and amplitude of the LH pulse (Forde et al. 2011). During the early luteal phase, LH pulsatility has a low amplitude (delta LH, 0.3 to 1.8 ng) and high frequency (20 to 30 pulses/24 h), and during the middle of the luteal phase, LH pulses have high amplitude (delta LH, 1.2 to 7.0 ng) and low frequency (6 to 8 pulses/24 h) (Rahe et al. 1980). This amplitude and frequency are still insufficient to promote final maturation and ovulation of the dominant follicle. Then, during the luteal phase of the estrous cycle, the dominant follicle undergoes atresia and all production of estrogen and inhibin decreases gradually, removing the negative feedback on the FSH secretion in the hypothalamus/pituitary. This event promotes new FSH secretion and the emergence of a second follicular wave (Atkins et al. 2013; Forde et al. 2011).

High estrogen production is an essential characteristic of the dominant follicle that largely depends on LH pulse frequency. In this phase, LH binding to its receptors on theca cells provides the conversion of cholesterol to testosterone through a series of catalytic reactions. Then, testosterone diffuses from theca cells to granulosa cells and is converted into estrogen by the aromatase enzyme (Hillier 1994; Sunderland et al. 1994). Estrogen has a local effect on follicular development and a systemic effect on the hypothalamus and hypophysis. During the follicular phase, when the P4 concentration is low, the estrogen produced by the preovulatory follicle induces an increase in the GnRH level in the hypothalamus, which promotes an increase in the amplitude and frequency of LH pulses that stimulates the final maturation and ovulation of the preovulatory follicle (Crowe and Mullen 2013; Sunderland et al. 1994). The maximum diameter of the preovulatory follicle is variable, being 10 to 14 mm in *Bos indicus* (Sales et al. 2012a; Figueiredo et al. 1997) and 14–20 mm in *Bos taurus* (Ginther et al. 1989a).

In addition, hormonal action intraovarian factors are also important to estrous cycle regulation, acting indirectly by altering estradiol synthesis or directly through negative feedback that controls the hypothalamus and pituitary gland (Forde et al. 2011). Although the acquisition of LH receptors by the granulosa cell layer is considered the main mechanism that promotes the follicular selection process (Lucy 2007), the bioavailability of the insulin-like growth factor (Rivera and Fortune 2003) and the presence of other growth factors also contributes to the development, proliferation, and steroidogenic capacity of the dominant follicle (Knight and Glister 2006).

1.4 Estrus Synchronization Program

Estrus synchronization is a hormonal strategy performed in a group of females with the aim of inducing all animals in the same phase of the estrous cycle. Using this hormonal strategy, management for estrus identification and insemination are facilitated. Two possible methods for estrus synchronization are available: (I) interrupting the activity of the luteal phase or (II) extending the CL length. A luteolytic agent such as PGF2α or its synthetic analogs (Table 1.1), frequentely are used to interruption of the luteal phase and then induce estrus. On the other hand, the duration of the luteal phase can be modified using a progestin treatment, which is similar to the endocrine activity of CL and prevents estrus behavior during treatment. In this case estrus synchronization occurs after removing the P4 source.

1.4.1 Use of Prostaglandins and Synthetic Analogs

Since the 1970s, a wide variety of studies have evaluated the effects of PGF2α and its synthetic analogs in domestic species (Odde 1990; Rajamahendran et al. 1976). A treatment with PGF2α in cattle decreases the P4 plasma concentration in 24 h

Table 1.1 Active principle, dose, and commercial name of main prostaglandins and syntheticanalogs

Active principle	Dose (mg)	Commercial name
Cloprostenol	0.526	Bio-Cio[a]
Cloprostenol	0.530	Ciosin[b]
Cloprostenol	0.500	Cioprostinn[c]
Cloprostenol	0.526	Cioton[d]
Cloprostenol	0.526	Clocio[e]
Cloprostenol	0.526	SincroCio[f]
Cloprostenol	0.500	Sincrosin[b]
Cloprostenol	26.30	Induscio[g]
D-cloprostenol	0.150	Croniben[h]
D-cloprostenol	0.150	Prolise[i]
D-cloprostenol	0.150	VeteGlan[j]
Dinoprosttromethamine	25.00	Lutalyse[k]

Source: [a]BioVet, [b]MSD Sáude Animal, [c]Boehringer Ingelheim, [d]JA Sáude Animal, [e]Bimeda, [f]Ourofino Saúde Animal, [g]GlobalGen Vet Science, [h]Biogénesis Bagó, [i]União Química Farmacêutica Nacional SA, [j]Hertape Calier Saúde Animal SA, [k]Zoetis Saúde Animal

after injection in 80 to 100% of the animals. However, morphological CL regression is gradual and can be observed by ultrasound approximately 24 to 48 h after treatment. Despite PGF2α inducing efficient luteolysis, estrus behavior can be observed for a long time due to the stage of development of dominant follicle at the time of application of PGF2α. If a dominant follicle is present at the time of application, the estrus behavior occurs in a short period, approximately 24 to 72 h after luteolytic agent application. Otherwise, only the dominant follicle of the next follicular wave may be considered, in this case there will be a longer interval until the occurrence of estrous behavior, up to 144 h (Kastelic et al. 1990). Therefore, the synchronization program with PGF2α requires careful practice for estrus observation in treated animals.

Despite the high efficiency of protaglandin causing luteolysis and inducing estrus, it is worth noting that its use only has an effect in the presence of a responsive CL. Therefore, for more efficient use of luteolytic agents, it is necessary to identify CL by transrectal palpation or ultrasound. A CL up to the 5th day of life after the ovulation is not responsive to luteolysis because the CL is still in formation. In the same way that a CL after the 15th day of the cycle also does not respond to the application because it may have already undergone natural luteolysis (Fernandes et al. 2006; Meidan and Levy 2002; Levy et al. 2000).

On the other hand, when CL is not evaluated, estrus synchronization with luteolytic agent can be performed by two applications of PGF2α with intervals of 11–14 days. The second application induces luteolysis in animals that do not respond to PGF2α at the first treatment, resulting in synchronization rates of 70 to 80% of females (Binelli et al. 2014; Machado et al. 2007; Baishya et al. 1980). Following this strategy, animals that manifest estrus after each application of PGF2α can be inseminated, or for greater concentration in the number of females in estrus it is recommended to inseminate all animals only after the second application.

Another option for application of PGF2α in AI programs is the detection of estrus for 5 to 7 consecutive days followed by AI, and in females that did not express estrus, CL identification and PGF2α application is perfomed for further estrus observation and insemination. The estrus observation can continue during the next 5 to 7 days, for a total of 12–14 days of work. These three protocols for the synchronization with PGF2α are illustrated in Fig. 1.4.

Currently the main synthetic analogs of PGF2α (Table 1.4) commercially available are dinoprosttromethamine, cloprostenol, and D-cloprostenol. These analogs are considered potent luteolytic agents used routinely in estrous cycle synchronization programs. Intramuscular administration of conventional doses (dinoprosttromethamine 25 mg, cloprostenol 500μg, D-cloprostenol 150–300μg) is the most used but also have tested other routes for reduction of doses. Half a dose by vulvar submucosa injection resulted in similar synchronization rates; however, application in the vulvar submucosa is a route less practiced, especially for a nonexperienced worker (Baryczka et al. 2018; Valldecabres-Torres et al. 2012; Chacur et al. 2010; Colazo et al. 2002).

Although estradiol esters (estradiol-17β, benzoate, cypionate and estradiol valerate) are capable of causing luteolysis, the luteolytic effect depends on the ester and

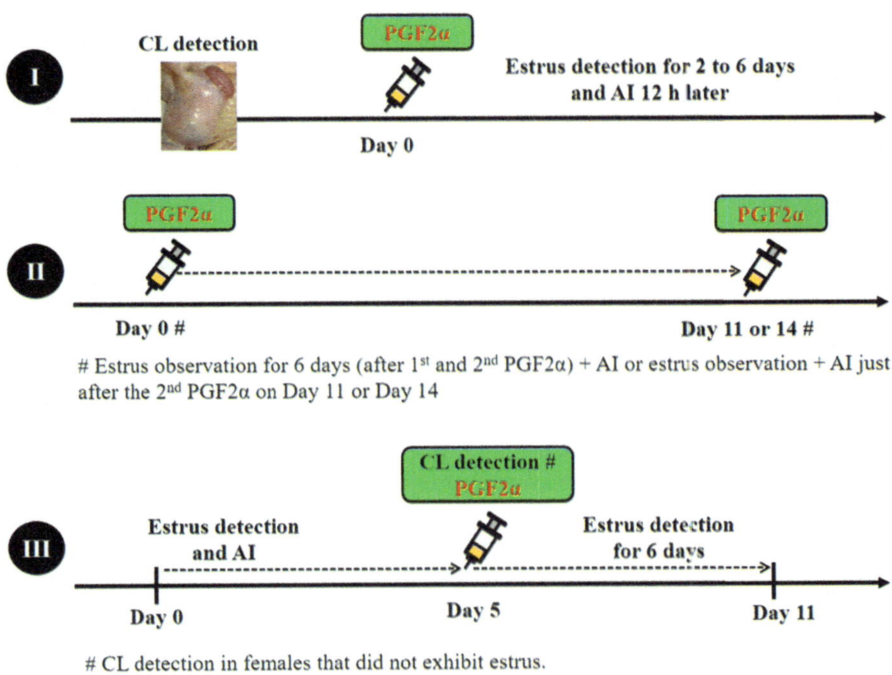

Fig. 1.4 Schematic representation of three treatments options for estrous synchronization with PGF2α and its synthetic analogs in cattle. *CL* corpus luteum, *AI* artificial insemination

dose used (Baruselli et al. 2004b). Of all available esters, estradiol valerate has the greatest potential for luteolysis; however, prostaglandins and their analogs are still more used to induce estrus in cyclic cows.

Commonly females in estrus are inseminated 12 h later, and conception rates are similar to those obtained in natural estrus. However, to facilitate the management of inseminations, females identified in estrus by the morning are inseminated in the late afternoon. Those already detected to be in estrus by the afternoon are inseminated in the morning of the next day. In addition to facilitating the management of the process, this strategy also results in better synchronization of oocyte and sperm viability.

1.4.2 Progesterone and Progestin

Since the 1950s, P4 and its synthetic analogs (progestins) have been used for estrous synchronization programs. The first studies involving P4 were carried out with formulations administered by an oral or injectable route for 14 to 21 days, producing an effective estrus synchronization rate but variable fertility (Hansel et al. 1966). These initial results were unsatisfactory, probably due to their use over a long time and the suboptimal concentration of P4, leading to the formation of persistent follicles containing low-quality oocytes (Fortune and Rivera 1999). In the last decade, the authors of this chapter have conducted a series of studies with injectable P4 formulations, showing promising results for ovarian follicular dynamics (Morotti et al. 2013a, b) and improved fertility of cows submitted to TAI program (Campos et al. 2016a, b). However, there is still no commercially available injectable formulation for use in estrous cycle synchronization programs. Natural P4 has been used in intravaginal devices, and synthetic P4 has been used in subcutaneous auricular implants (norgestomet) or oral use (melengestrol acetate). Despite the different forms of P4 presentation (Table 1.2), an intravaginal device is the most common for synchronization programs.

For the purpose of estrus synchronization, formulations containing P4 can be used to prolong the lutein phase of the estrous cycle or to simulate the effect of a CL, so the synchronization of the animals occurs when removing the source of P4. Although this strategy provides an effective synchronization of the females, in cyclic animals the result is more satisfactory when the application of a luteolytic agent is combined with the removal of the P4 source (Baruselli et al. 2004b). In addition to estrous cycle synchronization, P4 has also been used for cyclicity induction in prepubertal heifers (Gonzalez et al. 2020). This subject will be discussed further in another section of this chapter.

Table 1.2 Main forms of progesterone or progestin presentation (active principle) and their commercial names, administration routes, and concentrations

Active principle	Commercial name	Administration route	Concentration
Norgestomet	Crestar[a]	Auricular	3 mg
Melengestrol acetate	MGA, Premix[b]	Oral	0.5 mg/day
Progesterone	CIDR-B[b]	Intravaginal	1.9 g
Progesterone	DIB[b]	Intravaginal	0.5–1.0 g
Progesterone	Cronipres[c]	Intravaginal	0.5–1.0 g
Progesterone	PRIMER[d]	Intravaginal	0.5–1.0 g
Progesterone	Sincrogest[e]	Intravaginal	1.0 g
Progesterone	SincrogestInjetável	Injectable	150 mg/mL
Progesterone	Progecio[d]	Injectable	70 mg/mL
Progesterone	Repro One[f]	Intravaginal	0.5 g
Progesterone	Repro Neo[f]	Intravaginal	1.0 g
Progesterone	Repro Sync[f]	Intravaginal	2.0 g
Progesterone	Biprogest[g]	Intravaginal	1.25 g
Progesterone	Progestar[h]	Intravaginal	0.96 g
Progesterone	Betaproginn[h]	Injectable	25 mg/mL
Progesterone	Prociclar[i]	Intravaginal	0.75 g
Progesterone	Fertilcare 600[a]	Intravaginal	0.6 g
Progesterone	Fertilcare 1200[a]	Intravaginal	1.2 g

Source: [a]MSD Saúde Animal, [b]Zoetis Saúde Animal, [c]Biogénesis Bagó, [d]União Química Farmacêutica Nacional SA, [e]Ourofino Saúde Animal, [f]GlobalGen Vet Science, [g]Bimeda, [h]Boehringer Ingelheim, [i]Ceva Saúde Animal

1.5 Ovulation Synchronization Program for Timed Artificial Insemination

For the elaboration of an efficient ovulation synchronization program, the hormonal protocol must combine three basic principles: (I) synchronization of the emergence of the follicular growth wave, (II) synchronic control of the progesteronic phase, and (III) synchronized ovulation induction.

Based on these principles, protocols for ovulation synchronization aim to control the growth of the follicular wave to regulate the luteal phase and induce ovulation in a fixed amount of time, avoiding the necessity of estrus detection for AI. For this purpose, some hormones such as P4, progestin, PGF2α, estradiol esters (estradiol-17β, benzoate, cypionate and estradiol valerate), equine chorionic gonadotropin (eCG), human chorionic gonadotropin (hCG), and GnRH are combined to provide in estrus and ovulation synchronization.

1.5.1 Synchronization of Follicular Wave Emergence

Progesterone combined with estrogen and/or GnRH is an efficient hormonal combination to synchronization in the emergence of a new wave of follicular growth. Treatment with P4 and estradiol esters on a random day of the estrous cycle (day 0) has been effective for promoting follicular atresia and the emergence of a new follicular wave at a fixed time in cattle (Bó and Baruselli 2014). Exogenous estradiol induces atresia of small follicles by decreasing FSH release, whereas P4 is used to suppress the release of LH and inhibit the activity of large follicles, which are dependent on this gonadotropin. Therefore, application of estrogen and P4 at the same time promotes atresia of all of the follicles (large, medium, and small) present in the ovary and a new wave emergence 2 to 4 days later (Baruselli et al. 2004b; Bó et al. 1995, 2003).

In cyclic *Bos taurus* cows and heifers, the emergence of a new follicular wave occurred in 3 to 4 days after treatment with progestogen and 5 mg of estradiol-17β or progestin and 1 to 2.5 mg of estradiol benzoate (EB) (Martinez et al. 2005; Bó et al. 1995). Apparently, *Bos indicus* and *Bos taurus* females have similar intervals for the synchronization of the follicular wave. For example, in Brangus heifers treated with an intravaginal P4 device and 2 mg of EB the follicular wave emergence did not differ from wave emergence in Angus heifers (3.3 ± 0.6 versus 4.3 ± 0.2 days) (Bó et al. 2003). A similar time of emergence was also observed between Nelore and crossbred ½ Nelore x ½ Angus heifers (3.3 ± 0.6 versus 3.5 ± 0.1 days) (Carvalho et al. 2008). Heifers and Nelore cows receiving 2.0 mg of EB and insertion of an auricular implant of norgestomet presented intervals of 2.5 ± 0.2 days for a new wave of emergence (Sá Filho et al. 2011). However, doses of 2.5 and 5.0 mg of estradiol valerate (EV) resulted in intervals of wave emergence of 4.2 ± 0.3 and 6.1 ± 0.6 days in heifers and 3.1 ± 0.4 and 4.0 ± 0.5 days in cows, respectively (Bó and Baruselli 2014).

Another efficient alternative for the synchronization of follicular wave emergence is to associate P4 with GnRH analogs. A study carried with dairy cows (*Bos taurus*) in lactation evaluated the association of 2 mg of EB and CIDR versus 100μg of GnRH and CIDR and revealed follicular wave emergence at 4.8 ± 0.4 and 2.0 ± 0.2 days, respectively (Kim and Kim 2007). In high-yielding dairy cattle the emergence of a wave of follicular growth is a major challenge. In this context, studies show that the application of estrogen and GnRH analogs in association with the insertion of a P4 device appears to be more effective in synchronizing the emergence of the follicular growth wave than the isolated use of P4 and estrogen or P4 and GnRH. In addition, twice the dose of GnRH has been more effective in synchronizing the emergence of the wave in these animals (Silva 2020; Wiltbank and Pursley 2014). Despite the efficiency of both combinations, estrogen has a lower cost compared to that of GnRH.

1.5.2 Control of the Progesteronic phase

The progesteronic phase can be controlled efficiently by the use of a P4 device, by the use of luteolytic agents, and by the association of these, which is more effective for synchronization. The use of luteolytic agents is highly efficient for inducing estrus synchronization in cyclic females as previously discussed. However, the use of these agents in association with P4 devices has been more effective than their isolated uses. In this context, PGF2α or its synthetic analogs was proposed for increasing the fertility of cows submitted to synchronization with CIDR-B for 8–9 days associated with EB at the time of P4 device insertion (Macmillan and Peterson 1993a, b). Efficient control of the progesteronic phase has been reported with the removal of the P4 device between D6 and D9 (Bó and Baruselli 2014; Sá Filho et al. 2010a, b, 2011) and PGF2α application can be performed at any of the three times: (i) at the time of P4 removal, (ii) 24 to 48 h before P4 removal, or (iii) on day 0 at P4 device insertion (Bó et al. 2002; Macmillan and Peterson 1993a, b). Therefore, administration of luteolytic agents such as PGF2α before or at the end of P4 treatment is required to ensure luteolysis and adequate control of the progesteronic phase (Kastelic et al. 1999).

The time for application of the luteolytic agent has been widely studied in *Bos indicus* cattle. Initially, luteolysis was observed in 83.3% Nelore cows treated with 5 mg of EV between 2 and 7 days after treatment (Pinheiro et al. 1998). Therefore, due to the high percentage of induction of luteolysis using EV at the mentioned dose, it is not necessary to use PGF2α in protocols using progestin and EV at the beginning of hormonal treatment (Kastelic et al. 1999). The effects of the induction of luteolysiswith dinoprosttromethamine were investigated simultaneously or 48 h before P4 device removal. In this study, a higher P4 concentration (4.58 ± 0.21 ng/mL) was observed in cows treated simultaneously to the P4 device removal compared to those treated 48 h earlier (3.05 ± 0.21 ng/mL) (Peres et al. 2009). In addition, cows with early luteolysis improved the ovulation rate (77.0% versus 85.4%) and increased the pregnancy rate by aproximadely16%.

In summary, to obtain greater efficiency in TAI protocols, the addition of PGF2α 2 to 3 days before or at the end of P4 treatment is required to induce luteolysis in cyclic females. In a normal estrous cycle, this strategy increases the proestrous phase (low P4 and high estradiol), better preparing the uterus for pregnancy or allowing subsequent ovulation. However, it is possible that this anticipated luteolysis is not necessary when using more potent PGF2α analogs such as cloprostenol or D-cloprostenol.

1.5.3 Synchronized Induction of Ovulation

The use of ovulation inducers in association with P4 treatment has been proposed to reduce the dispersion of ovulation and improve the ovulation rate. In addition to the wave emergence, the use of estradiol esters [EB or estradiol cypionate (EC)] at the end of P4 treatment induces ovulation between 72 and 84 h after P4 removal in 75% of cows (Morotti et al. 2013a, b; Bó et al. 1995). Several hormones are used to induce ovulation in cattle, such as estradiol esters (estradiol-17β, EB, and EC), GnRH analogs, hCG, and LH. The main ovulatory mechanism associated with the use of ovulation inducers is related to the induction of an endogenous LH surge (Baruselli et al. 2004b). Estradiol esters are considered indirect inducers, and GnRH and LH are considered direct inducers. According to cost benefit analysis, the most used hormone for induction of ovulation in *Bos indicus* cattle is EB 24 h after P4 removal (1 mg) or EC along with device removal (0.5 mg in heifers to 1 mg in cows) (Torres et al. 2014). However, GnRH anologs can also be used for this purpose from the time of insemination up to 16 h before.

Frequently, within the TAI protocols the insemination is performed 48 h after P4 removal or approximately 8 to 12 h before the expected time for ovulation (Baruselli et al. 2004a, b; Hanlon et al. 1996). However, to reduce management practices, administration of estradiol esters was proposed to induce ovulation simultaneously or 24 h after removal of the P4 device. Similar results to those with EB for induction of ovulation were obtained with EC administration (Sales et al. 2012a). Similar intervals between CIDR removal and LH surge (54.6 ± 3.4 versus 59.3 ± 3.5 h) and between CIDR removal and ovulation (81.6 ± 5.0 versus 86.4 ± 4.8 h) were observed in heifers treated with 0.5 mg of EC 0 or 24 h after intravaginal device removal (Sales et al. 2012b). Application of EB or EC to induce ovulation after P4 removal resulted in similar ($P > 0.05$) follicular responses (ovulatory follicle diameter: 13.1 versus 13.9 mm), the interval from P4 device removal to ovulation (70.2 versus 68.5 h) and ovulation rate (77.8% versus 82.8%). In addition, the pregnancy per AI was similar ($P > 0.22$) between cows treated with EB (57.5%; 277/482) and EC (61.8%; 291/471) (Sales et al. 2012b). In another study, an interval of 45 h was observed between injection of 0.5 mg of EC simultaneously to P4 device removal and LH surge (Sales et al. 2015).

Other ovulation inducers, such as synthetic GnRH analogs, hCG, or LH, can be used. However, the use of LH is restricted to inducing ovulation in protocols for ovarian superovulation due to its high cost. The same limiting factor, cost, also has to be considered for hCG and GnRH, although GnRH has been widely used alone or in association with EC or EB in females that exhibit low or no estrus expression during TAI protocols (Rodrigues et al. 2019). Generally, indirect ovulation inducers are proconized in conventional TAI protocols, being that the ovulation is normally obtained by cheaper alternatives by the use of estrogens. On the oder hand, direct inductors are used either in specific situations, either for TAI or more often for superovulation protocols.

1.6 Strategies for Improving Fertility at TAI

TAI programs represent one of the most important advances in reproductive biotechniques in cattle. The main advantages of this reproductive tool are undoubtedly associated with ease of management, since the TAI protocol eliminated the practice of estrus identification and its low efficiency, contributed to a greater intensification in the use of AI, which made the technique better known and widespread in the world. In addition, TAI has increased the service rate to 100% of the synchronized animals, which significantly contributes to increasing the reproductive efficiency of the herd and the economic performance of the farm (Baruselli et al. 2018).

The goal of the TAI program is to inseminate all treated animals at the same time and to obtain pregnancy rates of approximately 40 to 60%. The best reproductive rates are achieved with well-managed herds, with high quality nutrition, good health, and semen quality and a well-trained team. Although good results can be obtained with TAI, a reduction of approximately 15 to 20% in the conception rate of *Bos indicus* cattle with a high rate of anestrous has been reported (Baruselli et al. 2004a; Fernandes et al. 2001).

Postpartum cows may undergo a prolonged anestrous period during which they do not show behavioral signs of estrus, which is decisive when reproduction is dependent on bull mating. In this context, anestrous postpartum represents one of the biggest challenges for beef breeding, mainly due to endocrine imbalance regarding adequate LH secretion during this period, which is considered the main factor for anestrous in *Bos indicus* cattle (Baruselli et al. 2004a, 2011). Additionally, many herds managed on tropical pastures may suffer nutritional restrictions due to the seasonality of forages, which may aggravate postpartum anestrus conditions. Therefore, due to characteristics of estrus expression, poor efficiency of estrus detection, and a high percentage of cows in postpartum anestrous (Stevenson et al. 2015; Lamb et al. 2010; Crowe 2008; Yavas and Walton 2000b), AI programs based on estrus detection present with low efficiency, mainly for tropical and subtropical herds (Baruselli et al. 2004a; Bó et al. 2003). On the other hand, TAI program represents one of the most efficient alternatives. However, even receiving a hormonal protocol for TAI, fertility may be affected if hormonal adjustments or management practices are not performed properly.

Administration of eCG and temporary calf removal (TCR) are strategies performed to provide appropriate gonadotropin support for growth and ovulation induction of the dominant follicle in females submitted to TAI (Barreiros et al. 2014; Campos et al. 2013; Sá Filho et al. 2014; Yavas and Walton 2000a, b). Practices such as this have been associated with TAI protocols and improved postpartum fertility because gonadotropin support provides an increase in follicular activity. However, the use of this gonadotropin support has been highly indicated mainly in animals with intense anestrus postpartum or that have a low body condition score (BCS) (Baruselli et al. 2004b).

eCG is a glycoprotein hormone produced by the corium of a pregnant mare with wide application in TAI protocols. This glycoprotein has a high molecular weight and has an excellent gonadotrophic support for follicular growth due its mixed bioactivity (at approximately 2/3 FSH and 1/3 LH) (Soumano et al. 1996; Murphy and Martinuk 1991). It can be used at the end of the hormonal protocol to provide adequate gonadotropic aid for dominant follicle growth, mainly in acyclic cows and heifers (Barreiros et al. 2014). A schematic representation of the pharmacological action and physiological variations of TAI protocols is shown in Fig. 1.5.

Many studies evaluated the effect of eCG administration or TCR use on the conception rate for TAI in cattle (Baruselli et al. 2004a, 2018; Prata et al. 2018; Barreiros et al. 2014; Campos et al. 2013; Sá Filho et al. 2014; Yavas and Walton 2000a, b). Initially, similar conception rate was observed after TAI between cyclic lactating Nelore cows that were submitted to TCR (50.5%) compared to control group (53.5%). However, there was a 22% increase in the conception rates of *Bos indicus*

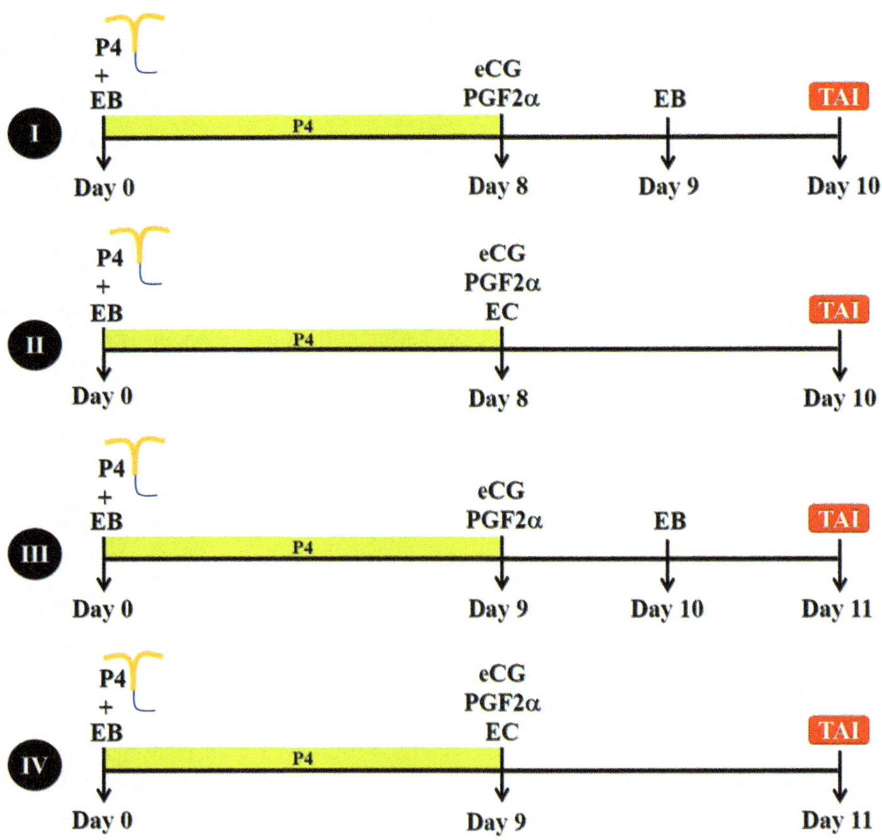

Fig. 1.5 Diagrammatic representation of the protocol for the synchronization of ovulation and timed artificial insemination (TAI) in cattle. Treatment with progesterone (P4) for 8 days [ovulation induced by estradiol benzoate (EB, I) and estradiol cypionate (EC, II)] or 9 days [ovulation induced by EB (III) and EC (IV)]. D, day; P4, progesterone (1.0–1.9 g); EB (D0), 2 mg; EB (D9), 1 mg; PGF2α, 12.5 mg of dinoprost or 125–250μg of cloprostenol; EC, 1 mg; eCG, 300–400 IU equine chorionic gonadotropin

cows with high anestrous rate after the same treatment (Ereno et al. 2007). In addition, after TAI programs, females who do not become pregnant have hormone treatment-induced cyclicity and may result in pregnancy rates of approximately 60 to 65% in the first 45 days of the breeding season (Baruselli et al. 2004a). Therefore, eCG administration for acyclic cows has been recommended to provide a pregnancy rate of around 50%, which is similar to that observed in cyclic cows (Baruselli et al. 2004a; Kastelic et al. 1999).

The use of eCG has been proposed in cows with a high postpartum period (intense anestrus) and BCS < 3 to obtain conception rates around 50% (Baruselli et al. 2004a). However, an interaction between BCS and the postpartum period can also influence conception rates to TAI. Therefore, administration of eCG in cows among 30 to 60 days postpartum is recommended regardless of the BCS (2 to 3.5). Additionally, management using TCR for 56 h was effective for increasing the conception rate of cows submitted to TAI with BCS < 2.5 compared to cows that were of control group (48.2% versus 28.2%; $P < 0.05$) (Peña 2007).

Some studies were conducted to evaluate the effect of eCG and TCR after P4 device removal on the growth of the dominant follicle in *Bos indicus* cattle. Studies have shown increased dominant follicle diameter and increased ovulation rate (85%) in Nelore cows treated with GnRH associated with 48 h of TCR compared to cows that were injected with GnRH without TCR (51%) (Meneghetti and Vasconcelos 2001). A common effect of eCG and TCR on TAI protocols is associated with increased ovulation rates due to a greater synchronization of the preovulatory LH surge (Sá Filho et al. 2006; Cavalieri et al. 1997; Williams et al. 1996). The effects on dominant follicle growth are likely partly related to the percentage of cyclicality and the postpartum period at the beginning of the TAI protocol. It was reported that follicles with a high maximum diameter promote a linear increase in both the follicular diameter and conception rates in *Bos indicus* and *Bos taurus* cattle (Baruselli et al. 2018; Sá Filho et al. 2010a; Borsato et al. 2004).

The luteotrophic effect of eCG was already shown by an in increase in the P4 concentration (8.6 ± 0.4 versus 6.4 ± 0.4 ng/mL; $P < 0.05$) 12 days after TAI in Brangus cows (Baruselli et al. 2004a). In addition, the TCR practice also results in a higher concentration of P4 after TAI and provide a higher pregnancy rate especially in acyclic females. Thus, the importance of these gonadotrophic stimuli is remarkable because the interaction between the P4 concentration and early embryonic development are key events during the maternal recognition process of gestation (Mann and Lamming 2001).

1.6.1 TAI in Heifers: Reaching Sexual Maturity and Induction of Puberty

The age of puberty is one of the most important parameters to determine the occurrence of the first calving both in beef and dairy cattle. Although puberty is often associated with endocrine mechanisms, gene-environment interactions can also affect reproductive physiology in cattle. In heifers, follicular growth in the prepubertal period is characterized by the FSH secretion, occurrence of the follicular wave emergence and dominance follicular, but without the occurrence of ovulation (Wiltbank et al. 2002). Until puberty, endocrine mechanisms promote negative feedback on the hypothalamus, which decreases GnRH and LH pulsation, blocking the final growth of the dominant follicle and ovulation (Day et al. 1987). In this context, an interruption of inhibitory factors on the hypothalamus occurs only when the animal has adequate weight and body development, which allows for full ovarian activity (Seneda et al. 2019). Therefore, it is observed that nutrition can greatly affect the occurrence of puberty.

In general, zebuine and crossbred heifers experience delayed puberty approximately 15 to 27 months old and body weight around 280 to 300 kg (Teodoro et al. 1993). Taurine heifers start puberty approximately with 10 to 15 months and weight from 220 to 250 kg (Patterson et al. 1992). In the last weeks of the prepubertal period and soon after puberty, there is a modification in the size of the female genital system, possibly by increasing the concentration of P4 and estradiol, signaling the acquisition of sexual maturity. Then, puberty is the beginning, but not the fullness, of the activity of the reproductive female trait (Honaramooz et al. 2004). It is important to note that sexual maturity occurs with the full development of the reproductive system, as evidenced by the increase in the uterine size and tone, besides regular ovarian activity.

A combined analysis of the uterus and ovaries should always be considered before using early hormonal treatment programs for the synchronization of estrus and ovulation in heifers. Heifers with regular ovarian activity, but presenting uterine dimensions characteristic of the prepubertal period, generally have low pregnancy rates after AI. A specific pattern of uterine diameter to start reproduction has been studied, but comparative analysis among animals before and after puberty can provide standardization for both the diameter and the tone of the uterus.

A five-point reproductive tract scoring system (RTS) was developed to estimate the pubertal status and reproductive potential of beef heifers by transrectal palpation of the uterine horns, ovaries, and ovarian structures (Table 1.3) (Rosenkrans and Hardin 2003; Andersen et al. 1991; LeFever and Odde 1987) system is practical and highly accurate, and females with score I, II, or III are considered prepubertal and

Table 1.3 Reproductive tract scoring system (RTS) used to evaluation of sexual maturity in heifers

Score	Ovarian and uterine characteristics
RTS I	Immature reproductive tract, without uterine tone, and without palpable ovarian structures
RTS II	Uterine horn diameter with 20 to 25 mm, without uterine tone, and follicles ≤8 mm
RTS III	Mild uterine tone and follicles 8 to 10 mm in size
RTS IV	Uterine horn diameter of 30 mm, presence of uterine tone, and follicles ≥10 mm
RTS V	Presence of a palpable corpus luteum

Adapted from Andersen et al. (1991)

those with score IV or V are considered pubertal and respond better to reproductive practices.

Prior to the breeding season a puberty-inducing hormone treatment can be used for heifers that have not yet reached sexual maturity. P4 and estradiol-based hormonal protocols have been proposed to induce cyclicity in prepubertal heifers (Rasby et al. 1998). Although some mechanisms are not fully known, it seems that P4 sensitizes the hypothalamus, reducing estradiol receptors and inhibiting GnRH and LH pulsatility (Andersen et al. 1991). In addition to possible effects on the hypothalamus, 17β-estradiol and P4 treatment increases the expression of receptors for GnRH in the pituitary and improves LH synthesis and secretion (Looper et al. 2003). Therefore, P4 treatment for 10 to 14 days mimics a functional CL and consequently induces puberty, possibly by reducing estradiol receptors in the hypothalamus, which after P4 removal increases LH secretion and promotes ovulation. For greater efficiency, at the end of treatment with P4, 1 mg of EB is applied to increase the ovulatory potential of the dominant follicle (Gonzalez et al. 2020; Rodrigues et al. 2013).

The induction of cyclicity with the P4 device (previously used for 24 days) was tested in prepubertalNelore heifers and resulted in 85.6% of cyclicity in P4-treated heifers and 81.1% in heifers injected with 1 mg of EB at the time of device removal (Sá Filho et al. 2006). Cyclicity induction in *Bos indicus* heifers has been tested with different sources of P4. Inductions using a P4 device (fourth use) versus P4 injectable (150 mg, i.m.) were efficient for cyclicity induction (81.5% versus 86.3%; respectively; Fig. 1.6) (Gonzalez et al. 2020). A similar cyclicity rate was obtained

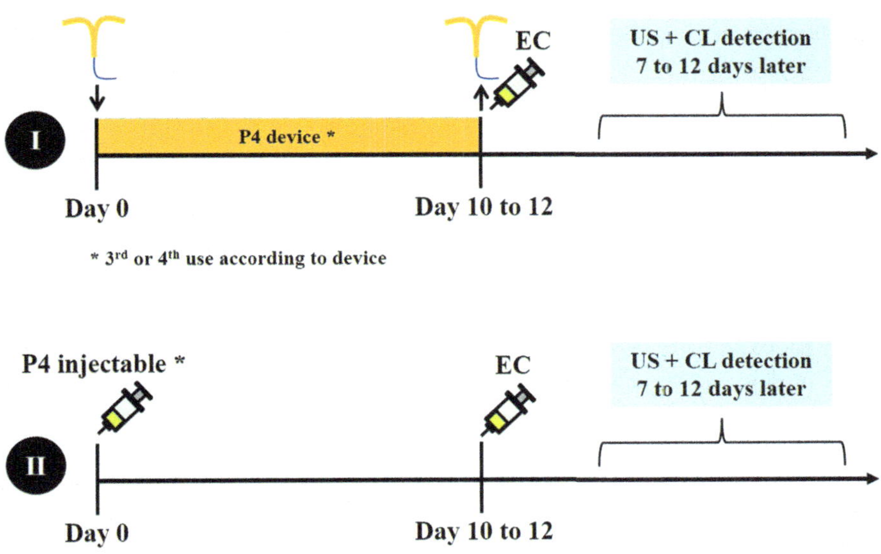

Fig. 1.6 Diagrammatic representation of cyclicity induction protocols in prepubertal heifers. Treatment with progesterone (P4) for 10–12 days [P4 intravaginal device at 3rd or 4th use (I) or P4 injectable (II)]. EC, estradiol cypionate (0.5 mg); * P4 injectable, 150 mg; US, ultrasound evaluation for identification of corpus luteum (CL)

in females treated with P4 and EB (64%) compared to those treated with P4 and EC (67.2%). However, after TAI, greater pregnancy/AI and pregnancy/treated were achieved with EC in relation to the EB (Sá Filho et al. 2015). It is worth highlighting that the authors observed maintenance of P4 concentrations above 1 ng/mL for 7 days in prepubertalNelore heifers in females treated with a new or reused intravaginal device for 24 days.

1.6.1.1 TAI in *Bos indicus* Heifers

TAI programs in heifers have resulted in high reproductive rates similar to those found in cows. However, *Bos indicus* heifers treated with a first-use P4 device (CIDR-B) and/or containing high P4 concentration can reduce the diameter of the preovulatory follicle and interfere with the conception rate for TAI (Dias et al. 2009; Carvalho et al. 2008). This occurs because a high P4 concentration can significantly suppress gonadotropin secretion, specifically LH, impairing the follicular growth rate (Dias et al. 2009; Kinder et al. 1996). Therefore, synchronization of ovulation in heifers is commonly performed with a lower circulating P4 concentration, employing a monodose P4 device, using a reused device (second or third use), by PGF2α anticipation (day 0 or 7), and/or employing eCG gonadotropic stimulation at device removal (Dias et al. 2009; Carvalho et al. 2008).

Although it provides a reduction in endogenous P4 concentration in cows, the reuse of intravaginal P4 devices in heifers increases the efficiency of TAI, reduces protocol costs, and results in similar conception rates to those of cows (Sales et al. 2011b; Peres et al. 2009). Another alternative for *Bos indicus* heifers is the use of norgestomet which does not compromise the follicular growth or pregnancy rate of TAI (Sá Filho et al. 2013, 2015). Figure 1.7 shows the principal hormone protocols for the synchronization of ovulation and TAI protocols for *Bos indicus* heifers.

1.7 Use of Sexed Semen for AI and TAI

The choice of the calf's sex in the livestock system is considered a determining factor in beef and dairy cattle. For example, a male calf has low or no zootechnical value in dairy farms compared to a female calf. However, a male calf has great importance for beef cattle farms due to its high potential production. In this context, sexed semen has been used in AI programs to increase efficiency in production systems and genetic improvement programs for better targeting sex-related characteristics (Baruselli et al. 2007; Weigel 2004).

Sexed semen has been highlighted in cattle breeding due to the benefits generated by its use (Seidel Jr et al. 1997, 1999). Among various advantages, sexed sperm can provide genetic improvements for milk and meat production, favoring expressions related to sex, decreased incidence of dystocia by the birth of female calves weighing around 2 kg less than males, better efficiency in the production of breeders

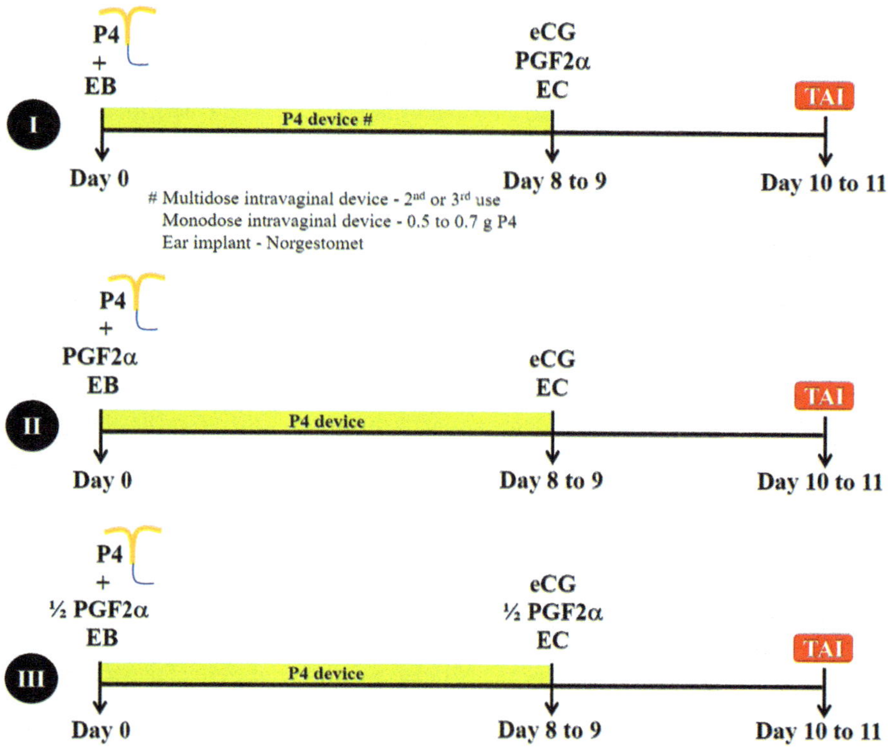

Fig. 1.7 Diagrammatic representation of protocols for the synchronization of ovulation and timed artificial insemination (TAI) in heifers. (I) Treatment with progesterone device (multidose intravaginal, 2nd or 3rd use; monodose intravaginal, 0.5–0.7 g; Ear implant, norgestomet) for 8–9 days. (II) Treatment with progesterone for 8–9 days with PGF2α injection at the beginning of the protocol. (III) Treatment with progesterone for 8–9 days with ½ PGF2α injected on day 0 and ½ PGF2α together with P4 device removal. P4, 0.5 and 0.7 g of progesterone; EB, 2 mg of estradiol benzoate; PGF2α, 12.5 mg of dinoprost or 125–250μg of cloprostenol; eCG, 300–400 IU of equine chorionic gonadotropin; EC, 1 mg of estradiol cypionate

for replacement (heifers or bulls), and increased number of calves when it is desired to increase herd size for production (Seidel Jr. 2007; Cerchiaro et al. 2007). In addition, advantages from the production of desired sex embryos (superovulation and *in vitro* fertilization) can be obtained, and the best performance using sexed semen is obtained with *in vitro* embryo production (Morotti et al. 2014; Pellegrino et al. 2016; Pontes et al. 2010).

Currently, flow cytometry is the most efficient technology for sexing sperm, and this approach is based on DNA differences between X and Y (Seidel Jr. 2007; Seidel Jr. and Garner 2002). In bulls, X sperm contains around 4% more DNA in their chromosome than Y sperm (Garner 2006), and such a difference allows sperm classification with an efficiency close to 95% for determining the desired sex (Morotti et al. 2014; Pontes et al. 2010). In general, the sexing process involves a preparation and adding the DNA-binding dye Hoechst 33342 in the ejaculated (Garner et al. 2013). In the sequence the mix of the ejaculate and DNA dye is analyzed in the flow cytometer that can make over 30,000 consecutive evaluations of individual sperm per second, sorting the ejaculate into three parts: X sperm, Y sperm, and waste (indiscriminate and dead sperm) (Seidel Jr. 2014). Therefore, even with high-speed spermatic sorting, producing sexed sperm straws with a conventional concentration (20 to 50 million sperm) is not viable or economical. For this reason, sexed semen was initially commercialized in straws with approximately 2 million sperm (Schenk et al. 2009; Weigel 2004).

In this context, many studies have proven that there is considerable variation in fertility among bulls (both sexed and non-sexed sperm) and that the use of low fertility bulls and low concentration dose may result in a marked reduction in fertility (DeJarnette et al. 2008, 2011; Andersson et al. 2004). Approximately 10 to 20% of bulls have low fertility with a conventional semen dose; therefore, in this case, a reduction in fertility of more than 10 to 20% is expected after using a dose of two million sperm (Schenk et al. 2009; Seidel Jr. and Schenk 2008a, b). However, there is evidence of a relationship between sexed and non-sexed semen. For example, bull fertility using low doses of non-sexed sperm is usually a good indicator of fertility using a low sexed sperm dose (Seidel Jr. 2014).

A comparison among conception rates in Holstein heifers and cows using sexed semen from different bulls and different insemination doses (2.1, 3.5, and 5.0 million sperm) recorded the best pregnancy rates in nulliparous heifers using the 5 million sperm dose (59.5% pregnancy rate) compared to those of the 2.1 and 3.5 million sperm doses (46.4% and 52.2%, respectively) (DeJarnette et al. 2008). Bull or sperm dose did not affect the conception rate among cows (27.0, 29.1, and 30.3% for the 2.1, 3.5, and 5.0 million sperm doses, respectively). This study indicated that using an increased sperm dose of some bulls could improve the conception rate in nulliparous heifers, whereas neither the bull nor the sexed sperm dose affected the conception rate in cows. Therefore, a high insemination dose is only advantageous for some bulls, and it is recommended to identify ahead of time the maximum bull fertility based on semen quality factors and perform tests with low insemination dose.

In addition, a lower sperm dose, sexed semen has lower fertility also due to a lot of damage caused during the sexing process (diluents, dye, classification pressure, freezing process, etc.). Using good management practice, pregnancy rates in cattle with sexed sperm are approximately 80% of those rates from a conventional semen dose, which represents a reduction of 10 to 20% in pregnancy rates compared to those of the conventional non-sexed semen dose (Butler et al. 2014; Seidel Jr. 2014). There is variation in the conception rates reported in the literature; however, there is a consensus regarding the conception rate of heifers after estrous detection. It is estimated that AI with sexed semen ranges from approximately 70 to 90% of those with conventional semen, depending mainly on the farm management conditions. Conception rates of 40 to 68% were reported for Holstein heifers after insemination with sexed semen and between 67 and 82% using conventional semen. Despite the lower fertility reported for sexed semen, adjustments of time of AI regarding the onset of estrous can improve conception rates (Seidel Jr. et al. 1999).

For AI programs using sexed semen frequentelyfemales are inseminated based on estrous detection after natural estrus or synchronization with PGF2α using two doses 14 days apart. Thus, insemination can be performed about 12 to 24 h after estrous detection (Seidel Jr. and Schenk 2008a, b) or about 80 to 82 h after the second administration of PGF2α (Kurykin et al. 2007). Using heat detectors (radiotelemetry, Heat Watch®) in Jersey heifers, AI with sexed semen was performed at different times considering the onset of estrous (12–16 h, 16–20 h, 20–24 h, and 24–30 h). In this study, higher conception rates were reported for females inseminated 16–24 h after the onset of estrous (average 53.7%) compared those inseminated after 12–16 h (37.7%); however, similar conception rates were observed compared to those of females inseminated 24–30 h after the onset of estrous (45.5%) (Sá Filho et al. 2010a, b). Based on estrous observation, an intravaginal P4 device can also be used from days 0 to 8 in association with PGF2α injection 24 h before device removal (Underwood et al. 2010a). This strategy facilitates the management of synchronization and estrus identification.

Other studies have also reported that using a high number of sexed sperm per insemination after estrous detection (2.1–4.2 million) in two stages with a 12 h interval, GnRH administration at the time of estrous detection (Sá Filho et al. 2010a, b), and the deposition site in the uterus (Kurykin et al. 2007) did not affect the conception rates in dairy heifers. On the other hand, conception rate is affected by number of services to which that female is submitted, first (55.3%a), second (46.1%a), or third services (34.8%b) (Sá Filho et al. 2010a, b), suggesting that fertility following AI with sexed semen tends to decrease with the number of services (DeJarnette et al. 2009). In this context, sexed semen is more appropriate for heifers (Butler et al. 2014; Seidel Jr. 2014), and its use is recommended in the first service postpuberty, followed by the use of conventional semen in the next AI. In addition, this strategy enables maximum fertility using sexed semen; it is possible to reduce dystocia (female calves cause fewer calving problems) and increase the proportion of females calves in heifers and primiparous around 65% (Weigel 2004).

Initially, low conception rates were reported for beef and dairy cattle submitted to TAI with sexed semen, but studies have contributed to promote satisfactory strategies. Heifers inseminated with sexed sperm between 55 and 56 h after CIDR removal and PGF2α injection had a 34% pregnancy rate, whereas those submitted to TAI between 67 and 68 h after CIDR removal presented a 49% pregnancy rate (Seidel Jr. and Schenk 2008a, b). Lactating dairy cows submitted to OvSynch protocol (GnRH, PGF2α, and GnRH) had similar pregnancy rates to TAI using a 10 million sexed sperm dose (43.9%), 2 million sexed sperm dose (40.5%), or 10 million non-sexed control sperm dose (55.6%) (Schenk et al. 2009).

Strategies have been developed to improve the reproductive performance of the herd after using sexed semen. The effect of time for AI (54 h after implant removal/16–18 h before ovulation versus 60 h after implant removal/10–12 h before ovulation) and semen [conventional (40 million sperm) versus sexed (2 million sperm)] were studied in 389 beef cows (*Bos indicus*) 30–60 days postpartum. In this study, conception rates were similar between TAI using sexed semen at 54 h (48.4%) and 60 h (55.1%) after norgestomet implant removal and among conventional

(58.9%), sexed X (52.0%), and sexed Y semen (49.0%). However, 6 hours delay in TAI increased by 9% the conception rate of animals inseminated with sexed semen [54 h (37.4%) versus 60 h (46.4%)]. These data suggest that the most appropriate time to perform TAI with sexed semen is 60 h after P4 device removal (10–12 h before ovulation) (Souza et al. 2008). The same experimental design was tested in dairy cattle (*Bos taurus*) using ovulation synchronization with CIDR-B insertion, 2 mg of EB and PGF2α on day 0; CIDR-B removal and PGF2α on day 8; 1 mg of EB on day 9; and TAI on day 10 (54 or 60 h after CIDR-B removal). Similar results were obtained, demonstrating that a 6 h delay in TAI with sexed semen also increased conception rates in dairy cattle (Sales et al. 2011a).

These increased conception rates observed when using sexed semen after 54 or 60 h of device removal or 10–12 h before ovulation is justifiable because sexed sperm require less time for capacitation due to the flow cytometry process (Lu and Seidel Jr. 2004). In addition, ultrasonography evaluation to determine the diameter of the dominant follicle at the time of TAI with sexed semen has improved pregnancy rates. Using sexed semen for TAI, an effect of the type of semen [conventional (45.4%) versus sexed (54.2%)] and of the diameter of the dominant follicle at TAI [≥ 11 mm (57.9%) versus <11 mm (44.1%)] was observed on the pregnancy rates (Sá Filho et al. 2012).

Currently, sexing sperm has achieved important advances and process has been performed based on next-generation technologies (SexedULTRATM - Sexing Technologies, Navasota, TX), which provides sex-sorted semen that is commercially available for dairy and beef cattle. This sex-sorted semen is presented at a concentration of 4×10^6 spermatozoa per straw. In addition, this semen includes adjustments to the composition of the medium that include the prestaining seminal treatment, modifications in the staining medium itself, and freezing extenders, which contribute to greater balance and pH maintenance for prolonged times (Vishwanath and Moreno 2018; Thomas et al. 2017; de Graaf et al. 2014).

Sex-sorting from next-generation technologies was compared with the different methodologies of sex-sorted semen in Nelore cattle submitted to the TAI protocol. For all cows, the TAI protocol was similar as demonstrated in Fig. 1.8. In this study, the pregnancy rate for TAI ($P < 0.0001$) is promising for the current sexing methodology used (Table 1.4) (Baruselli et al. 2017).

Another study conducted in Brazil (Marques et al. 2018) analyzed the conception rates after TAI with sex-ultra semen in suckled Nelore cows (n = 281). In addition to testing different doses (4, 6, and 8 × 106 sperm from a single Nelore bull), this study evaluated the rational use of semen according to estrus behavior (sex-sorted to cows in estrus and non-sorted to cows with no estrus expression) after P4 removal (Fig. 1.8). The conception rate was similar among the cows inseminated with 4 (59.5%, 47/79), 6 (58.4%, 45/77), and 8 x 106 sperm (51 9%, 40/77). Cows in estrus showed 56.7% (123/233) and those not in estrus showed 33.3% (16/48) conception rates. Interestingly, this study highlights a strategy with high efficiency for the use of sex-sorted semen in commercial field conditions. It emphasizes the positive results achieved (pregnancy rate higher than 50%), which can be attributed

Table 1.4 Pregnancy rate TAI in Nelore cows inseminated with conventional semen or different methodologies of sex-sorted semen

Nelore bulls	Methodology	Pregnancy rate % (n/N)
Conventional semen (20×10^6sptz)	Frozen semen without sexing	52.0[a] (112/199)
Sex-sorted (2.1×10^6sptz)	Previous sexing methodology	28.2[c] (58/206)
Sex-ultra (2.1×10^6sptz)	Current sexing methodology	37.6[b, c] (72/191)
Sex-ultra (4×10^6sptz)	Current sexing methodology with enhanced concentration	43.0[b] (86/200)

a–b: indicates statistical difference among different methodologies of sex-sorted semen
Adapted from Baruselli et al. (2017)

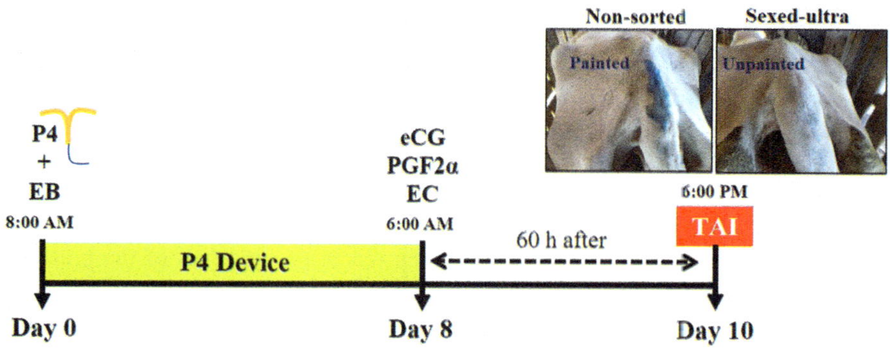

Fig. 1.8 Diagrammatic representation of the TAI protocol performed in Nelore cows inseminated with sexed-ultra (cows unpainted at the base of the tail/estrus expression) or conventional semen/non-sorted (cows painted at the base of the tail/no estrus expression). Treatment based on P4-estradiol and insemination performed 60 h after removal of the P4 device. P4, progesterone (0.6 g); EB, estradiol benzoate (2 mg); PGF2α, cloprostenol (0.526 mg); eCG, equine chorionic gonadotropin (300 IU); EC, estradiol cypionate (1 mg). (Adaptations from Marques et al. (2018))

to the new methodology of sexed semen and to the fact that this semen was only used in cows that expressed estrus, as indicated by unpainted tails.

Another important consideration about sexed semen is related to the time when sexing is performed based on freezing. Conventionally, sexed semen preparation begins with semen collection from bulls followed by sperm sex sorting (X and Y chromosomes) and freezing. Currently also is possible to obtaining sexed sperm from previously frozen doses (reverse-sorted semen, RSS), which represents a breakthrough in livestock management due to the possibility of its use in association with other biotechnologies. In addition, there is no need for the bull to be close to the sexing laboratory and allows for the sexing of freezing semen from bulls that have died. In this context, this technique can also be used as an alternative in AI (Underwood et al. 2010b) or IVF programs (Morotti et al. 2014) to preselect the sex of the offspring produced.

1.8 Resynchronization Program

Resynchronization is a reproductive strategy that aims to concentrate the breeding period from females that failed to become pregnant after receiving the AI procedure. Using resynchronization, non-pregnant females are quickly diagnosed and inseminated in the shortest possible time, reducing the service period and making the breeding season shorter. This strategy is very interesting because it is associated with a higher pregnancy rate from AI, reduction in the number of bulls to natural breeding after TAI or AI program, in addition to several benefits associated with the management of calves (Baruselli et al. 2018; Marques et al. 2015).

Insemination is a worldwide established biotechnology that provides economic returns by improving offspring production rates. However, this return is closely linked to strategies that can be associated with using this technique. Among the main factors that affect the productive performance of calves until weaning, there is the interaction between the month of birth and geographic location, such as variation in the calf weight at weaning. For example, in cattle from tropical countries Nelore calves born from August to October had high weaning weights (Bocchi and Albuquerque 2004), possibly due to a higher concentration of calving in the late winter and early spring after conception during months of high food availability. Therefore, a short breeding season aims to improve progeny performance. Furthermore, calves whose conceptions occurred in the first half of the breeding season had increased production rates, such as earlier slaughter, earlier reproductive life, and their mothers had a higher pregnancy rate at the end of the breeding season (Carneiro et al. 2012).

1.8.1 Estrus Resynchronization

A wide variety of hormonal treatments can be used for synchronization of estrus and ovulation in cattle, and this strategy allows for inseminating all treated animals, resulting in conception rates of approximately 50% and a service rate of 100% (Baruselli et al. 2004a, b, 2018). Using an ovulation synchronization protocol, synchronous estrus return is expected in those females who failed to conceive after the first TAI. Commonly, this estrus behavior is concentrated between 18 and 23 days later, and reproductive practices can be strategically combined at this time (Baruselli et al. 2004a, b; Cavalieri et al. 2004). In this estrus return, conventional AI can be used, or bulls can be rationally allocated together with cows for natural breeding from 10 to 15 days after TAI (Campos et al. 2013). This management results in high conception rates at the end of the breeding season (Torres-Júnior et al. 2009). Generally, visual estrous detection in *Bos indicus* cows results in low service rates (25%) due to a short estrous duration (<12 hours) and high incidence of estrous occurring at night (30%) in Nelore cows (Pinheiro et al. 1998; Galina et al. 1996). Furthermore, postpartum anestrous usually has a high incidence in *Bos indicus* cattle, especially if kept in regions with seasonal forages (Ayres et al. 2008). Therefore, difficulty in estrous detection in nonpregnant cows is a challenge for improving the number of pregnancies following AI because fewer than half of nonpregnant cows have become inseminated again (Campos et al. 2013).

The TAI program followed by estrous detection and subsequent AI usually results in 60 to 65% pregnancy rate in the first 45 days of the breeding season compared to 30% after natural mating and 20% after conventional AI. Failure in inseminating nonpregnant cows in a short period results in prolonged intervals among services and in delayed pregnancies (Baruselli et al. 2018; Marques et al. 2015; Campos et al. 2013; Galvao et al. 2007; Bartolome et al. 2005; El-Zarkouny, 2004). In this context, ovulation resynchronization is a strategy to increase the number of pregnancies in the first half of the breeding season. Therefore, the aim of ovulation resynchronization is to eliminate the estrous observation, increase the number of calves born by AI, and reduce the number of bulls needed.

1.8.2 Main Strategies for Ovulation Resynchronization

In general, hormonal treatments for ovulation resynchronization in cattle are similar to those previously described for the TAI protocol. One of the most common treatments for resynchronization involves the use of an intravaginal P4 device and EB, combined with application of PGF2α, EC and/or GnRH (Sá Filho et al. 2014; Campos et al. 2013; Galvao et al. 2007; El-Zarkouny, 2004). This hormonal combination allows for synchronization of the emergence of a new wave of follicular growth in nonpregnant females. The ovulation resynchronization can be started after confirming that the female is not pregnant (traditional resynch/30 days after 1st TAI) (Marques et al. 2015) or before even knowing the female's gestational status (early resynch/22–23 days after 1st TAI and super-early resynch/14 days after 1st TAI) (Baruselli et al. 2018; Sá Filho et al. 2014; Campos et al. 2013).

Although TAI has shown satisfactory results, reductions of 15 to 20% in the conception rate are reported for cattle with high anestrous rate (Ayres et al. 2008; Baruselli et al. 2004a, b). In this context, administration of eCG in anestrous suckled cows has been recommended to provide pregnancy rates of approximately 50%, which is similar to those observed in cyclic cows (Barreiros et al. 2014; Kastelic et al. 1999). Therefore, the use of gonadotropic support (equine chorionic gonadotropin) or temporary calf removal (48–72 h) has been recommended at this time to final growth of the dominant follicle in resynchronization protocols of postpartum cattle (Campos et al. 2013).

Resynchronization 30 days (traditional resynch) was one of the first resynchronization strategies used, which consists in performed a pregnancy diagnosis from 28 to 32 days after the first TAI (Fig. 1.9). Therefore, only nonpregnant females are submitted to the second hormonal protocol, resulting in a 40-day interval between TAI (Marques et al. 2015; Bartolome et al. 2005). This resynchronization results in

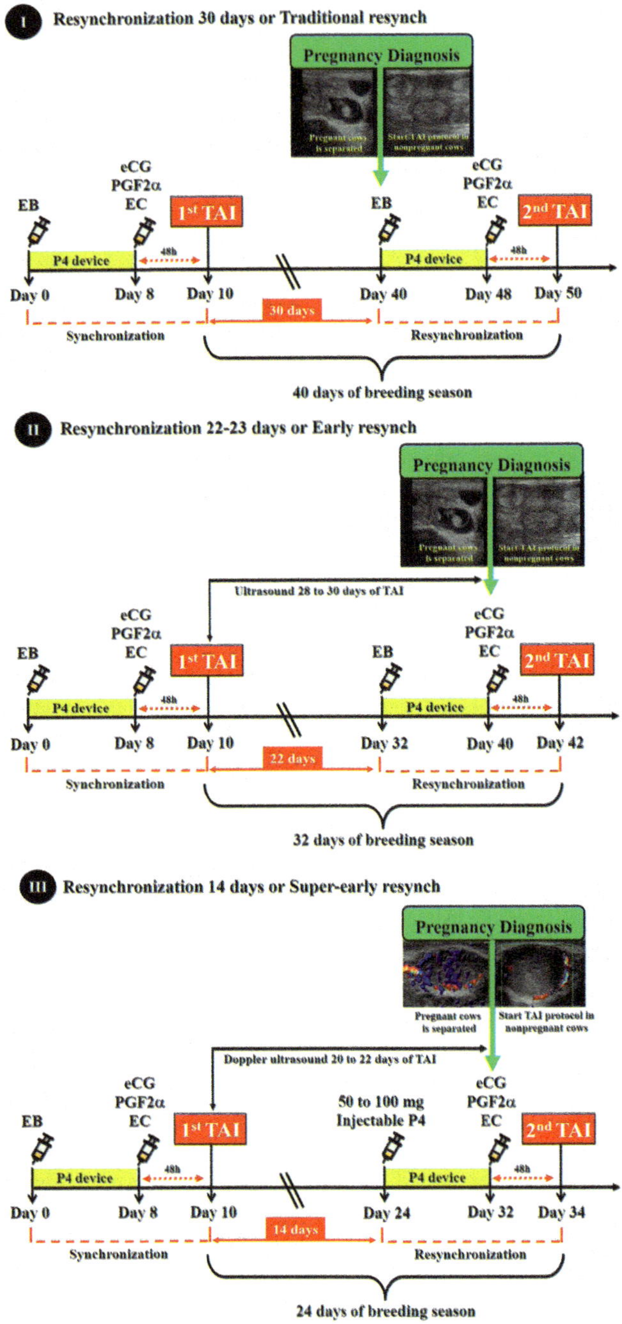

Fig. 1.9 Schematic representation of the protocol used for timed artificial insemination (TAI) and resynchronization of ovulation in nonpregnant cattle. I) Resynchronization 30 days/traditional resynch; II) Resynchronization 22–23 days/early resynch; III) Resynchronization 14 days/super-early resynch. *P4* progesterone (0.5–1.9 g), *EB* estradiol benzoate (2 mg), *EC* estradiol cypionate (0.5 mg for heifers and 1 mg for cows), *PGF2α* prostaglandin (125–500μg), *eCG* equine chorionic gonadotropin (300–400 IU)

a conception rate of 50 to 60% for each TAI, and the final cumulative pregnancy rate can be over 80%.

Another possibility for ovulation resynchronization (early resynch/22–23 days after 1st TAI) is to start a second hormonal protocol 22–23 days after TAI without a previous pregnancy diagnosis. Therefore, all females initially treated for 1st TAI protocol receive a P4 intravaginal device and EB injection between 22 to 23 days regardless of gestational status (Sá Filho et al. 2014; Campos et al. 2013; Chebel et al. 2003). Upon P4 device removal (8–9 days from insertion), the pregnancy ultrasound evaluation is performed, and nonpregnant females are injected with PGF2α and an ovulation inducer, such as EC (Fig. 1.9).

The resynchronization program 14 days after 1st TAI (super-early resynch) anticipates insemination at 8 days in relation to resynch 22–23 and 16 days in relation to resynch 30, making the breeding season extremely short and a 14-day interval between TAIs. This resynch is initiated 14 days after the first insemination, being that all females receive a combination of treatments with intravaginal P4 (device) and 50–100 mg of injectable P4, IM. Then, on day 22, the diagnosis of pregnancy is made by assessing CL vascularization with color Doppler. Cows with low or absent vascularization are considered nonpregnant and continue resynchronization treatment with device removal, PGF2α, EC, and eCG, receiving TAI 48 h later (day 24). Cows with moderate CL or strong vascularity are considered pregnant and do not receive any hormonal treatment except reassessment with B-mode ultrasound (Baruselli et al. 2018).

The use of color Doppler ultrasonography to characterize the CL vascularization is essential for classifying female's gestational status (Hassan et al. 2018). After the luteolysis process (approximately 14 to 17 days of the estrous cycle), the pregnant female maintains high vascularization in the CL, and therefore, the Doppler evaluation allows an indirect assessment of gestational status from the CL vascularization score. This technique allows for a much earlier pregnancy diagnosis than B-mode ultrasound identification. Nevertheless, females with vascularized CL need to be reevaluated with B-mode ultrasound 8 to 10 days later to visualize the gestational vesicle and detect eventual false positives (females classified as pregnant by the CL vascularization score but without a gestational vesicle) (Baruselli et al. 2017; Pugliesi et al. 2017; Siqueira et al. 2013).

Reproductive programs that allow insemination every 21 days would be a more ideal because they would have a 100% (21/21) service rate. In this context, resynch at 14 days is the strategy that most closely approximates this ideal service rate, being 87.5% (24/21) to resynch 14, 66% (32/21) to resynch 22–23, and 52.5% (40/21) to resynch 30 (Baruselli et al. 2018). However, the implementation of super-early resynch is more laborious and, for success, requires an ultrasound with Doppler function in addition to high professional experience.

1.8.3 Success Rate Using Resynchronization

Satisfactory results are achieved with resynchronization programs in cattle, which results with pregnancy rates of approximately 50% after each TAI. These strategies result in pregnancy rates of 80 to 90% during days 24 to 40 of the breeding season (according to resynch), contributing to greater genetic gain of the herd (more pregnant from AI) besides reducing the number of bulls on the farm. In some situations, natural breeding (use of bulls) may be associated with resynchronization for 10–15 days after the 2nd or 3rd TAI until the end of the breeding season (Baruselli et al. 2018; Marques et al. 2015).

On the other hand, there is variation in conception according to the number of calving. Primiparous and secundiparous cows showed a reduction in the conception rate from the 1st to 2nd TAI (20%) compared to heifers and multiparous cows. This reduction is possibly associated with nutritional limitations and not the category's own fertility. Conception rates among different categories of Nelore females were different (Fig. 1.10) in a study evaluating reproductive performance (Marques et al. 2015). At the end of the breeding season, heifers had a higher conception rate (85%) than those of primiparous (76%) and multiparous cows (78%). In this study, all animals received a source P4 (intravaginal device for the cows or ear implants for heifers) in combination with 2 mg of EB on day 0. All females received 250μg of cloprostenol, 300 IU of eCG, EC (1 mg to cows and 0.5 mg to heifers), and P4 removal on the eighth day. TAI was performed 48 hours after P4 removal. Thirty days after the 1st TAI, all females were evaluated by ultrasonography, and the nonpregnant bovines were resynchronized with the same hormonal treatment. The pregnancy rate was assessed by ultrasound 30 days after TAI.

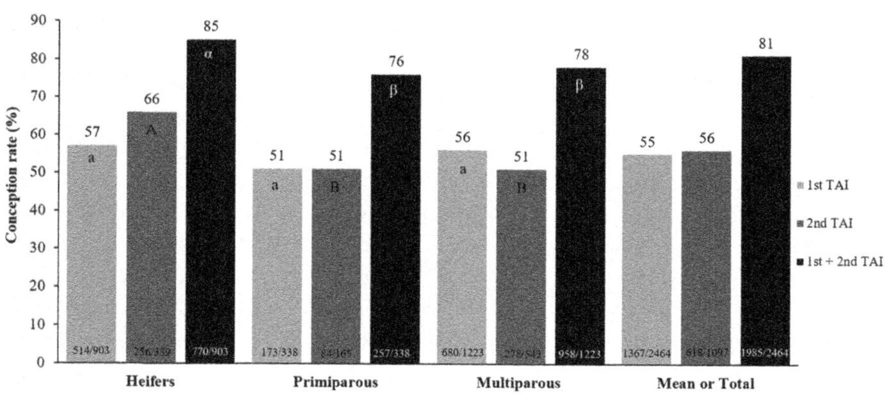

Fig. 1.10 Conception rates among different categories of Nelore females after the timed artificial insemination (1st TAI) and ovulation resynchronization (2nd TAI). (Adapted from Marques et al. (2015))
a-b: indicates statistical difference for categories in 1st TAI; A-B: indicates statistical difference for categories in 2nd TAI; and α-β: indicates statistical difference for categories in 1st and 2nd TAI

Early resynch anticipates the 2nd TAI in 8–9 days, but both pregnant and non-pregnant cows receive a P4 device and 1–2 mg of EB between 19 and 23 days after the first TAI. In addition, to promote follicular atresia and synchronize follicular growth, estradiol esters can induce luteolysis (Vieira et al. 2014; Pinheiro et al. 1998). However, studies with *Bostaurus* (El-Zarkouny 2004; Chebel et al. 2003) and *Bos indicus* (Sá Filho et al. 2014; Campos et al. 2013) females reported no interruption of pregnancies obtained in the 1st TAI using GnRH or EB in the early resynchronization. There was no reduction in the conception rates in Nelore cows treated (54%) or untreated (47.8%) with 2 mg of EB 23 days after TAI (Campos et al. 2013). In addition, pregnancy losses were similar in Nelore females treated with 1.0 mg of EB (4.1%) compared to untreated females (2.0%) and maintained with bulls for natural mating (Sá Filho et al. 2014). Conception rates are similar to those obtained with resynch 30 days, as shown previously, resulting in approximately 50% in the 2nd TAI and a total of 75% of conception in 32–33 days of service (Sá Filho et al. 2014; Campos et al. 2013). In this resynch, variations in conception rates were also observed according to number of calving. Nonsuckled cows, heifers, and postpartum cows have a decrease of 20% in conception rates. Furthermore, primiparous cows showed a similar conception rate (40%) in two subsequent (Marques et al. 2015; Crepaldi et al. 2014; Sá Filho et al. 2014). In cows that did not become pregnant after the 1st TAI, a 2nd TAI (resynchronization of ovulation) may be performed 32 days later, and a 3rd TAI may be repeated, resulting in a total conception rate of 90% after three services in 64 days of a breeding season (Baruselli et al. 2018).

A similar pregnancy rate using resynch 22 versus resynch 14 was observed in the 1st synchronization (48% versus 53%) and resynchronization (56% versus 51%), respectively. Cumulative pregnancy rates after 32 and 24 days of the breeding season were also similar for resynch 22 (77%, 97/126) and resynch 14 (75%, 89/118). However, resynch 14 improved the service rate every 21 days with 66% for resynch 22–23 and 87.5% for resynch 14 (Baruselli et al. 2017).

In general, reduced conception rates are observed in the 2nd and 3rd TAI. In addition to the number of calving, it is necessary to consider other factors, such as BCS and health status of the herd. Reduction in the period of service for 62 to 80 days using three TAI provides a short calving interval and a considerable improvement in reproductive efficiency (with an average calving interval of less than 12 months). Pregnancy rates between 75 and 90% were reported for cattle submitted to resynchronization, reducing the numbers of bulls needed for natural matting or even not requiring them at all.

1.9 Dominant Follicle Manipulation During the TAI Protocol

The size of the dominant follicle at the time of TAI has been extensively investigated due to its influence on the reproductive behavior and performance of animals submitted to synchronization of ovulation. Although the TAI protocol aims for ovulation synchronization and does not require estrus detection, many studies have investigated the relationship of this behavior, diameter of dominant follicle and the fertility of animals undergoing TAI programs (Moraes et al. 2019; Morotti et al. 2013a, b, 2018a; Pfeifer et al. 2012, 2015; Sales et al. 2012a, b; Sá Filho et al. 2010a, b, 2011). In this context, dominant follicle size at the time of insemination has been widely studied in which the largest follicle diameter at TAI is positively associated with higher estrus expression, higher ovulation rate, larger size of CL, higher concentration of P4, and higher pregnancy probability (Moraes et al. 2019; Morotti et al. 2018a; Pfeifer et al. 2015).

1.9.1 Estrus Expression and Fertility in TAI

Considering that the follicular diameters may differ at the time of TAI and that this may determine a higher or less ovulatory potential according to follicle size (Gimenes et al. 2008), block TAI has been suggested as a possibility to improve fertility of females undergoing a TAI program. Block TAI can be applied as a strategy to increase the pregnancy rate in Nelore cows (Pfeifer et al. 2015). On day 10 of the TAI protocol, the blocks are divided according to the size of the dominant follicle; in this way, TAI is performed at different times in each group (Table 1.5). Performing TAI in a block of Nelore cows increased the pregnancy rate by 16.7% (65.5%; 129/203) compared to the group subjected to conventional TAI on day 10 (48.8%; 102/209; Pfeifer et al. 2015).

For better understanding of the relationship of dominant follicle size at TAI, estrus expression, and fertility, an estrus score can be performed based on the painting intensity remaining at the base of the tail. Normally, at the removal of the P4 device, all animals have the sacral and tail base region painted with marker stick, and at the time of TAI, the animals are classified into different scores (Nogueira et al. 2019). According to the proportion of paint removed, animals can be classified into (Fig. 1.11): score I, no or low paint removal (no estrus expression), score II, partial paint removal (up to 75%; low estrus expression), and score III, complete or > 75% paint removal (high estrus expression). This practical strategy allows for usefully identifying cows with greater estrus expression and consequently

Table 1.5 Pregnancy rate in cows inseminated either in blocks or conventionally according to the diameter of the dominant follicle at timed artificial insemination (TAI)

Group	Follicular diameter#	TAI timing	Pregnancy rate* % (n/N)
Control	Conventional TAI	48 h after P4 removal	48.8 (102/209)
Block TAI	> 15 mm	0 h	63.5 (129/203)
	13.0–14.9 mm	6 h later	
	10.1–12.9 mm	24 h later	
	≤ 10 mm	30 h later	

#Diameter of dominant follicle 48 h after P4 device removal (day 10)
* P-value <0.01
Adapted from Pfeifer et al. (2015)

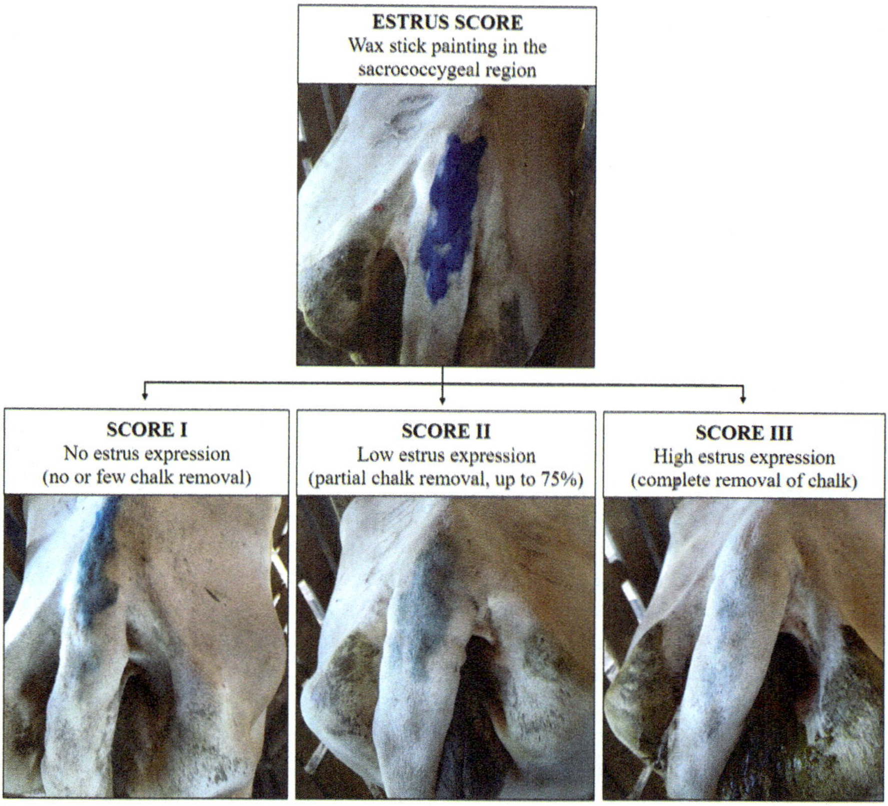

Fig. 1.11 Scheme of the estrus expression intensity score during timed artificial insemination (TAI) using wax stick painting in the sacrococcygeal region. Commonly, paint management is performed at the time of removal of the intrvaginal progesterone device and the estrus score classification performed immediately before insemination. Cows with high estrus expression are conventionally inseminated and low estrus expression or no estrus expression receive insemination together with GnRH analogue application to potentialize the occurrence of ovulation

improving the pregnancy rates in TAI (score I, 40.0%; score II, 49.7%; and score III, 60.9%), allowing the cows with low score to be targeted for additional treatments aimed at improving pregnancy.

Even using TAI, the classification of the estrus score is interesting because scores I and II present lower reproductive performance. Based on this it was suggested application of 100μg of gonadorelin (GnRH group; n = 470) against 1 mL of saline (control group; n = 467; Rodrigues et al. 2019) to animals with I and II scores. Cows with score III (estrus group; n = 1347) received no additional treatment. The pregnancy rate was greater in the estrus group (57.09%; 769/1347) than in the control (36.18%, 169/467) and GnRH groups (45.95%, 216/470). However, GnRH injection increased pregnancy by approximately 10% in relation to the control. Therefore, using P4-estradiol-based TAI protocols, estrus expression can be efficiently monitored with painting on the sacrococcygeal region. Thus, GnRH application in cows with a low estrus score is a simple strategy that can increase the pregnancy rate in cattle.

1.9.2 Antral Follicle Count and Fertility in TAI

The antral follicle count (AFC) is a highly variable reproductive characteristic among the females of a herd, but with very high repeatability throughout the productive female life (Lima et al. 2020; Moraes et al. 2019; Jimenez-Krassel et al. 2017; Burns et al. 2005). For this reason, the AFC has become an interesting reproductive tool since it is closely related to the efficiency of reproductive biotechniques, in addition to the great possibility of affecting female fertility (Garcia et al. 2020; Morotti et al. 2018a, b). In general, AFC can be determined in a simple way through the ultrasound exam to quantify the number of follicles ≥3 mm in diameter and usually is performed on a random day of the estrous cycle of the cows. Therefore, considering the AFC variability, cows can be classified as low, intermediate, or high AFC according to the number of antral follicles present in the ovary during ultrasound evaluation (Silva-Santos et al. 2014; Burns et al. 2005).

The evaluation of the AFC is inserted in a very current context of the selection of females and that seems to be better established for the production of embryos (Garcia et al. 2020; Seneda et al. 2019; Santos et al. 2016; Silva-Santos et al. 2014). For example, for both *in vivo* production and *in vitro* production, the selection of donors with high AFC is positively associated with better reproductive performance in these biotechniques (Zangirolamo et al. 2018; Santos et al. 2016). However, relation between AFC and efficiency of the TAI program is not yet fully understood (Morotti et al. 2018a, b). In taurine cattle there are studies that show that AFC does not exert any influence, or it has been observed that high AFC is responsible for better reproductive performance (Evans et al. 2012; Ireland et al. 2010). However, the authors of this chapter have conducted a series of studies on this subject in zebuine cattle which reveal that low AFC appears to have better reproductive performance when subjected to TAI.

In this context, evaluating the ovarian follicular dynamics and fertility of Nelore cows submitted to TAI programs, Morotti et al. (2018a) revealed higher follicular diameters (Fig. 1.12) and pregnancy rates for females with low AFC compared to those with high counts (Table 1.6). In addition to these findings, other studies have also revealed an interaction of AFC with body condition score in *Bosindicus* (Moraes et al. 2019) and higher reproductive longevity and greater reproductive performance in *Bos taurus* with low AFC (Jimenez-Krassel et al. 2017).

In a study on the relationship between AFC and reproductive performance, and the pattern of expression of important genes for various cellular functions in *Bos indicus* cattle (Lima et al. 2020) showed promising data with positive influence of low AFC on fertility. In this study we found that very low AFC in Nelore cows resulted in a large dominant follicle diameter, a tendency to have higher P4 concentration and greater pregnancy rate in TAI program. In addition, Nelore heifers with

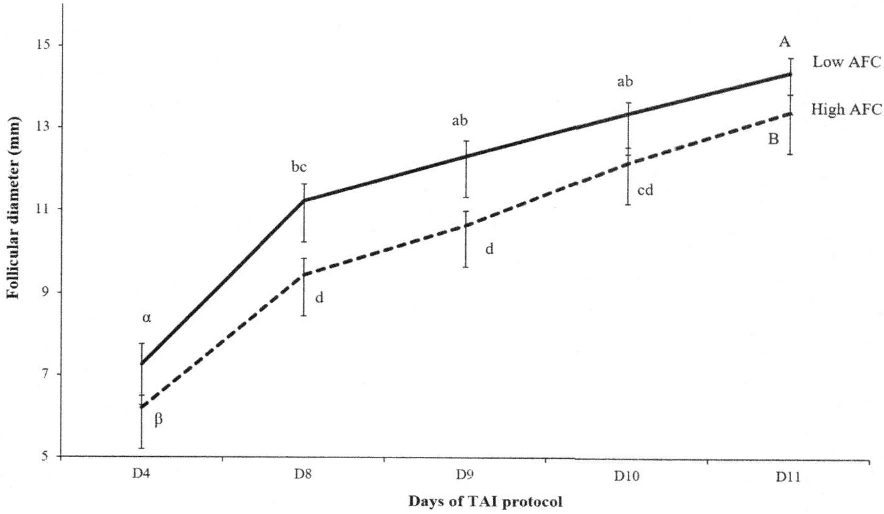

Fig. 1.12 Diameter of dominant follicle (mm) during ovarian follicular dynamics in Nelore cows with consistently high (≥45 follicles; dashed line) or low (≤15 follicles; continuous line) antral follicle count synchronized with a taimedartifical insemination (TAI) protocol. Values denoted using different Greek letters (α-β; punctual evaluation on day 4), lowercase letters (a-d; evaluations with 24-h intervals) or capital letters (A-B; evaluations with 12-h intervals) were different ($P < 0.05$). (Adapted from Morotti et al. (2018a))

Table 1.6 Mean (M) and standard deviation (SD) of the antral follicle count (AFC) and conception rates from Nelore cattle with consistently high, intermediate, or low AFCs following a TAI protocol in two different studies

Studies	AFC groups	AFC Mean ± SD	Conception rate % (n)
Morotti et al. (2018a)	Low (≤15 follicles)	11.3 ± 2.8^c	61.7^a (150/243)
	Intermediate (20–40 follicles)	29.4 ± 6.0^b	52.9^b (210/397)
	High (≥45 follicles)	52.8 ± 7.7^a	49.5^b (96/194)
	P-value	0.001	0.027
Moraes et al. (2019)	Low (≤10 follicles)	7.7 ± 2.6^c	57.7^a (176/305)
	Intermediate (11–29 follicles)	18.0 ± 3.4^b	49.7^b (155/312)
	High (≥30 follicles)	38.0 ± 12.4^a	47.9^b (57/119)
	P-value	<0.0001	0.008

Values with different superscripted letters (a, b) were different ($P \leq 0.05$) between the AFCs and conception rates

low AFC exhibited oocytes and cumulus cells with a better expression patterns of genes linked to intercellular communication, meiotic control, epigenetic modulation, adaptation and cellular stress response and follicular growth. However, studies on the ovarian follicular population are current, it is a subject that has been widely investigated, and many aspects are not yet fully understood.

1.10 Considerations

Currently, reproductive biotechniques have achieved great technological advances that have contributed greatly to increasing the development of livestock. Artificial insemination certainly represents one of the most popular assisted reproductive techniques that contributes significantly to genetic improvementis, being easy to apply and has been considered one of the greatest potentials for expansion in cattle. In this context, many strategies have been developed to stimulate the use of artificial insemination programs. Alternatives to pharmacological control of the estrous cycle, the use of timed artificial insemination, insemination with sexed semen, ovulation resynchronization, and dominant follicle manipulation, are highly effective strategies that are currently indicated as reproductive practices in bee and dairy cattle. Finally, maintaining these advances and utilizing the strategies discussed herein are great challenges but are necessary to increase the productive and reproductive efficiency of livestock.

References

Adams GP, Jaiswal R, Singh J, Malhi P (2008) Progress in understanding ovarian follicular dynamics in cattle. Theriogenology 69:72–80

Alfieri AA, Leme RA, Agnol AMD, Alfieri AF (2019) Sanitary program to reduce embryonic mortality associated with infectious diseases in cattle. Anim Reprod 16(3):386–393

Allrich RD (1994) Endocrine and neural control of estrus in dairy cows. J Dairy Sci 77:2738–2744

Amstalden M, Alves BR, Liu S, Cardoso RC, Williams GL (2011) Neuroendocrine pathways mediating nutritional acceleration of puberty: insights from ruminant models. Front Endocrinol 2:109

Anderson ML (2007) Infectious causes of bovine abortion during mid- to late-gestation. Theriogenology 68:474–486

Andersen KJ, Lefever DG, Brinks JS, Odde KG (1991) The use of reproductive tract scoring in beef heifers. Agri-Practice 12:19–26

Andersson M, Taponen J, Koskinen E, Dahlbom M (2004) Effect of insemination with doses of 2 or 15 million frozen-thawed spermatozoa and semen deposition site on pregnancy rate in dairy cows. Theriogenology 61:1583–1588

Atkins JA, Pohler KG, Smith MF (2013) Physiology and endocrinology of puberty in heifers. Vet Clin N Am Food Anim Pract 29:479–492

Ayres H, Martins CM, Ferreira RM, Mello JE, Dominguez JH, Souza AH, Valentin R, Santos IC, Baruselli PS (2008) Effect of timing of estradiol benzoate administration upon synchronization of ovulation in suckling nelore cows (*Bos indicus*) treated with a progesterone-releasing intravaginal device. Anim Reprod Sci 109:77–87

Baishya N, Ball PJH, Leaver JD, Pope GS (1980) Fertility of lactating dairy cows inseminated after treatment with cloprostenol. Br Vet J 136:227–239

Barreiros TR, Blaschi W, Santos GM, Morotti F, Andrade ER, Baruselli PS, Seneda MM (2014) D dynamics of follicular growth and progesterone concentrations in cyclic and anestrous suckling Nelore cows (*Bos indicus*) treated with progesterone, equine chorionic gonadotropin, or temporary calf removal. Theriogenology 81:651–656

Bartolome JA, Sozzi A, Mchale J, Melendez P, Arteche AC, Silvestre FT, Kelbert D, Swift K, Archbald LF, Thatcher WW (2005) Resynchronization of ovulation and timed insemination in

lactating dairy cows, II: assigning protocols according to stages of the estrous cycle, or presence of ovarian cysts or anestrus. Theriogenology 63:1628–1642
Baruselli PS, Teixeira AA (2018) Inseminação Artificial em Tempo Fixo (IATF) em Bovinos. In: Morani ESC, Rodrigues LH, Rancoletta M (eds) Manual de Reprodução nas espécies domésticas. Editora MedVet, pp 147–166
Baruselli PS, Reis EL, Marques MO, Nasser LF, Bó GA (2004a) The use of hormonal treatments to improve reproductive performance of anestrous beef cattle in tropical climates. Anim Reprod Sci 82:479–486
Baruselli PS, Marques MO, Reis EL, Carvalho NAT, Carvalho JB (2004b) Manipulação Hormonal do estro e da ovulação. Curso à Distância de Manipulação do Ciclo Estral em Bovinos de Corte (Módulo 3), 1
Baruselli PS, Martins SAH, Gimenes CM, Sales LU, Ayres JNS, Andrade H, Raphael AFC, Arruda RP (2007) Sexed semen: artificial insemination and embryo transfer. Revista Brasileira de Reprodução Animal 31:374–381
Baruselli PS, Ferreira RM, Sales JNS, Gimenes LU, Sá Filho MF, Martins CM,..... Bó GA, et al (2011) Timed embryo transfer programs for management of donor and recipient cattle. Theriogenology 76(9):1583–1593
Baruselli PS, Ferreira RM, Colli MHA, Elliff FM (2017) Timed artificial insemination: current challenges and recent advances in reproductive efficiency in beef and dairy herds in Brazil. Anim Reprod 14:558–571
Baruselli PS, Ferreira RM, Sá Filho MF, Bó GA (2018) Review: using artificial insemination v. natural service in beef herds. Animal 12:45–52
Baryczka A, Barański W, Nowicki A, Zduńczyk S, Janowski T (2018) Effect of single treatment with cloprostenol or dinoprost on estrus and reproductive performance in anestrous dairy cows after service. Pol J Vet Sci 21:383–387
Bergfeld EG, Kojima FN, Cupp AS, Wehrman ME, Peters KE, Garcia-Winder M, Kinder JE (1994) Ovarian follicular development in prepubertal heifers is influenced by level of dietary energy intake. Biol Reprod 51:1051–1057
Binelli M, Sartori R, Vasconcelos JLM, Monteiro Jr. PLJ, Pereira MHCV, Ramos RS (2014) Fertility of fixed-time protocols in dairy cattle evolution in fixed-time: from synchronization of ovulation to improved fertility, 494–506
Bó GA, Baruselli PS (2014) Synchronization of ovulation and fixed-time artificial insemination in beef cattle. Animal 8:144–150
Bo GA, Adams GP, Caccia M, Martinez M, Pierson RA, Mapletoft RJ (1995) Ovarian follicular wave emergence after treatment with progestogen and estradiol in cattle. Anim Reprod Sci 39:193–204
Bó GA, Cutaia L, Tribulo R (2002) Tratamientos hormonales para inseminación artificial a tiempo fijo en bovinos para carne: algunas experiencias realizadas en Argentina. Segunda Parte. Taurus 15:17–32
Bó GA, Baruselli PS, Martinez MF (2003) Pattern and manipulation of follicular development in *Bos indicus* cattle. Anim Reprod Sci 78:307–326
Bocchi ALTRA, Albuquerque LG (2004) I Idade da vaca e mês de nascimento sobre o peso ao desmame de bezerros nelore nas diferentes regiões brasileiras. Acta Scientiarum. Anim Sci 26:475–482
Borsato EALJHE, Rubin KCP, Saut JP, Barreiros TRR, Seneda MM (2004) Relação entre o tamanho do folículo ovulatório e taxa de concepção em novilhas *Bos taurus* X *Bos indicus* submetidas à inseminação artificial em tempo fixo. Revista Brasileira de Reprodução Animal 28:137–142
Boulton AC, Rushton J, Wathes DC (2017) An empirical analysis of the cost of rearing dairy heifers from birth to first calving and the time taken to repay these costs. Animal 11(8):1372–1380
Brickell JS, Wathes DC (2011) A descriptive study of the survival of Holstein-Friesian heifers through to third calving on English dairy farms. J Dairy Sci 94(4):1831–1838

Brumatti RC, Ferraz JBS, Eler JP, Formigonni IB (2011) Desenvolvimento de índice de seleção em gado corte sob o enfoque de um modelo bioeconômico. Archivos de zootecnia 60(230):205–213

Buckley F, Lopez-Villalobos N, Heins BJ (2014) Crossbreeding: implications for dairy cow fertility and survival. Animal 8:122–133

Butcher RL, Reber JE, Lishman AW, Breuel KF, Schrick FN, Spitzer JC, Inskeep EK (1992) Maintenance of pregnancy in postpartum beef cows that have short-lived corpora lutea. J Anim Sci 70:3831–3837

Butler ST, Hutchinson IA, Cromie AR, Shalloo L (2014) Applications and cost benefits of sexed semen in pasture-based dairy production systems. Animal 8(Suppl 1):165–172

Burns DS, Jimenez-Krassel F, Ireland JL, Knight PG, Ireland JJ (2005) Numbers of antral follicles during follicular waves in cattle: evidence for high variation among animals, very high repeatability in individuals, and an inverse association with serum follicle-stimulating hormone concentrations. Biol Reprod 73(1):54–62

Campos JT, Marinho LS, Lunardelli PA, Morotti F, Seneda MM (2013) Resynchronization of estrous cycle with eCG and temporary calf removal in lactating *Bos indicus* cows. Theriogenology 80:619–623

Campos JT, Morotti F, Costa CB, Bergamo LZ, Seneda MM (2016a) Evaluation of pregnancy rates of Bos indicus cows subjected to different synchronization ovulation protocols using injectable progesterone or an intravaginal device. Semina: Ciências Agrárias 37(6):4149–4155

Campos JT, Morotti F, Bergamo LZ, Costa CB, Seneda MM (2016b) Pregnancy rate evaluation in lactating and non-lactating Nelore cows subjected to fixed-time artificial insemination using injectable progesterone. Semina: Ciências Agrárias 37(4):1991–1996

Carneiro LC, Silva JCC, Mendes GP, Ferreira IC, Santos RM (2012) Efeito do mês de parição na taxa de gestação subsequente e no peso ao desmamedos bezerros de vacas Nelore. Acta Sci Vet 40(2):1030

Carvalho JB, Carvalho NA, Reis EL, Nichi M, Souza AH, Baruselli PS (2008) Effect of early luteolysis in progesterone-based timed AI protocols in *Bos indicus*, *Bos indicus* x *Bos taurus*, and *Bos taurus* heifers. Theriogenology 69:167–175

Cavalieri J, Rubio I, Kinder J, Entwistle K, Fitzpatrick L (1997) Synchronization of estrus and ovulation and associated endocrine changes in *Bos indicus* cows. Theriogenology 47:801–814

Cavalieri J, Hepworth G, Fitzpatrick LA (2004) Comparison of two estrus synchronization and resynchronization treatments in lactating dairy cows. Theriogenology 62:729–747

Cerchiaro I, Cassandro M, Dal Zotto R, Carnier P, Gallo L (2007) A field study on fertility and purity of sex-sorted cattle sperm. J Dairy Sci 90:2538–2542

Chacur MGM, Ferreira Aurelio PT, Scalon O Jr, Inague L, Do Nascimento Kronka S, Scalon LF (2010) Effects of lower doses of cloprostenol intramuscular or into vulvar submucosa on estrus induction and pregnancy rates in Nelore cows. Semina: Ciencias Agrarias 31:451–458

Chebel RC, Santos JEP, Cerri RLA, Galvão KN, Juchem SO, Thatcher WW (2003) Effect of resynchronization with GnRH on day 21 after artificial insemination on pregnancy rate and pregnancy loss in lactating dairy cows. Theriogenology 60:1389–1399

Colazo MG, Mapletoft RJ (2014) A review of current timed-AI (TAI) programs for beef and dairy cattle. Can Vet J 55:772–780

Colazo MG, Martínez MF, Kastelic JP, Mapletoft RJ (2002) Effects of dose and route of administration of cloprostenol on luteolysis, estrus and ovulation in beef heifers. Anim Reprod Sci 15:47–62

Crepaldi GAFBG, Vieira LM, Sá Filho MF, Guerreiro BM, Baruselli PS (2014) Reproductive efficiency of Nelore females submitted to three consecutive FTAI programs with 32 days of interval between inseminations. Anim Reprod 11:355

Crowe MA (2008) Resumption of ovarian cyclicity in post-partum beef and dairy cows. Reprod Domest Anim 43(Suppl 5):20–28

Crowe MA, Mullen MP (2013) Regulation and function of gonadotropins throughout the bovine oestrous cycle

Cuervo-Arango J, Garcia-Rosello E, Garcia-Munoz A, Valldecabres-Torres X, Martinez-Ros P, Gonzalez-Bulnes A (2011) The effect of a single high dose of PGF2alpha administered to dairy cattle 3.5 days after ovulation on luteal function, morphology, and follicular dynamics. Theriogenology 76:1736–1743

Day ML, Imakawa K, Wolfe PL, Kittok RJ, Kinder JE (1987) Endocrine mechanisms of puberty in heifers. Role of hypothalamo-pituitary estradiol receptors in the negative feedback of estradiol on luteinizing hormone secretion. Biol Reprod 37:1054–1065

de Graaf SP, Leahy T, Vishwanath R (2014) Biological and practical lessons associated with the use of sexed semen. Reproduction in Domestic Ruminants, 8, 9th

DeJarnette JM, Nebel RL, Marshall CE, Moreno JF, Mccleary CR, Lenz RW (2008) Effect of sex-sorted sperm dosage on conception rates in holstein heifers and lactating cows. J Dairy Sci 91:1778–1785

DeJarnette JM, Nebel RL, Marshall CE (2009) Evaluating the success of sex-sorted semen in US dairy herds from on farm records. Theriogenology 71:49–58

DeJarnette JM, Leach MA, Nebel RL, Marshall CE, Mccleary CR, Moreno JF (2011) Effects of sex-sorting and sperm dosage on conception rates of Holstein heifers: is comparable fertility of sex-sorted and conventional semen plausible? J Dairy Sci 94:3477–3483

Demetrio DGB, Santos RM, Demetrio CGB, Vasconcelos JLM (2007) Factors affecting conception rates following artificial insemination or embryo transfer in lactating holstein cows. J Dairy Sci 90:5073–5082

Dias CC, Wechsler FS, Day ML, Vasconcelos JL (2009) Progesterone concentrations, exogenous equine chorionic gonadotropin, and timing of prostaglandin F(2alpha) treatment affect fertility in postpuberal Nelore heifers. Theriogenology 72:378–385

Eaglesome MD, Garcia MM (1997) Disease risks to animal health from artificial insemination with bovine semen. Rev Sci Tech 16:215–225

El-Zarkouny SZSJS (2004) Resynchronizing estrus with progesterone or progesterone plus estrogen in cows of unknown pregnancy status. J Dairy Sci 87:3306–3321

Ereno RL, Barreiros TRR, Seneda MM, Baruselli PS, Pegorer MF, Barros CM (2007) Taxa de prenhez de vacas Nelore lactantes tratadas com progesterona associada à remoção temporária de bezerros ou aplicação de gonadotrofina coriônica eqüina. Rev Bras Zootec 36:1288–1294

Evans AC, Adams GP, Rawlings NC (1994) Follicular and hormonal development in prepubertal heifers from 2 to 36 weeks of age. J Reprod Fertil 102:463–470

Evans ACO, Mossa F, Walsh SW, Scheetz D, Jimenez-Krassel F, Ireland JLH,..... Ireland JJ, et al (2012) Effects of maternal environment during gestation on ovarian folliculogenesis and consequences for fertility in bovine offspring. Reprod Domestic Animals 47:31–37

Farnworth PG (1995) Gonadotrophin secretion revisited. How many ways can FSH leave a gonadotroph? J Endocrinol 145:387–395

Fernandes P, Teixeira AB, Crocci AJ, Barros CM (2001) Timed artificial insemination in beef cattle using GnRH agonist, PGF2alpha and estradiol benzoate (EB). Theriogenology 55:1521–1532

Fernandes CAC, Figueiredo ACS, Oliveira ER, Oba E, Vasconcelos TD (2006) Eficiency of different pgf2α analogous in the pospartum perido of dairy cows. In: World Buiatrics Congress, 2006, Nice, France. WBC, Nice

Ferraz MVC, Pires AV, Santos MH, Silva RG, Oliveira GB, Polizel DM, Biehl MV, Sartori R, Nogueira GP (2018) A combination of nutrition and genetics is able to reduce age at puberty in Nelore heifers to below 18 months. Animal 12(3):569–574

Figueiredo RA, Barros CM, Pinheiro OL, Soler JM (1997) Ovarian follicular dynamics in Nelore breed (*Bos indicus*) cattle. Theriogenology 47:1489–1505

Forde N, Beltman ME, Lonergan P, Diskin M, Roche JF, Crowe MA (2011) Oestrous cycles in *Bos taurus* cattle. Anim Reprod Sci 124:163–169

Fortune J, Rivera G (1999) Folículo dominante persistente em bovinos: aspectos básicos e aplicados. Arquivos da Faculdade de Veterinária UFRGS 27:22–34

Foster DL, Jackson LM, Padmanabhan V (2006) Programming of GnRH feedback controls timing puberty and adult reproductive activity. Mol Cell Endocrinol 25:254–255

Frandson RD, Wilke WL, Fails AD (2009) Anatomy and physiology of farm animals. Wiley
Galina CS, Calderon A, Mccloskey M (1982) Detection of signs of estrus in the charolais cow and its Brahman cross under continuous observation. Theriogenology 17:485–498
Galina CS, Orihuela A, Rubio I (1996) Behavioural trends affecting oestrus detection in zebu cattle. Anim Reprod Sci 42:465–470
Galvao KN, Santos JE, Cerri RL, Chebel RC, Rutigliano HM, Bruno RG, Bicalho RC (2007) Evaluation of methods of resynchronization for insemination in cows of unknown pregnancy status. J Dairy Sci 90:4240–4252
Gama LT, Bressan MC, Rodrigues EC, Rossato LV, Moreira OC, Alves SP, Bessa RJB (2013) Heterosis for meat quality and fatty acid profiles in crosses among *Bos indicus* and *Bos taurus* finished on pasture or grain. Meat Sci 93:98–104
Garcia SM, Morotti F, Cavalieri FLB, Lunardelli PA, de Oliveira Santos A, Membrive CMB,..... Seneda MM et al (2020) Synchronization of stage of follicle development before OPU improves embryo production in cows with large antral follicle counts. Animal Reprod Sci 221:106601
Garner DL (2006) Flow cytometric sexing of mammalian sperm. Theriogenology 65:943–957
Garner DL, Evans KM, Seidel GE (2013) Sex-sorting sperm using flow cytometry/cell sorting. Methods Mol Biol 927:279–295
Gimenes LU, Sá Filho MF, Carvalho NA, Torres-Junior JR, Souza AH, Madureira EH, Trinca LA, Sartorelli ES, Barros CM, Carvalho JB, Mapletoft RJ, Baruselli PS (2008) Follicle deviation and ovulatory capacity in *Bos indicus* heifers. Theriogenology 69:852–858
Ginther OJ, Kastelic JP, Knopf L (1989a) Composition and characteristics of follicular waves during the bovine estrous cycle. Anim Reprod Sci 20:187–200
Ginther OJ, Knopf L, Kastelic JP (1989b) Temporal associations among ovarian events in cattle during oestrous cycles with two and three follicular waves. J Reprod Fertil 87:223–230
Ginther OJ, Bergfelt DR, Kulick LJ, Kot K (1999) Selection of the dominant follicle in cattle: establishment of follicle deviation in less than 8 hours through depression of FSH concentrations. Theriogenology 52:1079–1093
Ginther OJ, Bergfelt DR, Beg MA, Kot K (2002) Role of low circulating FSH concentrations in controlling the interval to emergence of the subsequent follicular wave in cattle. Reproduction 124:475–482
Goff AK (2004) Steroid hormone modulation of prostaglandin secretion in the ruminant endometrium during the estrous cycle. Biol Reprod 71:11–16
Gonzalez SM, Assunção IEG, Duarte J, Oliveira CHA, Guedes TA, Morotti F, Seneda MM, Andrade ER (2020) Induction of puberty in Bos indicus heifers in the western Amazon region. Semina: Ciências Agrárias 41(5 Suppl):2153–2162
Hanlon DW, Williamson NB, Wichtel JJ, Steffert IJ, Craigie AL, Pfeiffer DU (1996) The effect of estradiol benzoate administration on estrous response and synchronized pregnancy rate in dairy heifers after treatment with exogenous progesterone. Theriogenology 45:775–785
Hansel W, Donaldson LE, Wagner WC, Brunner MA (1966) A comparison of Estrous cycle synchronization methods in beef cattle under feedlot conditions. J Anim Sci 25:497–503
Hassan M, Arshad U, Bilal M, Sattar A, Avais M, Bollwein H, Ahmad N (2018) Luteal blood flow measured by Doppler ultrasonography during the first three weeks after artificial insemination in pregnant and non-pregnant Bos indicus dairy cows. J Reprod Dev 65(1):29–36
Haworth GM, Tranter WP, Chuck JN, Cheng Z, Wathes DC (2008) Relationships between age at first calving and first lactation milk yield, and lifetime productivity and longevity in dairy cows. Vet Rec 162(20):643–647
Hillier SG (1994) Current concepts of the roles of follicle stimulating hormone and luteinizing hormone in folliculogenesis. Hum Reprod 9:188–191
Honaramooz A, Aravindakshan J, Chandolia RK, Beard AP, Bartlewski PM, Pierson RA, Rawlings NC (2004) Ultrasonographic evaluation of the pre-pubertal development of the reproductive tract in beef heifers. Anim Reprod Sci 80:15–29
Ireland JJ, Smith GW, Scheetz D, Jimenez-Krassel F, Folger JK, Ireland JLH,..... Evans ACO, et al (2010) Does size matter in females? An overview of the impact of the high variation in

the ovarian reserve on ovarian function and fertility, utility of anti-Müllerian hormone as a diagnostic marker for fertility and causes of variation in the ovarian reserve in cattle. Reprod Fertil Dev 23(1):1–14

Jainudeen MR, Hafez ESE (2013) Reproduction in farm animals. Raven Press, Somerset

Jimenez-Krassel F, Scheetz DM, Neuder LM, Pursley JR, Ireland JJ (2017) A single ultrasound determination of≥ 25 follicles≥ 3 mm in diameter in dairy heifers is predictive of a reduced productive herd life. J Dairy Sci 100(6):5019–5027

Kastelic JP, Knopf L, Ginther OJ (1990) Effect of day of prostaglandin F2α treatment on selection and development of the ovulatory follicle in heifers. Anim Reprod Sci 23:169–180

Kastelic JP, Olson WO, Martinez M, Cook RB, Mapletoft RJ (1999) Synchronization of estrus in beef cattle with norgestomet and estradiol valerate. Can Vet J 40:173–178

Keri G, Horvath A, Nikolics K, Teplan I (1985) In the central role of GnRH in the regulation of reproductive functions: mechanism of action and practical results. J Steroid Biochem 23:719–723

Kim IH, Kim UH (2007) Comparison of the effect of estradiol benzoate plus progesterone and GnRH on the follicular wave emergence and subsequent follicular development in CIDR-treated, lactating dairy cows with follicular cysts. Anim Reprod Sci 98:197–203

Kinder JE, Bergfeld EGM, Wehrman ME, Peters KE, Kojima FN (1995) Endocrine basis for puberty in heifers and ewes. J Reprod Fertil 49:393–407

Kinder JE, Kojima FN, Bergfeld EG, Wehrman ME, Fike KE (1996) Progestin and estrogen regulation of pulsatile LH release and development of persistent ovarian follicles in cattle. J Anim Sci 74:1424–1440

Knight PG, Glister C (2006) TGF-β superfamily members and ovarian follicle development. Reproduction 132:191–206

Knopf L, Kastelic JP, Schallenberger E, Ginther OJ (1989) Ovarian follicular dynamics in heifers: test of two-wave hypothesis by ultrasonically monitoring individual follicles. Domest Anim Endocrinol 6:111–119

Kurykin J, Jaakma U, Jalakas M, Aidnik M, Waldmann A, Majas L (2007) Pregnancy percentage following deposition of sex-sorted sperm at different sites within the uterus in estrus-synchronized heifers. Theriogenology 67:754–759

Lamb GC, Mercadante VR (2016) Synchronization and artificial insemination strategies in beef cattle. Vet Clin N Am Food Anim Pract 32(2):335–347. https://doi.org/10.1016/j.cvfa.2016.01.006. Epub 2016 Apr 29

Lamb GC, Dahlen CR, Larson JE, Marquezini G, Stevenson JS (2010) Control of the estrous cycle to improve fertility for fixed-time artificial insemination in beef cattle: a review. J Anim Sci 88:E181–E192

LeFever DG, Odde KG (1987) Predicting reproductive performance in beef heifers by reproductive tract evaluation before breeding. CSU Beef Res Rep:13–15

Levy N, Kobayashi S, Roth Z, Miyamoto A, Meidan R (2000) Administration of prostaglandin F2α during the early bovine luteal phase does not alter the expression of ET-1 and its type a receptor: a possible cause for corpus luteum refractoriness. Biol Reprod 63:377–382

Lima MA, Morotti F, Bayeux BM, Rezende RG, Botigelli RC, De Bem THC, Fontes PK, Nogueira MFG, Meirelles FV, Baruselli PS, Silveira JC, Perecin F, Seneda MM (2020) Ovarian follicular dynamics, progesterone concentrations, pregnancy rates and transcriptional patterns in *Bos indicus* females with a high or low antral follicle count. Sci Rep 10:19557

Looper ML, Vizcarra JA, Wettemann RP, Malayer JR, Braden TD, Geisert RD, Morgan GL (2003) Influence of estradiol, progesterone, and nutrition on concentrations of gonadotropins and GnRH receptors, and abundance of mRNA for GnRH receptors and gonadotropin subunits in pituitary glands of beef cows. J Anim Sci 81:269–278

Løvendahl P, Chagunda MGG (2010) On the use of physical activity monitoring for estrus detection in dairy cows. J Dairy Sci 93:249–259

Lu KH, Seidel GE Jr (2004) Effects of heparin and sperm concentration on cleavage and blastocyst development rates of bovine oocytes inseminated with flow cytometrically-sorted sperm. Theriogenology 62:819–830

Lucy MC (2007) The bovine dominant ovarian follicle. J Anim Sci 85(13 Suppl):89–99

Machado R, Barbosa RT, Bergamaschi MA, Figueiredo RA (2007) A inseminação artificial em tempo fixo como biotécnica aplicada na reprodução dos bovinos de corte. In Embrapa Pecuária Sudeste-Artigo em anais de congresso (ALICE). In: SEMANA DO ESTUDANTE, 18, 2007, São Carlos, SP. Palestras. Embrapa Pecuária Sudeste, São Carlos, p 2007

Mackey DR, Sreenan JM, Rochet JF, Diskin MG (2000) The effect of progesterone alone or in combination with estradiol on follicular dynamics, gonadotropin profiles, and estrus in beef cows following calf isolation and restricted suckling. J Anim Sci 78:1917–1929

Macmillan KL, Peterson AJ (1993a) Current advances in the manipulation of reproductive function in domestic animals a new intravaginal progesterone releasing device for cattle (CIDR-B) for oestrous synchronisation, increasing pregnancy rates and the treatment of post-partum anoestrus. Anim Reprod Sci 33:1–25

Macmillan KL, Peterson AJ (1993b) A new intravaginal progesterone releasing device for cattle (CIDR-B) for oestrous synchonization, increasing pregnancy rates and the treatment of post-partum anoestrus. Anim Reprod Sci 33:1–25

Manikkam M, Rajamahendran R (1997) Progesterone-induced atresia of the proestrous dominant follicle in the bovine ovary: changes in diameter, insulin-like growth factor system, aromatase activity, steroid hormones, and apoptotic index. Biol Reprod 57:580–587

Mann GE, Lamming GE (2001) Relationship between maternal endocrine environment, early embryo development and inhibition of the luteolytic mechanism in cows. Reproduction 121:175–180

Marques MO, Morotti F, Silva CB, Ribeiro-Júnior M, da Silva RCP, Baruselli PS, Seneda MM (2015) Influence of category–heifers, primiparous and multiparous lactating cows–in a large-scale resynchronization fixed-time artificial insemination program. J Vet Sci 16:367–371

Marques MO, Morotti F, Lorenzetti E, Bizarro-Silva C, Seneda MM (2018) Intensified use of TAI and sexed semen on commercial farms. Anim Reprod 15(3):197–203

Martinez MF, Kastelic JP, Bo GA, Caccia M, Mapletoft RJ (2005) Effects of oestradiol and some of its esters on gonadotrophin release and ovarian follicular dynamics in CIDR-treated beef cattle. Anim Reprod Sci 86:37–52

Mee JF (2008) Prevalence and risk factors for dystocia in dairy cattle: a review. Vet J 176:93–101

Meidan R, Levy N (2002) Endothelin-1 receptors and biosynthesis in the corpus luteum: molecular and physiological implications. Domest Anim Endocrinol 23:287–298

Menegassi SRO, Barcellos JOJ (2015) Aspectos Reprodutivos do Touro: Teoria e Prática. Agrolivros, Guaíba, 280p

Meneghetti MVER, Vasconcelos JLM (2001) Efeito da remoção temporária dos bezerros nos folículos dominantes e na taxa de ovulação ao primeiro GnRH em protocolos de sincronização em vacas Nelore em anestro. Revista Brasileira de Reprodução Animal 25:286–288

Mizuta K (2003) Estudo comparativo dos aspectos comportamentais do estro e dos teores plasmáticos de LH, FSH, progesterona e estradiol que precedem a ovulação em f^emeas bovinas Nelore (*Bos taurus* indicus), Angus (*Bos taurus* taurus) e Nelore × Angus (*Bos taurus* indicus × *Bos taurus* taurus). Tese de Doutorado em Reprodução Animal – Faculdade de Medicina Veterinária e Zootecnia, Universidade de São Paulo, São Paulo., 98 p

Moenter SM, Brand RC, Karsch FJ (1992) Dynamics of gonadotropin-releasing hormone (GnRH) secretion during the GnRH surge: insights into the mechanism of GnRH surge induction. Endocrinology 130:2978–2984

Moraes FLZ, Morotti F, Costa CB, Lunardelli PA, Seneda MM (2019) Relationships between antral follicle count, body condition, and pregnancy rates after timed-AI in Bos indicus cattle. Theriogenology 136:10–14

Moran C, Quirke JF, Roche JF (1989) Puberty in heifers: a review. Anim Reprod Sci 18:167–182

Morotti F, Campos JT, Seneda MM (2013a) Fixed-time artificial insemination using injectable progesterone: ovarian follicular dynamics and pregnancy rates of Nelore cows (*Bos indicus*) with and without a corpus luteum. Semina: Ciências Agrárias 34:3867–3875

Morotti F, Campos JT, Oliveira RE, Seneda MM (2013b) Ovarian follicular dynamics of Nelore (*Bos indicus*) cows subjected to a fixed-time artificial insemination protocol with injectable progesterone. Semina: Ciências Agrárias 34:3859–3866

Morotti F, Sanches BV, Pontes JH, Basso AC, Siqueira ER, Lisboa LA, Seneda MM (2014) Pregnancy rate and birth rate of calves from a large-scale IVF program using reverse-sorted semen in *Bos indicus*, *Bos indicus*-taurus, and *Bos taurus* cattle. Theriogenology 81:696–701

Morotti F, Moretti R, Santos GM, Silva-Santos KC, Cerqueira PHR, Seneda MM (2018a) Ovarian follicular dynamics and conception rate in *Bos indicus* cows with different antral follicle counts subjected to timed artificial insemination. Anim Reprod Sci 188:170–177

Morotti F, Zangirolamo AF, da Silva NC, da Silva CB, Rosa CO, Seneda MM (2018b) Antral follicle count in cattle: advantages, challenges, and controversy. Anim Reprod 14(3):414–420

Murphy BD, Martinuk SD (1991) Equine chorionic gonadotropin. Endocr Rev 12:27–44

Nogueira GP (2004) Puberty in south American *Bos indicus* (zebu) cattle. Anim Reprod Sci 82-83:361–372

Nogueira E, Silva MR, Silva JCB, Abreu UPG, Anache NA, Silva KC,.... Rodrigues WB, et al (2019) Timed artificial insemination plus heat I: effect of estrus expression scores on pregnancy of cows subjected to progesterone–estradiol-based protocols. Animal 13(10):2305–2312

Odde KG (1990) A review of synchronization of estrus in postpartum cattle J Anim Sci 68:817–830

Orihuela AN (2000) Some factors affecting the behavioural manifestation of oestrus in cattle: a review. Appl Anim Behav Sci 70:1–16

Patterson DJ, Corah LR, Brethour JR, Higgins JJ, Kiracofe GH, Stevenson JS (1992) Evaluation of reproductive traits in *Bos taurus* and *Bos indicus* crossbred heifers: relationship of age at puberty to length of the postpartum interval to estrus. J Anim Sci 70:1994–1999

Pellegrino CA, Morotti F, Untura RM, Pontes JH, Pellegrino MF, Campolina JP, Seneda MM, Barbosa FA, Henry M (2016) Use of sexed sorted semen for fixed-time artificial insemination or fixed-time embryo transfer of in vitro-produced embryos in cattle. Theriogenology 86:888–8893

Peña DM (2007) Estrategias de destete temporario y tratamientos con ecg en protocolos de inseminación artificial a tiempo fijo (IATF) con dispositivos con progesterona en vacas cruza cebú con cria. Magíster en Ciencias Agropecuárias – Faculdad de Ciências Agropecuárias, Universidad Nacional de Córdoba., 37 p

Peres RF, Claro I Jr, Sá Filho OG, Nogueira GP, Vasconcelos JL (2009) Strategies to improve fertility in *Bos indicus* postpubertal heifers and nonlactating cows submitted to fixed-time artificial insemination. Theriogenology 72:681–689

Pfeifer LFM, Leal SDCBDS, Schneider A, Schmitt E, Corrêa MN (2012) Effect of the ovulatory follicle diameter and progesterone concentration on the pregnancy rate of fixed-time inseminated lactating beef cows. Revista Brasileira de Zootecnia 41(4):1004–1008

Pfeifer LFM, Castro NA, Melo VTO, Neves PMA, Cestaro JP, Schneider A (2015) Timed artificial insemination in blocks: a new alternative to improve fertility in lactating beef cows. Animal Reprod Sci 163:89–96

Pinheiro OL, Barros CM, Figueiredo RA, Do Valle ER, Encarnação RO, Padovani CR (1998) Estrous behavior and the estrus-to-ovulation interval in nelore cattle *Bos indicus* with natural estrus or estrus induced with prostaglandin F2α or norgestomet and estradiol valerate. Theriogenology 49:667–681

Pontes JHF, Silva KCF, Basso AC, Rigo AG, Ferreira CR, Santos GMG, Sanches BV, Porcionato JPF, Vieira PHS, Faifer FS, Sterza FAM, Schenk JL, Seneda MM (2010) Large-scale in vitro embryo production and pregnancy rates from *Bos taurus*, *Bos indicus*, and indicus-taurus dairy cows using sexed sperm. Theriogenology 74:1349–1355

Prata AB, Drum JN, Melo LF, Araujo ER, Sartori R (2018) Effect of different chorionic gonadotropins on final growth of the dominant follicle in Bos indicus cows. Theriogenology 111:52–55

Pugliesi G, Rezende RG, Silva JCB, Lopes E, Nishimura T K, Baruselli PS,..... Binelli M, et al (2017) Uso da ultrassonografia Doppler em programas de IATF e TETF em bovinos. Rev Bras Reprod Anim 41(1):140–150

Rahe CH, Owens RE, Fleeger JL, Newton HJ, Harms PG (1980) Pattern of plasma luteinizing hormone in the cyclic cow: dependence upon the period of the cycle. Endocrinology 107:498–503

Rajamahendran R, Bedirian KN, Lague PC, Baker RD (1976) Luteolytic activity of a synthetic prostaglandin and PGF2alpha in heifers. Prostaglandins 11:143–153

Rasby RJ, Day ML, Johnson SK, Kinder JE, Lynch JM, Short RE, Wettemann RP, Hafs HD (1998) Luteal function and estrus in peripubertal beef heifers treated with an intravaginal progesterone releasing device with or without a subsequent injection of estradiol. Theriogenology 50:55–63

Rawlings NC, Evans AC, Honaramooz A, Bartlewski PM (2003) Antral follicle growth and endocrine changes in prepubertal cattle, sheep and goats. Anim Reprod Sci 78:259–270

Richards JS, Russell DL, Robker RL, Dajee M, Alliston TN (1998) Molecular mechanisms of ovulation and luteinization. Mol Cell Endocrinol 145:47–54

Rivera G, Fortune J (2003) Proteolysis of insulin-like growth factor binding proteins-4 and-5 in bovine follicular fluid: implications for ovarian follicular selection and dominance. Endocrinology 144:2977–2987

Roche JF (1996) Control and regulation of folliculogenesis – a symposium in perspective. Rev Reprod 1:19–27

Rodrigues ADP, Peres RFG, Lemes AP, Martins T, Pereira MHC, Day ML, Vasconcelos JLM (2013) Progesterone-based strategies to induce ovulation in prepubertal Nellore heifers. Theriogenology 79(1):135–141

Rodrigues WB, Silva AS, Silva JCB, Anache NA, Silva KC, Cardoso CJT, Garcia WR, Sutovsky P, Nogueira E (2019) Timed artificial insemination plus heat II: gonadorelin injection in cows with low estrus expression scores increased pregnancy in progesterone/estradiol-based protocol. Animal 13:2313–2318

Rosenkrans KS, Hardin DK (2003) Repeatability and accuracy of reproductive tract scoring to determine pubertal status in beef heifers. Theriogenology 59(5–6):1087–1092

Sá Filho M, Reis E, Ayres H, Gimenes L, Peres A, Carvalho C, Carvalho J, Araújo C, Baruselli P (2006) Effect of oestradiol valerate or benzoate on induction of a new follicular wave emergence in *Bos indicus* cows and heifers treated with norgestomet auricular implant. Reprod Fertil Dev 18:289

Sá Filho MF, Crespilho AM, Santos JE, Perry GA, Baruselli PS (2010a) Ovarian follicle diameter at timed insemination and estrous response influence likelihood of ovulation and pregnancy after estrous synchronization with progesterone or progestin-based protocols in suckled *Bos indicus* cows. Anim Reprod Sci 120:23–30

Sá Filho MF, Ayres H, Ferreira RM, Nichi M, Fosado M, Campos Filho EP, Baruselli PS (2010b) Strategies to improve pregnancy per insemination using sex-sorted semen in dairy heifers detected in estrus. Theriogenology 74:1636–1642

Sá Filho MF, Baldrighi JM, Sales JN, Crepaldi GA, Carvalho JB, Bó GA, Baruselli PS (2011) Nduction of ovarian follicular wave emergence and ovulation in progestin-based timed artificial insemination protocols for *Bos indicus* cattle. Anim Reprod Sci 129:132–139

Sá Filho MF, Girotto R, Abe EK, Penteado L, Campos Filho EP, Moreno JF, Sala RV, Nichi M, Baruselli PS (2012) Optimizing the use of sex-sorted sperm in timed artificial insemination programs for suckled beef cows. J Anim Sci 90:1816–1823

Sá Filho MF, Penteado L, Siqueira GR, Soares JG, Mendanha MF, Macedo GG, Baruselli PS (2013) Timed artificial insemination should be performed early when used norgestomet ear implants are applied for synchronizing ovulation in beef heifers. Theriogenology 80:642–647

Sá Filho MF, Marques MO, Girotto R, Santos FA, Sala RV, Barbuio JP, Baruselli PS (2014) Resynchronization with unknown pregnancy status using progestin-based timed artificial insemination protocol in beef cattle. Theriogenology 81:284–290

Sá Filho MF, Nasser LFT, Penteado L, Prestes R, Marques MO, Freitas BG, Monteiro BM, Ferreira RM, Gimenes LU, Baruselli PS (2015) Impact of progesterone and estradiol treatment before the onset of the breeding period on reproductive performance of *Bos indicus* beef heifers. Anim Reprod Sci 160:30–39

Sales JN, Neves KA, Souza AH, Crepaldi GA, Sala RV, Fosado M, Campos Filho EP, De Faria M, Sá Filho MF, Baruselli PS (2011a) Timing of insemination and fertility in dairy and beef cattle receiving timed artificial insemination using sex-sorted sperm. Theriogenology 76:427–435

Sales JNS, Neves KAL, Souza AH, Crepaldi GA, Sala RV, Fosado M, Filho EPC, De Faria M, Filho MFS, Baruselli PS (2011b) Timing of insemination and fertility in dairy and beef cattle receiving timed artificial insemination using sex-sorted sperm. Theriogenology 76:427–435

Sales JN, Carvalho JB, Crepaldi GA, Cipriano RS, Jacomini JO, Maio JR, Souza JC, Nogueira GP, Baruselli PS (2012a) Effects of two estradiol esters (benzoate and cypionate) on the induction of synchronized ovulations in *Bos indicus* cows submitted to a timed artificial insemination protocol. Theriogenology 78:510–516

Sales JNS, Carvalho JBP, Crepaldi GA, Cipriano RS, Jacomini JO, Maio JRG, Souza JC, Nogueira GP, Baruselli PS (2012b) Effects of two estradiol esters (benzoate and cypionate) on the induction of synchronized ovulations in *Bos indicus* cows submitted to a timed artificial insemination protocol. Theriogenology 78:510–516

Sales JNS, Carvalho JBP, Crepaldi GA, Soares JG, Girotto RW, Maio JRG, Souza JC, Baruselli PS (2015) Effect of circulating progesterone concentration during synchronization for fixed-time artificial insemination on ovulation and fertility in *Bos indicus* (Nelore) beef cows. Theriogenology 83:1093–1100

Santos GMG, Silva-Santos KC, Barreiros TRR, Morotti F, Sanches BV, de Moraes FLZ,..... Seneda MM, et al (2016) High numbers of antral follicles are positively associated with in vitro embryo production but not the conception rate for FTAI in Nelore cattle. Animal Reproduct Sci 165:17–21

Sartorelli ES, Carvalho LM, Bergfelt DR, Ginther OJ, Barros CM (2005) Morphological characterization of follicle deviation in Nelore (*Bos indicus*) heifers and cows. Theriogenology 63:2382–2394

Savio JD, Thatcher WW, Morris GR, Entwistle K, Drost M, Mattiacci MR (1993) Effects of induction of low plasma progesterone concentrations with a progesterone-releasing intravaginal device on follicular turnover and fertility in cattle. J Reprod Fertil 98:77–84

Schams D, Berisha B (2004) Regulation of corpus luteum function in cattle-an overview. Reprod Domest Anim 39:241–251

Schenk JL, Cran DG, Everett RW, Seidel GE Jr (2009) Pregnancy rates in heifers and cows with cryopreserved sexed sperm: effects of sperm numbers per inseminate, sorting pressure and sperm storage before sorting. Theriogenology 71:717–728

Seidel GE Jr, Allen CH, Johnson LA, Holland MD, Brink Z, Welch GR, Graham JK, Cattell MB (1997) Uterine horn insemination of heifers with very low numbers of nonfrozen and sexed spermatozoa. Theriogenology 48:1255–1264

Seidel GE Jr (2007) Overview of sexing sperm. Theriogenology 68:443–446

Seidel GE Jr (2014) Update on sexed semen technology in cattle. Animal 8(Suppl 1):160–164

Seidel GE Jr, Garner DL (2002) Current status of sexing mammalian spermatozoa. Reproduction 124:733–743

Seidel GE Jr, Schenk JL (2008a) Pregnancy rates in cattle with cryopreserved sexed sperm: effects of sperm numbers per inseminate and site of sperm deposition. Anim Reprod Sci 105:129–138

Seidel GE Jr, Schenk JL (2008b) Pregnancy rates in cattle with cryopreserved sexed sperm: effects of sperm numbers per inseminate and site of sperm deposition. Anim Reprod Sci 105:129–138

Seidel GE Jr, Schenk JL, Herickhoff LA, Doyle SP, Brink Z, Green RD, Cran DG (1999) Insemination of heifers with sexed sperm. Theriogenology 52:1407–1420

Seneda MM, Morotti F, Zangirolamo AF, Silva NC, Sanches TK, Blaschi W, Barreiros TRR (2019) Antral follicle population in prepubertal and pubertal heifers. Reprod Fertil Dev 31:10–16

Senger PL (1997) Pathways to pregnancy and parturition. Raven Press, Redmond

Silva LO (2020) Physiological responses of Bos taurus and Bos indicus cows submitted to hormonal treatments for estrous cycle synchronization. Master's dissertation, Escola superior de Agricultura Luiz de Queiroz, University of São Paulo, Piracicaba. https://doi.org/10.11606/D.11.2020.tde-06102020-183501. Retrieved 2021-01-23, from www.teses.usp.br

Silva-Santos KC, Santos GMG, Koetz Júnior C, Morotti F, Siloto LS, Marcantonio TN,..... Seneda MM, et al (2014) Antral Follicle Populations and Embryo Production–In Vitro and In Vivo–of B os indicus–taurus Donors from Weaning to Yearling Ages. Reproduct Domestic Animals 49(2):228–232

Siqueira LGB, Areas VS, Ghetti AM, Fonseca JF, Palhao MP, Fernandes CAC, Viana JHM (2013) Color Doppler flow imaging for the early detection of nonpregnant cattle at 20 days after timed artificial insemination. J Dairy Sci 96(10):6461–6472

Smith MF, Geisert RD, Parrish JJ (2018) Reproduction in domestic ruminants during the past 50 yr: discovery to application. J Anim Sci 96:2952–2970

Soumano K, Silversides DW, Doize F, Price CA (1996) Follicular 3 beta-hydroxysteroid dehydrogenase and cytochromes P450 17 alpha-hydroxylase and aromatase messenger ribonucleic acids in cattle undergoing superovulation. Biol Reprod 55:1419–1426

Souza AHSJNS, Crepaldi GA, Teixeira AA, Baruselli PS (2008) Effect of type of semen (sexed vs non-sexed) and time of AI (60h vs 64h) on pregnancy rates of postpartum Nelore cows inseminated in a fixed time. Anim Reprod 6:224

Stevenson JS, Hill SL, Bridges GA, Larson JE, Lamb GC (2015) Progesterone status, parity, body condition, and days postpartum before estrus or ovulation synchronization in suckled beef cattle influence artificial insemination pregnancy outcomes. J Anim Sci 93:2111–2123

Sunderland SJ, Crowe MA, Boland MP, Roche JF, Ireland JJ (1994) Selection, dominance and atresia of follicles during the oestrous cycle of heifers. J Reprod Fertil 101:547–555

Taylor C, Rajamahendran R (1991) Follicular dynamics and Corpus luteum growth and function in pregnant versus nonpregnant dairy cows. J Dairy Sci 74:115–123

Teodoro RLMJC, Regazzi AJ, Lemos AM, Freitas AF (1993) Características associadas com a maturidade sexual em vacas mestiças Holandês:Gir e vacas do cruzamento tríplice Jersey ou Suíço x Holandês:Gir. Revista da Sociedade Brasileira de Zootecnia 22:488–494

Thomas JM, Locke JWC, Vishwanath R, Hall JB, Ellersieck MR, Smith MF, Patterson DJ (2017) Effective use of SexedULTRA™ sex-sorted semen for timed artificial insemination of beef heifers. Theriogenology 98:88–93

Torres JR Jr, Penteado L, Sales JN, Sá Filho MF, Ayres H, Baruselli PS (2014) A comparison of two different esters of estradiol for the induction of ovulation in an estradiol plus progestin-based timed artificial insemination protocol for suckled *Bos indicus* beef cows. Anim Reprod Sci 151:9–14

Torres-Júnior JRS, Melo WO, Elias AKS, Rodrigues LS, Penteado L, Baruselli PS (2009) Considerações técnicas e econômicas sobre reprodução assistida em gado de corte. Revista Brasileira de Reprodução Animal 33(1):53–58

Underwood SL, Bathgate R, Ebsworth M, Maxwell WM, Evans G (2010a) Pregnancy loss in heifers after artificial insemination with frozen-thawed, sex-sorted, re-frozen-thawed dairy bull sperm. Anim Reprod Sci 118:7–12

Underwood SL, Bathgate R, Maxwell WM, Evans G (2010b) Birth of offspring after artificial insemination of heifers with frozen-thawed, sex-sorted, re-frozen-thawed bull sperm. Anim Reprod Sci 118:171–175

Vailes LD, Washburn SP, Britt JH (1992) Effects of various steroid milieus or physiological states on sexual behavior of Holstein cows. J Anim Sci 70:2094–2103

Valldecabres-Torres X, García-Roselló E, García-Muñoz A, Cuervo-Arango J (2012) Effects of d-cloprostenol dose and corpus luteum age on ovulation, luteal function, and morphology in nonlactating dairy cows with early corpora lutea. J Dairy Sci 95:4389–4395

Vieira LMSFMF, Pugliesi G, Guerreiro BM, Cristaldo MA, Batista EOS, Freitas BG, Carvalho FJ, Guimaraes LHC, Baruselli PS (2014) Resynchronization in dairy cows 13 days after TAI followed by pregnancy diagnosis based on corpus luteum vascularization by color doppler. Anim Reprod 11:378

Vishwanath R, Moreno JF (2018) Semen sexing–current state of the art with emphasis on bovine species. Animal 12(s1):s85–s96

Vizcarra JA, Wettemann RP, Morgan GL (1999) Influence of dose, frequency, and duration of infused gonadotropin-releasing hormone on secretion of luteinizing hormone and follicle-stimulating hormone in nutritionally anestrous beef cows. Domest Anim Endocrinol 16:171–181

Wathes DC, Pollott GE, Johnson KF, Richardson H, Cooke JS (2014) Heifer fertility and carry over consequences for life time production in dairy and beef cattle. Animal 8(s1):91–104

Webb R, Garnsworthy PC, Gong JG, Armstrong DG (2004) Control of follicular growth: local interactions and nutritional influences. J Anim Sci 82(E-Suppl):E63-74

Weigel KA (2004) Exploring the role of sexed semen in dairy production systems. J Dairy Sci 87:120–130

Williams GL, Gazal OS, Vega GAG, Stanko RL (1996) Mechanisms regulating suckling-mediated anovulation in the cow. Anim Reprod Sci 42:289–297

Wiltbank MC, Pursley JR (2014) The cow as an induced ovulator: timed AI after synchronization of ovulation. Theriogenology 81(1):170–185

Wiltbank MC, Gümen A, Sartori R (2002) Physiological classification of anovulatory conditions in cattle. Theriogenology 57:21–52

Wiltbank M, Lopez H, Sartori R, Sangsritavong S, Gumen A (2006) Changes in reproductive physiology of lactating dairy cows due to elevated steroid metabolism. Theriogenology 65:17–29

Xu Z, Garverick HA, Smith GW, Smith MF, Hamilton SA, Youngquist RS (1995) Expression of follicle-stimulating hormone and luteinizing hormone receptor messenger ribonucleic acids in bovine follicles during the first follicular wave. Biol Reprod 53:951–957

Yavas Y, Walton JS (2000a) Induction of ovulation in postpartum suckled beef cows: a review. Theriogenology 54:1–23

Yavas Y, Walton JS (2000b) Postpartum acyclicity in suckled beef cows: a review. Theriogenology 54:25–55

Yin W, Gore AC (2010) The hypothalamic median Eminence and its role in reproductive aging. Ann N Y Acad Sci 1204:113–122

Yoo HS (2010) Infectious causes of reproductive disorders in cattle. J Reprod Dev 56(Suppl):S53–S60

Zangirolamo AF, Morotti F, da Silva NC, Sanches TK, Seneda MM (2018) Ovarian antral follicle populations and embryo production in cattle. Anim Reprod 15(3):310–315

Chapter 2
Reproduction Management and Artificial Insemination in Dromedary Camel

Djallel Eddine Gherissi and Ramzi Lamraoui

Abstract The dromedary camel is a unique multipurpose species under extreme arid conditions. It is used for production, leisure, transport and agricultural work. Its profile as a production animal has given it major importance in ensuring the sustainability of marginalized communities in the poorest regions of the world, but also because of the nutritional and therapeutic properties of its milk and meat. Nowadays, the exploitation of this animal resource known an orientation towards intensification under the pretext of the expanding demand on camel products. The productivity of camel herds is conditioned by its reproductive potential. Often, camel herd performances are limited by low fertility rate, high inter-calving intervals, late puberty leading to low longevity, low milk production and unavailability of young calves for herd renew and fattening farms. To reach an optimal level of camel herd's numerical productivity and therefore accelerate the genetic progress, it is imperative to reduce the duration of the unproductive periods, namely the waiting period and the reproduction period. This involves through the choice of a suitable farming system, improving breeding practices, improving data recording and reproductive monitoring, application of strategic reproduction control and new reproduction techniques.

Here we review strategies of camel herd reproduction management and means to improve camel herd reproductive performances using sexual activity control in males and females and artificial insemination practice. Controlling ovarian cycles of the female camel consists of controlling follicular growth and timing of ovulation. Therefore, this technique can be used to induce and synchronize the ovarian function during a favorable reproduction period or even during the non reproductive

---

D. E. Gherissi (✉)
Institute of Agronomic and Veterinary Sciences, University of Souk-Ahras, Souk Ahras, Algeria

Laboratory of Animal Productions, Biotechnologies and Health, University of Souk-Ahras, Souk Ahras, Algeria

R. Lamraoui
Institute of Agronomic and Veterinary Sciences, University of Souk-Ahras, Souk Ahras, Algeria

Department of Biology of Living Organisms, Faculty of Natural and Life Sciences, University of Batna, Batna, Algeria

V. K. Yata et al. (eds.), *Sustainable Agriculture Reviews 54*, Sustainable Agriculture Reviews 54, https://doi.org/10.1007/978-3-030-76529-3_2

season. Zootechnical means do not really make possible to synchronize the ovarian cycle in camels. They are used as complementary tools to the pharmacological methods. The different hormonal based protocols are: melatonin, progesterone, prostaglandin F2α and GnRH. Artificial insemination in dromedary camel is a topic of contemporary research that tends to develop standard sperm collection protocols, monitoring of sperm quality, short-term preservation, cryopreservation and semen thaw with considerable interest to increase the pregnancy rate and therefore to market this technology in camels for large-scale use. Various genetic, sanitary and economic advantages could be insured by this biotechnology and overcome many problems in regard with camel fertility efficiency such as short breeding season, long gestation period, traditional reproductive management and widespread of genital and venereal infectious diseases.

Keywords Artificial insemination · Oestrus synchronization · Dromedary camel · Camel bull · Breeding · Reproduction performance · Sperm · Follicular cycle · Female camel · Calving interval

Abbreviations

°C	*degrees Celsius*
AF	*annual herd fertility*
AI	*Artificial insemination*
AV	*artificial vagina*
BLUP	*Best Linear Unbiased Prediction*
CASA	*Computer Assisted Sperm Analysis*
CFM	*Calving to first mating interval*
CI	*Calving interval*
CIDR	*Controlled Internal Drug Releasing*
COI	*Calving-oestrus interval*
DC	*dromedary camel*
eCG	*equine chorionic gonadotropin*
EDTA	*Ethylenediaminetetraacetic acid*
FAO	*Food and Agriculture Organization*
GnRH	*gonadotropin releasing hormone*
GPG	*GnRH-PGF2 α-GnRH*
h	*hour*
hCG	*human chorionic gonadotropin*
IIED	*International Institute for Environment and Development*
ILRI	*International Livestock Research Institute*
IMV	*Institut de médecine vétérinaire*
INRA	*Institut national de la recherche agronomique*
LH	*luteinizing hormone*
M1, M2	*Maiting1, Maiting 2…*

mA	*ampere*
min	*minutes*
ml	*millilitre*
NGF	*nerve growth factor*
OD	*Open days*
OIF	*ovultion inductor factor*
PGF2α	*Prostaglandin F2α*
PHT	*Progeny History Testing*
PRID	*Progesterone Relasing Intravaginal Device*
RP	*reproduction period*
SeNPs	*Selenium nanoparticles*
SSO	*small and smooth ovaries*
V	*volt*
WP	*waiting period*
ZnONPs	*Zinc oxide nanoparticles*

2.1 Introduction

The dromedary camel (DC: *Camelus dromedarius*) is the best livestock animal well adapted to arid ecological conditions. It fights against desertification by using a high floristic biodiversity with high digestibility and better food efficiency. In addition, its low gregarious outpatient pasture and long intestinal transit contribute for a better germination of the seeds after excretion. Camels are multipurpose animals well adapted to harsh environment. They are used for nutritional, dietetic, economic, social and cultural purposes. They offer milk and meat production, hair and are used for racing, transportation and tourism (Al Abri and Faye 2019). The nutritional value of camel meat has been shown to be much higher than beef, lamb and goat because it contains less fat with a comparable index of essential amino acids (Raiymbek et al. 2015). Camel meat is rich in essential amino acids such as leucine and lysine, essential fatty acids such as omega 3, minerals, vitamins and bioactive compounds such as carnosine, anserin, glutathione, etc. (Kadim Isam et al. 2018). Camel milk is 3–10 times richer than cow's milk in vitamin C. (Faye 1997; Konuspayeva et al. 2009; Al Abri and Faye 2019). Its hypo-allergic property is due to its low content on β-casein and β-lactoglobulin (Konuspayeva et al. 2009). It's whey protein fractions are higher (El-Agamy 2006). So, dromedary camel contributes to the subsistence of transhumant and nomadic Saharan communities, often with restricted access to various protein resources of animal origin. Thus, it constitutes for them an element of food security in the face of climate change, a long-term capital ensuring the well-being of the family and an element of social prestige. Finally, the dromedary participates in sports and leisure activities and represents a cultural heritage in many countries and contributes in socio-economic activities and diversification of many marginalized area of arid areas worldwide (Al Abri and Faye 2019).

Poor reproductive performance in camel species is mainly due to a low fertility rate and increased inter-calving interval. The low numerical productivity of camel herds is often discussed among cameleers with adversarial opinions in regard to its aspects and its risk factors. However, the veterinarian should provide advice based only on specific information with reference to herd reproductive history. The serious difficulty often arises is about the reproduction evaluation in DC herds is often confronted to the following constraints: a pastoral breeding system in continuous displacement, traditional breeding practices and lack of recording information related to reproduction, absence of state veterinary services interventions on reproduction associated to difficulties to reach remote breeding areas for regular monitoring by private veterinarians, lack of appropriate training for health and management of camel herds. The diagnosis and treatment in this animal are generally approached in the same way as for other farm animals.

The reproduction control in camel species implements technical procedures or herd management strategies that make possible optimizing reproductive performances, particularly during anoestrus period (seasonal, post-partum) taking into the count the breeding systems of the species. It aims to reduce the rate of infertility, choice of calving period, calving synchronization, and decrease of unproductive periods (advancement/synchronization of puberty, decrease of anoestrus length, mating and conception during the non rutting season). Efficient methods to reproduction control in the camel are required as the interest grows in the potential application of artificial insemination and embryo transfer as tools for improving genetic traits, such as milk, meat and wool production and racing ability (Al Eknah 2000).

Compared to other farm animals, little research on camel reproduction control and artificial insemination has been found in the camel side. Oestrous synchronisation solves oestrous detection problems and makes artificial insemination more effective (Helmy 1991; Minoia et al. 1992). In female camels, ovulation may happen through intrauterine semen deposition without the need for coitus (Chen et al. 1985; Musa et al. 1990). This makes AI a reliable technology for continuous genetic advancement using males of high genetic potential and inseminating female groups that have undergone synchronization of follicular development. However, the large-scale use of this biotechnology faces many difficulties, such as semen selection, male camel sexual activity, heterogeneity of recovered semen quality (small volume, low sperm concentration and high viscosity) (Skidmore et al. 2013), limited knowledge of semen storage, optimum insemination time and sperm dose (Al-Bulushi et al. 2018).

2.2 Reproduction and Genetic Improvement of Dromedary Camels

The DC (*Camelus dromedarius*) is known for its rather long reproductive cycle characterized by a long prepubertal period, long gestation period, long interval between generations and a low fertility rate (Gherissi et al. 2020a). Likewise, the incidence of genital pathologies, abortions and neonatal mortality are high (Gherissi et al. 2019, 2020a, b). All these factors together lead to a low numerical productivity of camel herds, particularly in extensive breeding system (Faye 2018; Brigitte 2005).

Genetic improvement according to the classical method involves the animal identification, assessment of their performance, selection of breeders and use of breeders in order to achieve genetic progress. In fact, genetic improvement can only take place if performance and pedigree are recorded. Selection would be effective when it is carried out following estimation of breeding value using progeny (BLUP: Best Linear Unbiased Prediction) widely practiced in various animal species.

$$BV_p = 2\left(\frac{n}{n+k}\right)\left[P(progeny) - CA\right]$$

$$k = \frac{4-h^2}{h^2} \qquad r^2 = \frac{n}{n+k}$$

BVp: breeding value of the parents, r^2: reliability, n: number of offspring (informants), P (progeny): Mean value of the performance of the progeny, CA: Comparison average which characterizes the environment.

According to the previous formula, the precision of the selection depends closely on the quality and quantity of the individual controls of the available informants and the herd size used to calculate comparison average of the performance. Thereby, the low reproduction levels within camel herds can generate serious difficulties in order to give to appreciate with great reliability the genetic potential of the parents who will serve as reproducers (Emami Mibody et al. 2016). Contrariwise, good herd reproduction statue make easier to appreciate genetic value and animals with the highest predicted value can be selected as parents with minimum error.

On the other hand, genetic progress level depends on the variance of the genetic value in the population, the intensity of selection, the reliability of the estimated breeding value and the generation interval. In this regard, Al-Sobayil et al. (2006) and Almutairi et al. (2010) showed that camels species have a high genetic variability reflected in the heritabilities of various traits (body weight and growth rates were moderate to high; $h^2 = 0.24–0.40$, birth weight $h^2 = 0.37$, Daily gain ranged $h^2 = 0.25$ and 0.49, milk yield at 305 days $h^2 = 0.24$ and test day yields $h^2 = 0.22$). This reflects a potential for ample genetic gain if systematic selection is to be implemented (Al Abri and Faye 2019; Bahbahani et al. 2019).

$$GP\ \ per\ \ year = \frac{S_A \cdot i \cdot r_{VE,\widehat{VE}}}{t}$$

GP: genetic progress, S_A: variance of the genetic value in the population, i: the intensity of selection, $r_{VE,\widehat{VE}}$: reliability of the estimated breeding value t: generation interval.

The selection intensity was indicated by the proportion of animals required for the next generation to be parents. Reproductive efficiency and management of camel herds therefore have an important influence on the rate of parents necessary for the next generation to be produced and, therefore, on the rate of genetic progress. The generation interval in this species is long (approximately 8 years) and is highly dependent on genetic potential, environmental conditions, husbandry systems and breeding methods. A higher reproduction rate for a given population size means a lower number of breeding animals and, therefore, a higher selection intensity. A greater number of offspring per breeder also promotes a more accurate estimate of genetic values (Emami Mibody et al. 2016). Another benefit of increased reproductive rates is the faster dissemination of superior genetic material. In this context, reproduction control, in particular using reproductive biotechnologies, can have a direct effect on increasing the rate of genetic progress of camel herds.

Artificial insemination (AI) leads to high reduction in the use of breeding males. In dromedaries, male and female camels with high genetic values could constitute a nucleus of selection of less than 1% of the entire population. AI increases the available information on siblings and therefore increases the genetic value accuracy estimated in their parents (Van Arendonk and Bijma 2003). Furthermore, AI allows increasing the selection intensity and precision of males based on the progeny test especially for traits with low heritability (such as functional traits). The semen exchange between different herds allows for the establishment of genetic links between them and reduces the risk of venereal diseases transmission between herds. Sperm freezing gives organizations and AI centers ability to create camel gene banks as back-up stores of camel genetic diversity. In addition, cryopreservation would facilitate the exchange and transport of semen and the international exchange of camel's genetic material.

AI requires availability of technical skills in AI centers and camel farms with effective means of communication between these two stakeholders. However, in many camel breeding countries, financial means are lacking and the structuring of camel breeding is inadequate for setting up successful AI operations. In addition, most camel herders are reluctant to camel breeding intensification and prefer mobile and pastoral systems which make difficult to follow the follicular cycle and implement timed AI at large scale. Therefore, AI is probably not predictable for livestock in extensive grazing systems.

2.3 Reproductive Cycle of Female Camels

The female camel is strict monotocus and seasonal polyoestral breeder (Akral and Khanna 1995). During the non breeding period the sexual behavior disappears but the follicular waves persist at low growth levels (Gherissi et al. 2018) similar to what happen in ewe (Dogan et al. 2020) but different to mare who undergoes a deep ovarian rest in the non breeding season (Boeta et al. 2006). Ovulation is induced by GnRH-like factor or Ovulation inductor factor (OIF) contained in camel bull seminal plasma which is identified as nerve growth factor (bNGF) (Sanjay et al. 2012; Bogle et al. 2012) also with luteotrophic effect on the corpus luteum (Silva et al. 2014). This molecule induces pre-ovulatory luteinizing hormone (LH) surge 2–3 h after mating (Marie and Anouassi 1986) followed by ovulation (32–40 h after copulation) and subsequent formation of corpus luteum (El-Allali et al. 2017). Without mating at the breeding period (mature follicle of 11–19 mm); female's exhibit repeated cycles of follicular growth and regression without a cyclic corpus luteum. Three main phases have been described for follicular wave in female camel (Fig. 2.1): growth, maturity and regression. After emergence of small follicles

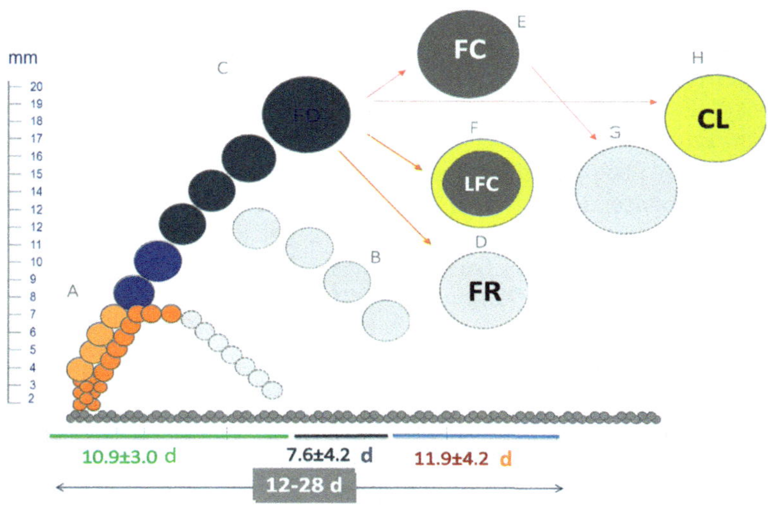

Fig. 2.1 The main follicular cycle phases in female camel (*FC* follicular cyst, *LFC* luteinized follicular cyst, *FR* dominant follicle regression, *CL* corpus luteum)
Following this first recruitment phase, several follicles (3–6) continue to grow (5–17 mm) until the emergence of one or two large dominant follicle which becomes mature; including 5–10 mm medium size follicles and 11–17 mm pre-ovulatory follicles. The dominant follicle undergoes a stagnation stage for 5–7 days. In 50% of unmated females, regression occurs until disappearance in the ovarian stroma. In other unmated females (50%) the largest follicle continues to grow attaining a large size (30–65 mm); this cyst-like follicle cannot ovulate spontaneously however, it regresses thereby permitting the expression of the next follicular growth waves in an average of 18 days later. In mated females, following ovulation, the mature follicle is usually transformed into a functional corpus luteum. (Adapted from Zarrouk et al. 2003)

(<4 mm in diameter) at the periphery of the ovaries during the recruitment phase, several follicles (3–6) continue to progress as medium size follicles (5–10 mm) until the dominance of one or two large follicles (11–19 mm) that become(s) mature pre-ovulatory follicle(s). The growth phase (<4 mm–10 mm) extend over 10.9 ± 3.0 days and dominance phase lasts over 7.6 ± 4.2 days (Skidmore et al. 1994).

Follicular regression occurs in 50% of unmated females until total desperation of these follicles in the ovarian stroma. The largest follicle tends to develop in other uncoupled females (50%) to reach an oversized follicle (>25 mm), which constitutes a follicle cyst that decreases ovulatory response (Tibary 2018). Presence of 2 or more co-dominant follicles in camelids is not rare and can occur in up to 40% of the follicular wave (Manjunatha et al. 2015; Campbell et al. 2015). A combination of stimuli, including the chemical factor in the seminal plasma, neurohormonal responses to the mechanical stimuli of coitus and the male effect, could result in the ovulatory response in the camel (Al Eknah 2000).

The anovulatory hemorrhagic follicles cysts seems to be the most frequent among ovarian cysts in female camels (Gherissi et al. 2019; Tibary 2018). The pathophysiology of anovulatory hemorrhagic follicles is poorly understood and their presence seems to not disturb follicular wave patterns (Tibary 2018). The follicular cyst or the corpus luteum of the non pregnant females regresses 18 days and 11 days later, respectively (Fig. 2.1), allowing the next waves of follicular growth to begin (Zarrouk et al. 2003; Marie and Anouassi 1987). The total length and the seasonal frequency of follicular wave in the female camel was found to vary considerably (17–28 days and maintain or reduced frequency of follicular waves) according to the geographic location, nutrition statute, start and end of the breeding season, body condition, photoperiod (Abdoon 2001; Vyas et al. 2004; El-Allali et al. 2005; Sghiri and Driancourt 1999; Gherissi et al. 2018, 2020b). In mated females, the ovulated follicle is usually transformed into a functional corpus luteum and its maintenance is conditioned by the suppression of PGF2α release before day 10 after ovulation (Skidmore et al. 1998).

2.4 Camel Herd Reproductive Performance Evaluation

2.4.1 Data Collection for Reproduction Analysis

Obtaining individual reproductive history is integral to an accurate evaluation of reproductive herd performances and to explain eventual difficulties that arise there. Regarding the reproductive particularities under pastoral camel breeding conditions, a method based on data collection from breeder's offspring history called "Progeny History Testing" (PHT), has been suitably adapted to camel reproduction investigations. This technique was developed firstly in Ethiopia by a team from the Ministry of Agriculture, and then widely used by the International Livestock Research Institute (ILRI). PHT method was approved by the International Institute for Environment and Development and the World Food and Agriculture Organization (IIED and FAO) in their data collection guides for different farming systems.

Under semi intensive management system, identification of all the animals in the herd using the most relevant means to camel species is essential. The breeding management in general and reproduction in particular must be done in compliance with certain preliminary conditions relating to the animal and its environment leading to monitoring and data collection by the breeder himself and stakeholders, in particular by taking observations, clinical examinations and preventive or curative treatments. The observations recording in semi-intensive camel farms is a concept that is absent in almost all camel breeders. This preliminary practice for the evaluation of the reproductive status of farms is fundamental and must be encouraged by using clear or even cryptic language. A mutually agreed system of abbreviation ratings can be very useful. Such information is manifold; they are particularly linked to calving (day, number, nature, type, interventions, etc.), sexual cycle (postpartum cycle number, number of cycles used, etc.) mating (days, duration of sexual receptivity, mating time, mating after induced or natural oestrus), pregnancy diagnosis (day, nature, result), body condition score, pathologies (absence of oestrus, ovarian and uterine pathologies, infections, embryonic mortality, mastitis, etc.). All observations must always refer to the identity of the animal. It will be accompanied by the date or even the time of the observation as well as the nature of the observation. The use of recording, scoring and organizing observations for processing is highly recommended. The most modest would be a simple farm management camel breeding calendar. Computerized systems adapted to camel breeding could emerge, which would effectively improve the management of several other aspects related to reproduction such as milk production, nutrition, etc.

2.4.2 *Reproduction Indicators of Camel Herds*

2.4.2.1 Calving Interval (CI)

The CI is the most important criteria to look at when talking about reproduction of the camel. The lengthening of this parameter leads to increase unproductive period (even drying period). It also decreases the herd numerical productivity by decreasing offspring production per year. Finally, it increases reform rate and herd replacement cost. Therefore, camel herders have every interest in the fact that their female camels give births often with the shortest intervals throughout their whole life.

In terms of herd performance; the interest is to measure the average of this parameter over the entire present females in order to identify the multiple individual and collective reasons for its lengthening/shortening. It is possible to shorten the CI of females by acting on waiting period (WP: time between calving and first mating) and reproductive period (RP: time between the first mating and the conception) (Fig. 2.2). WP can be shortened by rigorous peri-partum management with particular attention to late pregnancy and parturition (preparation, obstetric assistance and

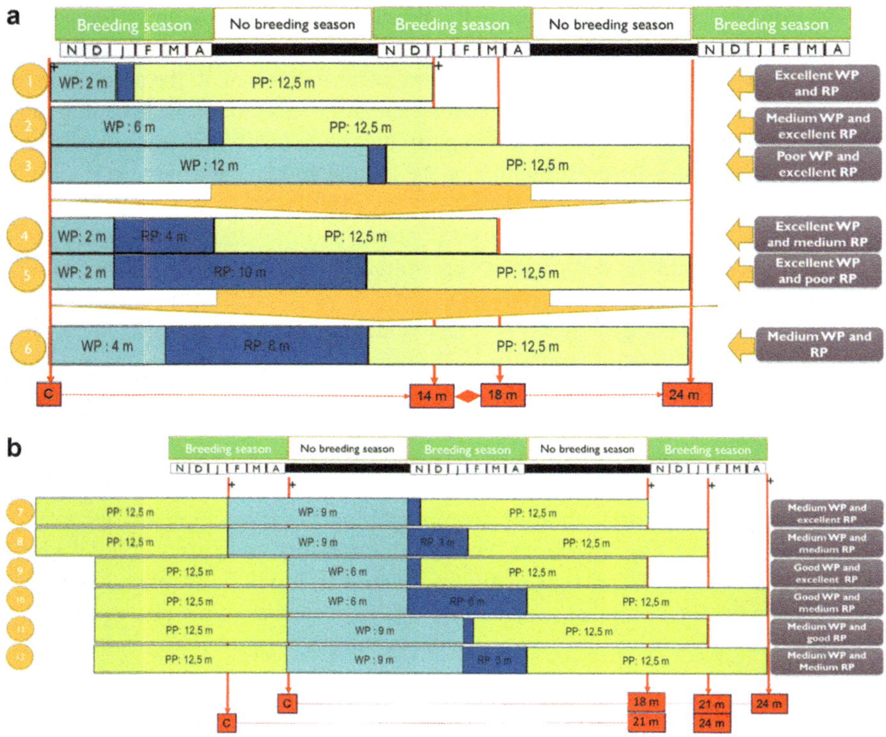

Fig. 2.2 Potential factors responsible for calving interval (CI): 18 months as objective and 24 months as threshold when calving occur at the beginning (**a**) or the middle/end (**b**) of the breeding season. WP: Waiting Period, RP: Reproduction Period, PP: Pregnancy Period. Calving at the beginning of the breeding season results on 2 to 12 months period of open days (OD) and females can achieve CI equal or less than 24 months. Excellent WP and RP lead to OD of about 2 months and CI of 14 months (box 1). Medium WP or RP lead to OD of about 6 months and CI of 18 months (boxes 2 and 4). Females with poor WP or RP will have an OD of 12 months and CI of 24 months (boxes 3 and 5). This last situation (CI of 24 months) could occur when moderate reproductive difficulties affect both WP and RP (Medium WP and RP, box 6). Animals with difficult reproductive status show association of anoestrus and infertility that lead to poor WP and RP and lead to OD and CI exceeding 12 months and 24 months (out of threshold), respectively. However, when the ovarian resumption and/or uterine involution periods overlaps with the seasonal anoestrus, the CI would be approximately 18–24 months. This is especially true when calving occurs on the middle or at the end of the breeding season. In this case, medium WP with excellent to good RP would give CI of 21–24 months (boxes 7, 8, 11 and 12). Similarly, the calving interval would be 21–24 months if there is a good WP with excellent to medium RP (boxes 9 and 10). In case of prolonged anoestrus associated with subfertility, the CI would exceed 24 months (out of threshold). (Adapted from Gherissi et al. 2020a)

balanced nutrition) and using good management practices (inputs and offspring's weaning). Calving at the beginning of the breeding season coincides with the wet period and food availability, which promotes a good body condition and early resumption of ovarian activity with mating and pregnancy during of the same season (Fig. 2.2). Regular checkups of postpartum animals to diagnose and treat related health problems (including mastitis, uterine infections, intercurrent diseases) and stillbirths are also recommended during WP. The RP management must be implemented by a regular balanced supply diet during the late pregnancy and postpartum period, camel bull flushing and sexual behavior and fertility control (male/female ratio, reform due to 'infertility …), as well as the specific diagnosis and treatment of genital diseases (metritis, bursitis …) (Tibary et al. 2005; Ali et al. 2010).

2.4.2.2 Calving-Oestrus Interval (COI)

Healthy female camels show first postpartum oestrus between 14 and 42 days under pastoral livestock systems (Elias et al. 1984). When female camels were mated between 5 and 6 weeks post partum, adequate conception rates were obtained (Derar et al. 2014). The COI could, however, be extended to the next breeding season when calving occurs in the middle or at the end of the breeding season. Taking into account these factors, the COI target was set at 2 to 6 months and the threshold at 9 months (Gherissi et al. 2020a; Table 2.1). Open days (OD).

High rates (70–80%) of cyclic females were reported at 30 days postpartum in well-fed dromedaries, therefore the conception rates at 45 days postpartum were about the average scores (Tibary and Anouassi 1997a; Derar et al. 2014), but ovarian cyclicity can resume after 8–10 months when the diet is inappropriate for ovarian functioning (Skidmore et al. 1994). Therefore; when female camels conceive within the same calving season OD would be less than 6 months. However, if the dam is mated during the next breeding season OD period could last, instead form 7 to 12 months. Accordingly, the objective for this parameter was set at 180 days and the threshold at >360 days (Gherissi et al. 2020a, Table 2.1).

Table 2.1 Standard objectives and thresholds of reproductive parameters of female camels

Variable	Unit	Objective	Threshold
Age at first oestrus	Months	24–36	>42
Age at first mating	Months	24–36	>48
Birth to conception interval	Months	36–48	>54
Age first calving	Months	48–54	>60
Calving to oestrus interval	Months	2–6	>9
Open days	Days	≤180	>360
Calving interval	Months	≤18	>24
Annual fertility	%	>85	<75
Mating-pregnancy diagnosis	Days	21	30
Pregnancy length	Months	12.5	13.5
Culling age	Years	15–17	> 17
Calving number	Integer	5	$7 > n > 3$
Culled females per year	%	<5	>30

Adapted from Gherissi et al. (2020a)

2.4.2.3 The Annual Herd Fertility AF (Pregnancy Rate)

It reflects the ratio within a year between the number of pregnant female camels and the number of mated females. The fertility rate for camel herds looking to improve camel numerical productivity should rages between 70% and 95% (Faye 2018). The agreed objective and threshold for this parameter are >85% and <75%, respectively (Gherissi et al. 2020a).

2.4.2.4 Female Camel's Calving Interval Objective and Threshold

The CI depends waiting period (postpartum to first mating), reproduction period (first mating to conception) and the female's fertility. Calving to first mating interval (CFM) in pastoral conditions occurs within 14 to 42 days (Elias et al. 1984). Derar et al. (2014) showed that camels could be mated between the fifth and sixth week postpartum with satisfactory open days and conception rates. The time to first oestrus (associated to first mating) after calving varies from 1 to 38 months with mean value of 10.3 ± 5.77 months (Gherissi et al. 2020a). High levels of CFM would be due to poor nutritional conditions leading to late resumption of follicular activity that takes up 8–10 months (Gherissi et al. 2018). The pregnancy duration in female camel is about 12.5–13 months (Gherissi et al. 2017).

The prompt ovarian resumption and occurrence of first mating earlier in the same calving season allow considering CI objective of 14–18 months and alarm threshold of 24 months (Fig. 2.2a). If calving occurs in the middle or late breeding season, the CI target would be 18–21 months and the alarm threshold would be 24 months (Fig. 2.2b). In fact, the objective CI is set at ≤18 months and the alarm threshold is set at >24 months (Table 2.1).

2.5 Reproduction Control in Dromedary Camel

Reproductive control tools can reduce unproductive periods by extending breeding season length and stimulating follicular growth and inducing ovulation during the non breeding season. They facilitate allow rational herd management to improve animal's resilience to harsh livestock condition under arid environment (calves care, fodder availability and postpartum control) and to adapt herd production according to market needs (milk, fattening and calves). In addition, they allow accelerating genetic progress, firstly by facilitating the widespread dissemination of AI and secondary by developing ovarian stimulation protocols and embryonic transplantation.

Synchronisation of ovarian activity principles applied in cows are used to synchronize ovarian activity in female camels, but methods are adjusted to physiological and behavioral characteristics of camelids, resulting in varying degrees of performance. The reproduction control in female dromedary camels is possible

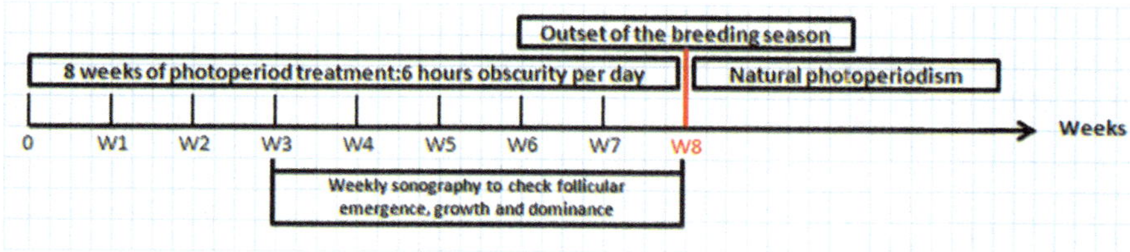

Fig. 2.3 Photoperiodic treatment of female camels 2 months before the beginning of the breeding season

using zootechnical means such as photoperiodic control (Vyas et al. 2008; Swelum et al. 2018a), food supplementation (Hammadi et al. 2001), male effect (Hafez and Hafez 2001) or hormonal exogenous supply such as melatonin (Dholpuria et al. 2012; Swelum et al. 2018b; Ainani et al. 2018; El-Allali et al. 2018) or progesterone (Monaco et al. 2013; Swelum and Alowaimer 2015; Swelum et al. 2018c), GnRH and PGF (Quzy et al. 2013; Al-Sobayil 2003, 2008; Manjunatha et al. 2018). The different applied protocols would be followed by blind natural mating or AI in fixed time using fresh or frozen semen.

2.5.1 Photoperiodic Control

In female camels, the photoperiodic control consists in applying a blindfolded eyes using black mask 6 h per day beginning 2 months before the start of the breeding season (November) (Vyas et al. 2008). Darkness perception by female camels using photoperiodic control leads to melatonin synthesis and release which advances the date of the sexual season. This remarkably induced a good follicular growth activity from the third week of treatment and the development of the ovulatory follicles between fifth and seventh week of treatment (Fig. 2.3; Vyas et al. 2008). Likewise, the ovulation, mating and pregnancy rates are quite considerable (Vyas et al. 2008).

In camel bulls, significant increase in testicular measurements and volume, libido, melatonin and testosterone concentrations were achieved by blindfold application during the non-breeding season to induce a shortened daily photoperiod of approximately 2.55 h relative to the natural day cycle (Swelum et al. 2018a).

2.5.2 Male Effect

Just one study was found in literature, documenting the pheromonal "male effect" causing female camels to cycle earlier in the breeding season (Hafez and Hafez 2001). In fact, the use of the male effect in camel species is rarely considered. Behavioral and endocrine processes of this practice to control reproduction in camels are by far to be compared with other species such as sheep and goats. These two species are spontaneous ovulation while the camel is characterized by induced ovulation through mating.

2.5.3 Flush Feeding

The daily feed supplementation using 4 kg of commercial concentrated food during the late pregnancy (about 2 months before calving) and at the start of the postpartum period (5 kg/day for 3 months) helps to improve the female camels overweight at the time of calving and allows a rapid and significantly higher rate of ovarian activity resumption (Hammadi et al. 2001), short waiting period (PA) (around 40 days), high rate (approximately 70%) of females mated for the first time at 60 days after parturition (Hammadi et al. 2001) and high conception rate (Mostafa et al. 2016a, b). In addition, this practice helps to better growth of camels up to the age of 90 days (Hammadi et al. 2001).

2.5.4 Ultrasound-Guided Aspiration of the Follicle

The elimination of the dominant follicle using ultrasound-guided fine needle aspiration makes possible to end the current follicular wave and emergence of new follicular pool in an average time of 2.3 ± 0.5 days. The dominant follicle deviation occurs after 8.8 ± 1.1 days (Skidmore et al. 2009).

2.5.5 Exogenous Melatonin Based Protocols

The effectiveness of treatment based on exogenous melatonin consists of mimicking short days either at the beginning or at the end of the breeding season. For female dromedary camels; two melatonin treatment methods are possible. The first option is to administer it 2 months ahead the natural breeding season to support the follicular waves emerging and ultimate growth of the dominant follicle(s) at the start of the breeding season (Dholpuria et al. 2012). For example, in Algerian local condition it carries out at early September and continues until the end of October over 8-week period (Fig. 2.4). The second option is to start treatment at the end of the breeding season to allow an ovarian follicular activity extension for the rest of the year (non breeding season) (El-Allali et al. 2018). This treatment begins at early April and ends in late May (Fig. 2.5). At the end of these two proposed treatments, monitoring of ovarian activity is recommended to ensure the quality of ovarian cycles.

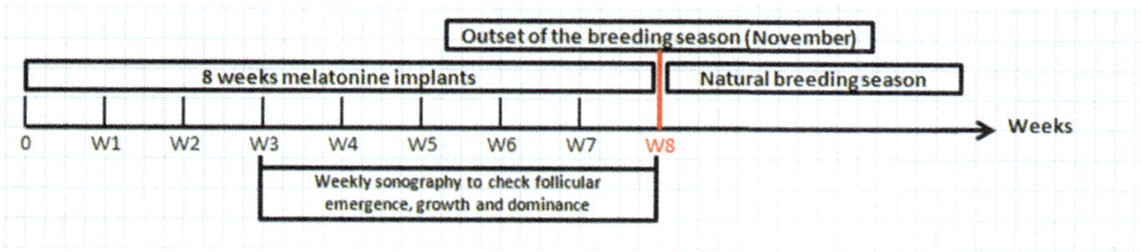

Fig. 2.4 Exogenous melatonin treatment of female camels 2 months before the beginning of the breeding season

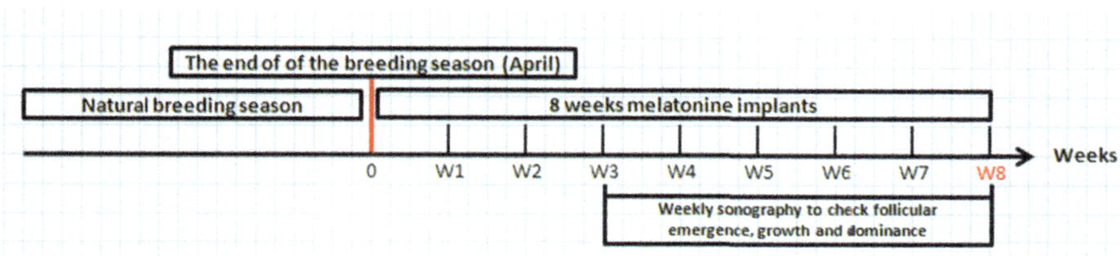

Fig. 2.5 Exogenous melatonin treatment of female camels 2 months after the end of the breeding season

To administer exogenous melatonin for camels, Dholpuria et al. (2012) recommend to use commercial implants containing 18 mg of melatonin at a dosage of 1 implant per 28 kg bodyweight (approximately 20 implants per female camel), which to be inserted subcutaneously by a 12-gauge applicator, preferably sterilized beforehand (Fig. 2.6).

For camel bull; the treatment during the non breeding season with melatonin implants (18-mg per 28 kg of live bodyweight) or approximately 28–32 implants per bull two times at an interval of 35 days helps to improve significantly their reproductive performance (Plasma testosterone, testicular measurements, sexual behavior and spermogram) (Swelum et al. 2018b).

2.5.6 Exogenous Progesterone Based Protocols

Daily progesterone injection (100–150 mg) and various devices secreting progesterone (CIDR 1.38 g: Controlled Internal Drug Releasing Device, new PRID Δ 1.55 g: Progesterone Relasing Intravaginal Device) are experimented to synchronize follicular waves in female camels (Cooper et al. 1990, 1992; Monaco et al. 2013; Swelum and Alowaimer 2015; Hussein et al. 2015; Swelum et al. 2018c; Tibary 2018; Abo El-Maaty et al. 2019). Cleaning of perineum and flushing of vagina with iodopovidone solution before device insertion is very important since it clinically improves vaginal environment at the end of the treatment (Monaco et al. 2013). Female camels receive CIDR for 7 days (Abo El-Maaty et al. 2019), 10 days

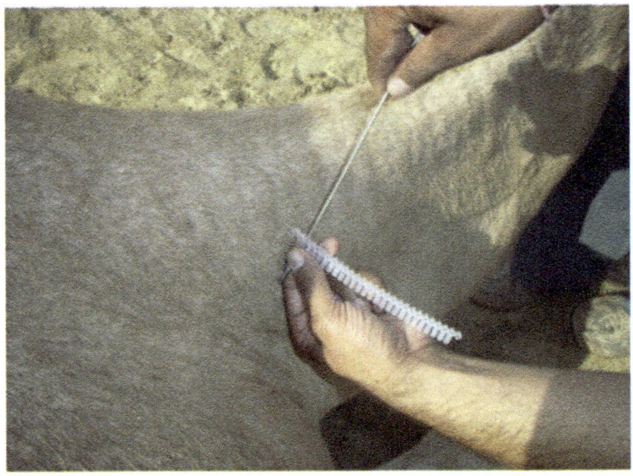

Fig. 2.6 The technique of application of subcutaneous melatonin implants in camel. (From Dholpuria et al. 2012)

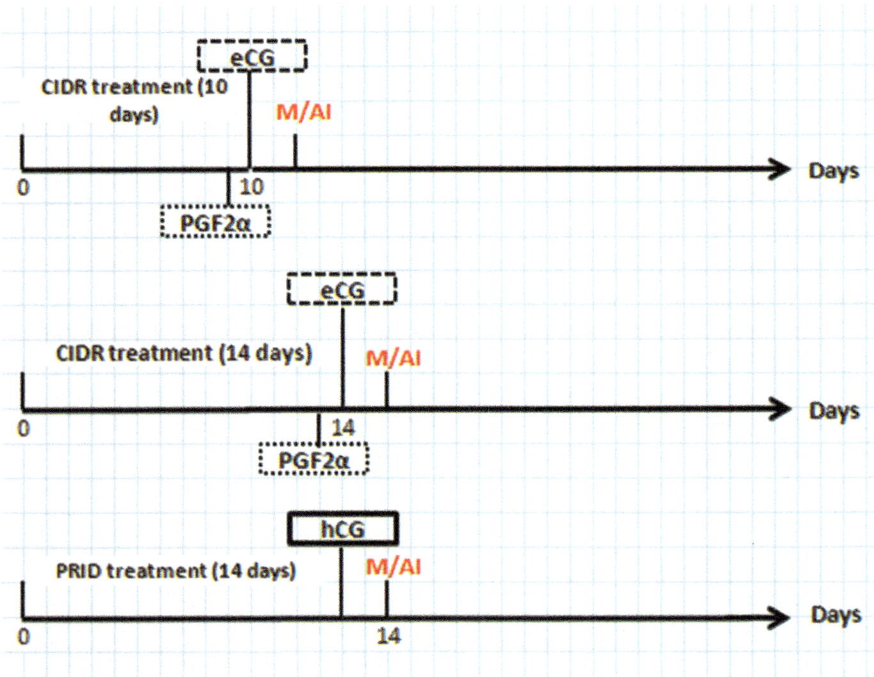

Fig. 2.7 Exogenous progesterone treatments during intra-vaginal devises (CIDR, PRID) with eCG or hCG and timed artificial insemination

(Monaco et al. 2013, Fig. 2.7), 14 days (Swelum and Alowaimer 2015; Al-Fatlawi and Al-Hamedawi 2017; Fig. 2.7) or 17 days (Al-Sobayil 2008), while new PRIDΔ is held 14 days for Swelum et al. (2018c; Fig. 2.7). The progesterone plasmatic concentration increases earlier after device insertion (2 days) but its plasmatic peak is observed after 12 days (Swelum and Alowaimer 2015).

There are conflicting studies on the effectiveness of PRIDs and CIDRs (Monaco et al. 2013; Tibary 2018). The authors conclude that the treatment of exogenous progesterone could fully prevent sexual receptivity and suppress the development of large follicles without suppressing follicular wave's growth. These devices have sometimes been related with increasing spontaneous ovulation (Tibary 2018).

There are conflicting reports on the efficacy of PRIDs and CIDRs (Monaco et al. 2013; Tibary 2018). The authors opinions converge that exogenous progesterone treatment could completely prevent sexual receptivity and suppresses the growth of large follicles but could not suppress the follicular wave. Sometimes, these devices have been associated with increased spontaneous ovulation (Tibary 2018). Injection of PGF2α (500μg) one day before CIDR withdraw (Monaco et al. 2013), eCG (2000 IU) at the day of CIDR withdraw (Al-Sobayil 2008) or hCG (3000 IU) at the last day of PRID treatment (Al-Sobayil 2008) were also practiced in female camels (Fig. 2.7). The daily injection of progesterone (100 mg, IM) for 10–16 days provided promising results in multiple ovulated female camels (McKinnon et al. 1994; Tibary 2018). More comprehensive studies are necessary to better understand the effectiveness of different associated treatments to exogenous progesterone administration on follicular wave synchronization in female camel. A recent study found that long-acting injections of progesterone can be used in camels and can be more effective than daily injections (Chhaibi et al. 2016). Synchronizing follicular waves using progesterone devices reminds a controversial issue between the authors particularly on their effects at different breeding seasons (Tibary 2018). Their effectiveness at the beginning and in the middle of the breeding season has been reported vain due to its action in reducing the follicular diameter but not follicular number (Monaco et al. 2013). During the breeding season Swelum and Alowaimer (2015) reported a high percentage of camels presenting ovulatory follicles 2–4 days after CIDR withdrawal showing therefore its significant interest in follicular wave synchronization. In another more study; Swelum et al. (2018c) confirmed the interest of new PRIDΔ in the synchronization of follicular waves during the breeding season with a high rate of females which carry ovulatory follicles 2 days after the end of treatment. The ovulation rate, the non-return rate are significantly increased, the open days period reduced mainly and the fertility rate is higher when the blind natural mating or timed AI are practiced 48 h after PRID withdraw (Al-Sobayil 2008; Al-Fatlawi and Al-Hamedawi 2017; Swelum et al. 2018c).

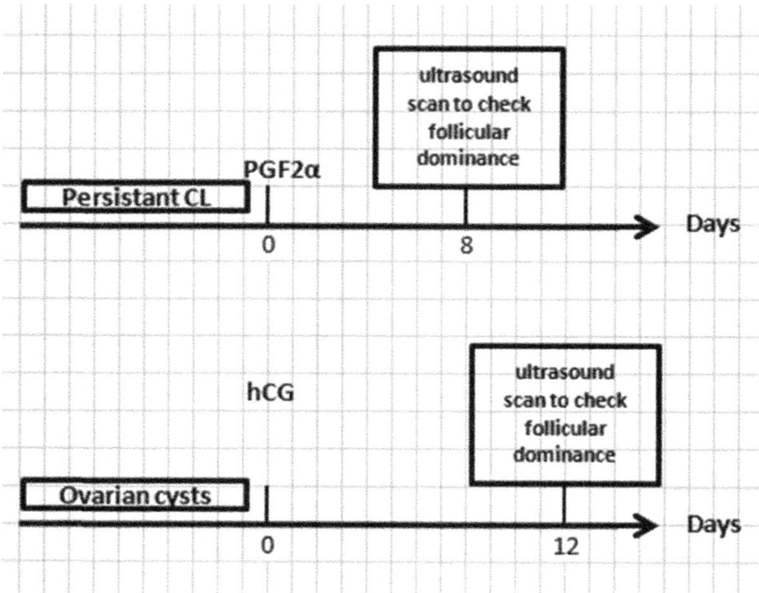

Fig. 2.8 Prostaglandine (PGF2α) and human chorionic gonadotrphin (hCG) treatments of persistent corpus luteum and ovarian cysts, respectively

2.5.7 Prostaglandin and Human Chorionic Based Protocols

PGF2alpha (500μg) administration for female camels with a persistent CL and hCG (4500 IU) for those with cystic ovaries were reported to have high efficacy on the number of females with ovaries showing ovulatory follicles 8 and 12 days post-treatment, respectively, but the pregnancy rate for both protocols remained low (Quzy et al. 2013; Fig. 2.8).

2.5.8 GnRH-Based Protocols

"*Select synch*" treatment 2 months ahead the natural breeding season with GnRH (250μg) on day 0 and PGF2α (5 ml) on day 7 (Skidmore et al. 2009) or day 10 (Mostafa et al. 2016a, b) did not improve the quality of follicular wave at the end of the treatment (Fig. 2.9). GnRH alone may be effective in synchronizing follicular waves to mitigate the harmful effects of prostaglandin on the antioxidant potential of camels and their response to breeding (Abo El-Maaty et al. 2019). Follicular wave emergence and follicular deviation occur about 3 days (range 2.5 and 3.5 days) and 5.5 days (range 4 and 11 days) respectively, after GnRH administration

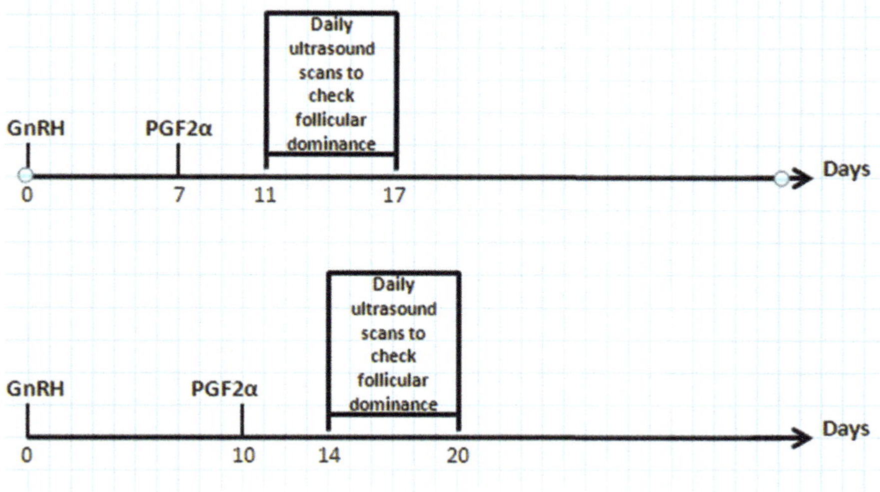

Fig. 2.9 Select synch (GnRH-PG) protocol for oestrus induction in female dromedary camels

(Manjunatha et al. 2015; Skidmore et al. 2009) and 4.5 days and 10.7 days for GnRH-PGF2α treatment.

The Ovsynch (GnRH-PGF2α-GnRH: 0–7-9) and eCG (2000 IU) treatments for female camels with small and smooth ovaries (SSO) 2 months ahead the natural breeding season induced high effective rates of animals with matures follicles at day 8 and day 12 post treatment, respectively (Quzy et al. 2013; Fig. 2.10). The GPG (0–7-14) treatment showed high proportion (≈73%) of female camels had a total of 22 dominant follicles (1.3–1.9 cm) of which 21 ovulated after GnRH injection day 14 (Skidmore et al. 2009; Fig. 2.10).

Ovsynch (0-7-10- 22:M1) protocol with timed natural mating followed by pregnancy diagnosis 21 days later and ovarian ultrasonographic monitoring of non pregnant females for second timed mating was examined by Manjunatha et al. (2018) with high rate of follicular wave synchronization and consistent pregnancy rate (Fig. 2.11).

Ovsynch -Resynch (GPG-M1-GP-M2; Fig. 2.12) treatments were experimented recently by Manjunatha et al. (2018) in dromedary camels during the breeding season. Firstly these authors used Ovsynch protocol (GnRH-PGF2α-GnRH: 0-7-10) followed by timed mating 12 days later (day 22). Then, systematic GnRH injection was received by all female camels at day 36 regardless to their pregnancy statue; followed by pregnancy diagnosis at day 43. PGF2α (day 43) was administrated to non pregnant females which were resubmitted to second timed mating at day 48. This protocol showed high proportion of ovulated female camels after Ovsynch protocol. The GnRH in Resynch protocol resynchronized effectively the follicular development for second mating of non pregnant females with high early and late pregnancy rates and low rate of pregnancy loss (Manjunatha et al. 2018).

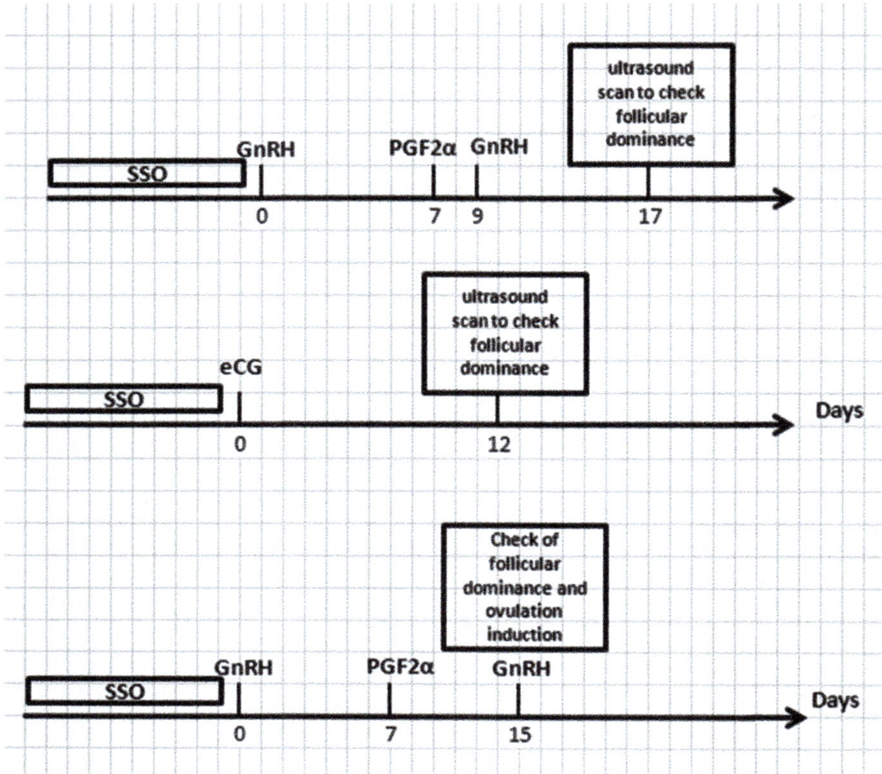

Fig. 2.10 Ovsynch (GnRH-PG-GnRH) and eCG (equine Chorionic Gonadotropins) protocols for oestrus induction in female dromedary camels with small and smooth ovaries (SSO) 2 months ahead the natural breeding season

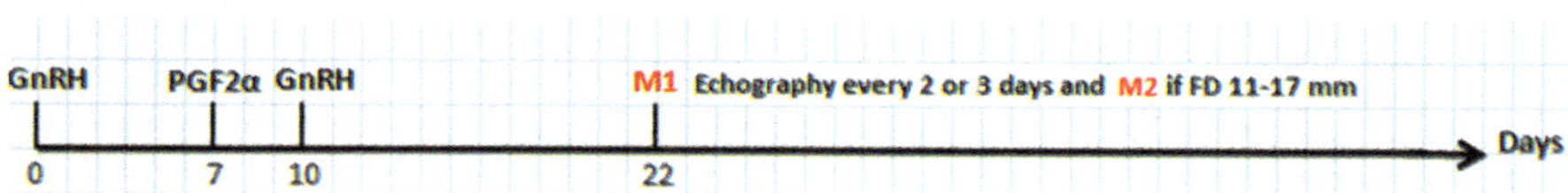

Fig. 2.11 Ovsynch (GnRH-PG-GnRH) protocol followed by blind natural mating and ultrasonographic monitoring of non pregnant females for second timed mating

The repetitive follicular wave resynchronization (GPG-M1-GP-M2-GP-M3; Fig. 2.12) after each mating was also explored by Manjunatha et al. (2018). They reported high synchronization rate at first mating and also high resynchronization rate at second and third timed natural mating. The pregnancy rate and the pregnancy loss rate after Ovsynch-first mating and Resynch protocols-second and third mating showed very interesting results (Manjunatha et al. 2018). An encouraging pregnancy cumulative rate is favorable to adopt this protocol in camels under farm and field conditions (Manjunatha et al. 2018).

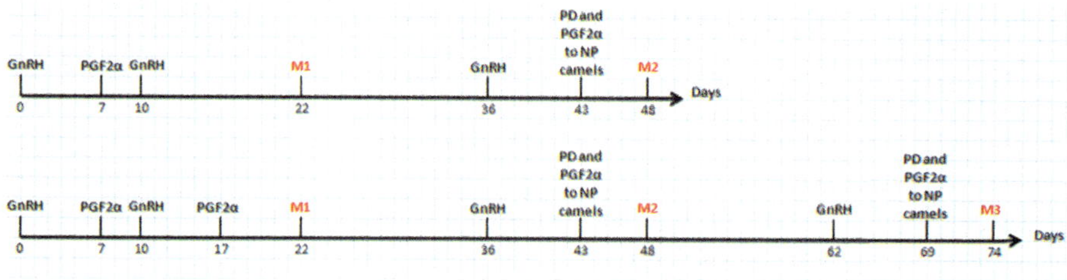

Fig. 2.12 Ovsynch -Resynch (GnH-PGFGnRH-M1-GnRH-PGF-M2) protocol for oestrus synchronization of female dromedary camel

The listed reproductive control methods could help to improve the CI in camel herds and therefore decrease the unproductive periods, overcome the constraint of the short breeding season, facilitate herd management practices (vaccination of mothers, maintenance and growth of camels, weaning, postpartum control …) and thus improve the birth rate and the numerical productivity, milk and meat production and in the long term decrease the generation interval for accelerating genetic progress.

2.6 Artificial Insemination

2.6.1 Definition

Artificial insemination (AI) is the act of depositing sperm by means of an instrument, at the most opportune time and in the most suitable place in the female genital tract. Since the 1960s, the use of AI as a breeding technique in camelidae has been reported with first camelid offspring from Bactrian camel obtained by this technique in 1961 (Elliot 1961). However, this technique has gained popularity over the last 25 years as interest is increasing to improve camel genetic characteristics such as milk, meat and wool quality, as well as racing ability. AI in DC faces several problems, including complications related to semen collection, bull sexual activity, semen quality (low volume, low sperm concentration and high viscosity), limited awareness of optimal insemination time and sperm dosage per insemination session, and lack of a standard storage technique (Elliot 1961; Tibary and Anouassi 1997c; Al-Bulushi et al. 2019b).

2.6.2 Importance of Artificial Insemination in Camels

The method offers many advantages: on the one hand, it facilitates the management of reproduction in the camel herds and increases the reproductive capacity of the males (ratio male/females) and therefore to obtain the most offsprings which helps rapid genetic improvement. On the other hand, it is a preventive measure against sexually transmitted diseases. Moreover, the use of AI would prevent the need to transport male or female animals for mating, thus reducing the charges and danger of moving valuable animals. The short and long term sperm conservation allows the international transportation of semen and conservation of the genetic material of males with high genetic level. It helps eliminate behavioral problems so AI would minimize the risks of injury. Finally, it facilities the application of cross breeding programs between different camelid's species.

Table 2.2 Advantages and disadvantages of sperm collection using artificial vagina

Advantages	Disadvantages
High sperm concentration	Decreased sperm mobility with prolonged contact of sperm with the inner membrane of the AV
High frequency of sperm collection during periods of sexual activity	Requires a trained camel bull
Not very aggressive technique, preserves the well-being and the natural sexual instinct	Requires high technicality of the operator
Allows observation of sexual behavior to assess male reproductive ability	Injuries risk of the animal and the operator
Can be replaced by candom device or held between the female's hind legs	Risk of refusal mating and especially when using dummy or with AV poorly suited
Can be used in the presence of female teaser in estrus or a camel dummy	High cost associated with the use of teaser females
Collection can be facilitated by creating an underground room or small laboratory underneath the dummy	
Method adapted to the insemination center: high collection rate, high sperm quality, less stressful manipulation of animals with high genetic potential, collection hygiene	
More adequate for short term and long term sperm preservation	
Higher pregnancy rate with fresh raw sperm	

Table 2.3 Advantages and disadvantages of sperm collection using an electroejaculator

Advantages	Disadvantages
Sperm collection during non breeding season	Low sperm concentration
More safe for the animal and operator	Requires pharmacological tranquilization of the camel bull
Lower cost	Aggressive technique that does not take into account the sexual instinct and comfort of the male camel
Collection of males with certain conditions that do not allow normal mating (locomotor diseases)	Low frequency and cadence of sperm collection
	More difficult sperm conservation
	Lower pregnancy rate

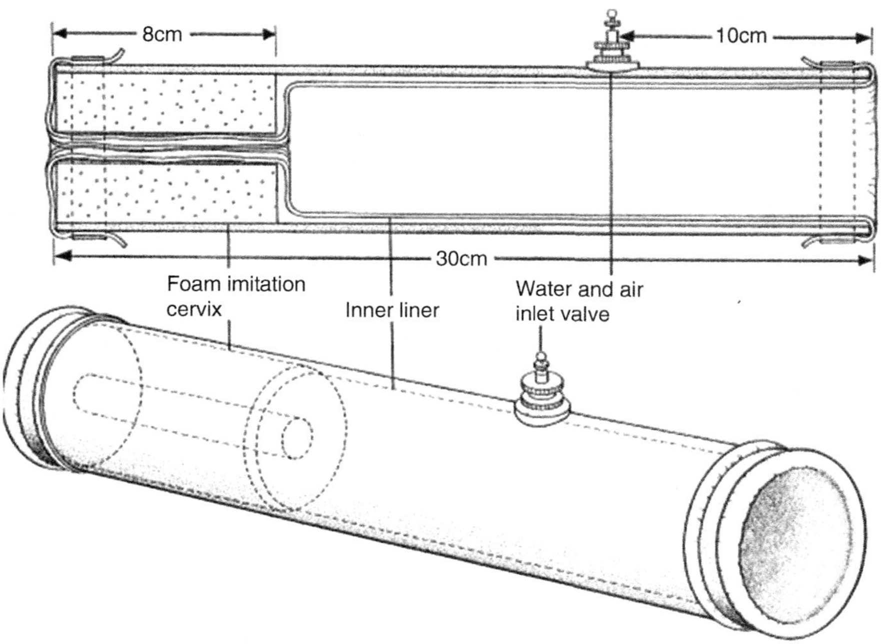

Fig. 2.13 Diagram of an artificial vagina used for DC. (Reprinted with permission of editor: Aurich Christine; from Bravo et al. 2000)

2.6.3 Semen Collection

The semen collection is the first step in AI and/or its examination. The primary objective of a semen collection is to obtain a large number of clean and optimum quality ejaculates (Monaco and Lacalandra 2019). The collection method has mighty effects on the physical and possibly biochemical properties of the ejaculated material (Tables 2.2 and 2.3) (Tibary et al. 2014). The artificial vagina (AV) and

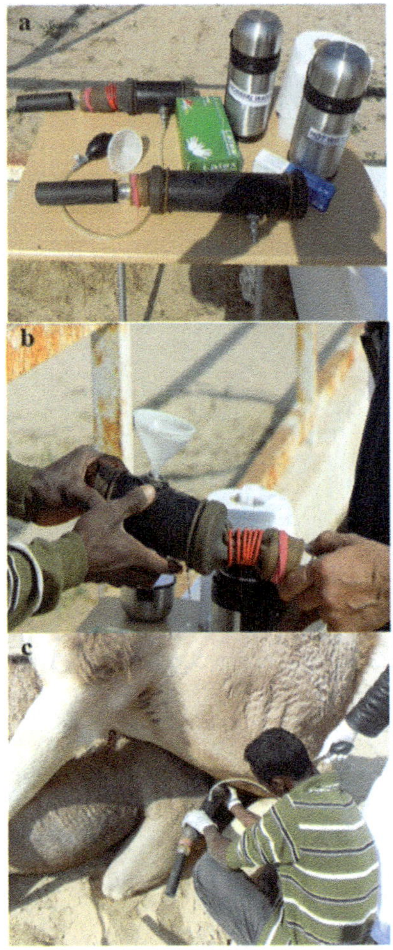

Fig. 2.14 Semen collection of male DCbull using modified bull-artificial vagina and female camel teaser with operator who direct penis by hand manipulation of sheath towards the opening of the artificial vagina. (**a**) Modified artificial vagina for collection of semen, (**b**) Coil on short artificial vagina to simulate cervical rings and (**c**) Collection of semen on a receptive female mount. (Photos reprinted with permission of the editor: Ahmed Bamouh, from Tibary and Anouassi 2018)

electroejaculation are the accepted methods for semen collection in camels (Bravo et al. 2000). The AV constitutes the means widely accepted by several authors who have worked on the dromedary using a female camel as a teaser or dummy. It's simple and practical device which has two parts. An outer cylinder made of rigid material, most often hard and thick rubber (thermal insulation) or plastic with an opening closed by a stopper. The inner latex or artificial rubber is introduced into the outer cylinder and its ends folded down and held in place by an elastic band. The cavity thus is formed by the outer cylinder and the inner liner. One end of the artificial vagina is lubricated: it will be used to introduce the penis; on the other is fixed a rubber cone at the end of which is fitted a graduated glass or better plastic tube to collect the sperm (Figs. 2.13 and 2.14).

Electroejaculation is rarely applied in the camel species (Tibary et al. 2014; Tibary and Anouassi 2018). This procedure requires the tranquillization or even the

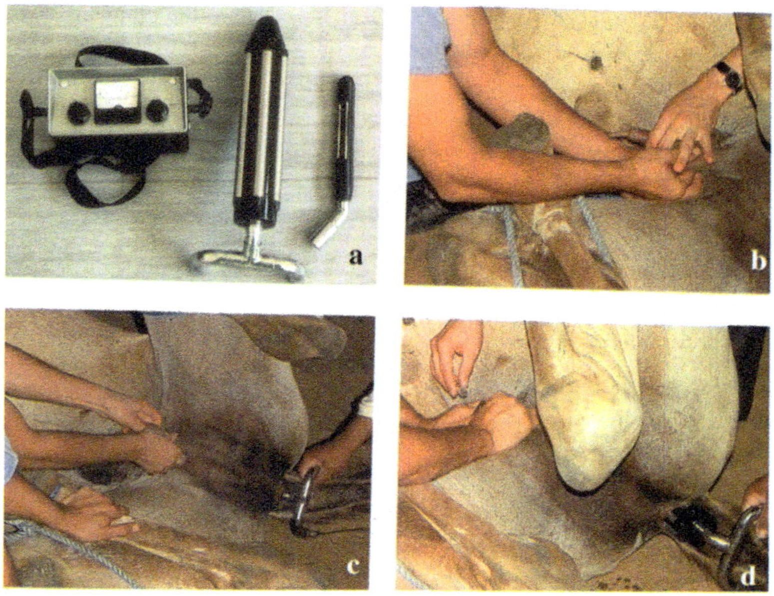

Fig. 2.15 Semen collection of male dromedary camel bull using an electroejaculator. (**a**) Electroejaculatro, (**b**) exteriorization of part of the penis outside the sheath, (**c**) placement of the electroejaculator after animal preparation, (**d**) semen collection step. (Photos reprinted with permission of the authors: Ahmed Tibary and Abdelhak Anouassi, from Tibary and Anouassi 1997b)

sedation of the animal, which would present a health risks (Fig. 2.15). The degradation of the physico-chemical quality and the concentration of the sperm collected by electroejaculation make it difficult its conservation and freezing (Tingari et al. 1986). So, electroejaculation (12V, 180 mA) can be applied with exceptional cases with camel bulls presenting highly vigorous and aggressively behavior, with low libido or with certain physical problems and for collection of semen during the non breeding season (Skidmore et al. 2013; El-Hussanein 2003). In this chapter, we don't consider this technique because its use does not meet the requirements of animal welfare.

Old methods of collecting semen such as collecting sperm directly from the vagina and collecting sperm by massaging the genital pool glands have also been reported in dromedaries (Tibary and Anouassi 1997b).

The modified bull-AV was commonly used for camel bull with good results (Bravo et al. 2000; Tibary and Anouassi 2018; Fig. 2.13). Some precautions must be taken into consideration in order to adapt the AV and avoid causing injuries to male bulls leading to the refusal of the animal for subsequent samples. The short modified AV allows also keeping away sperm from contact with the rubber liner of the AV and avoiding soiling the sample that adversely affect sperm quality. Therefore, it is recommended to:

- Shortening AV to allow the semen to pass directly into the collection flask. The internal part would have a diameter of 5 cm and a length of 30 cm (Bravo et al. 2000). The internal sleeve ends with a rubber receiving cone on which the collecting tube will be fixed. The whole will be secured with two rubber straps.
- To prevent contact with the rubber material, an additional removable plastic inner liner can be added, but the authors have found that males do not tolerate this well (Skidmore et al. 2013).
- For a perfect stimulation of ejaculation in dromedaries, the modified bull-AV should be equipped by foam imitation cervix of about 8 cm in length at its distal part. This provide a narrowing simulating cervical to receive the highly mobile urethral process of the camel penis to stimulate ejaculation during the extended copulatory (Bravo et al. 2000; Monaco et al. 2016; Monaco et al. 2018; Skidmore 2019). The filling water (55–60 °C) must be sufficient and it is necessary to blow air between the inner liner and the outer rigid wall to achieve an inner temperature of 38–40 °C and a pressure equal to that of the female vagina (Skidmore et al. 2013).
- The filling water (55–60 °C) must be in sufficient quantity and blowing air between the inner liner and outer rigid wall are important to obtain an internal temperature of 38–40 °C and a pressure equivalent to that of the female vagina (Skidmore et al. 2013).
- Keeping the vagina in an oven at 45 °C and fill it only in the minutes preceding the sample. Overpressure is not recommended because it may not leave enough space for the penis during ejaculation and cause the internal wall to burst. Correct filling is when thr upright opening of AV simulates a vaginal slit.

The semen collection should be done as much as possible in a quiet place to avoid any stressful situation for the animal and on non-pulverulent soil for the comfort of the operator and to avoid contamination of the sample as camels mate in a couched position. The use of a female teaser is essential. The impact of time of female teaser presentation and number of mating attempts before going to effective semen collection seem to have important effect on collected sperm volume and concentration (Al-Bulushi et al. 2018. These factors were reported to have a significant effect by increasing male excitation and improving sperm quality in other species (Hanzen 2015).

When the animal is in mating position, the penis will be directed by means of the operator hand applied to the sheath towards the opening of the artificial vagina, which directed slightly outwards while avoiding any excessive deviation of the penis. The sensory arousal of moist heat drives the intromission reflex. Before ejaculation is done, the male will make multiple thrusts, interspersed by periods of rest. The ejaculate typically happens in fractions and whole process can be prolonged between 5 and 50 min (Bravo et al. 2000). Thus, it is very difficult to know if ejaculation has occurred; collection continues until the male stands up. The effective ejaculation (intromission into the artificial vagina until the end of ejaculation) takes place after a variable duration ranging from 5 min to 20 min (Anouassi et al. 1992; Hassan et al. 1995) without yet a correlation between the duration of collection and the quality of the sperm. Soon after, the male stands up. Since the process of semen collection might be quite long, addition of about 1–2 ml of extender into the collection vessel before collection may be beneficial (Bravo et al. 2000). Immediately after the male stands up, the vagina is turned over so as to collect the sperm in the collection tube. The latter will, if necessary, be protected from any thermal shock by wadding or another isothermal envelope.

On the other hand, sexual peculiarities of the camel species (Tibary and Anouassi 1997a, b, c; Bravo et al. 2000; Monaco et al. 2018); such as the copulatory behavior ie, mating in sternal recumbency, lengthy ejaculation throughout copulation (from 5 min to 50 min), aggressive sexual behavior, the highly viscous nature of the semen and intrauterine semen deposition constitute major constraints to sperm collection and have made it mandatory to adapt sperm collection techniques with various degrees of success. We particularly mention:

- Using some specific hosing and welfare conditions to prevent injuries, stress and stereotypical behaviours (oral, locomotor, self-ejaculation) associated with poor
- reproductive performance (Monaco and Lacalandra 2019).
- Enhancing endocrine pathways and ejaculation quality of the male sexual by acting on the herd structure in which is maintained the camel bull, photoperiodic treatments and/or pharmacological (exogenous melatonin) protocols (Monaco et al. 2018; Monaco and Lacalandra 2019).
- Using female camel teaser and placed in sternal recumbency position with the operator in an underground room beneath the collection area which allows full observation of mating behavior and ejaculatory pattern (Hemeida et al. 2001; Fig. 2.16).
- Using artificial vagina devices implanted under the female teaser (Tibary and Anouassi 1997a, b, c; Al-Eknah et al. 2001; Fig. 2.17) or mounted on a dummy (El-Hussanein 2017; Fig. 2.18).
- Replace teaser female by a dummy designed in the similar position for natural mating, inside which is inserted an AV. The dummy is fixed on the floor of the collection yard and as a ceiling for a small laboratory to exchange AV from one

Fig. 2.16 Semen collection of male dromedary camel bull using artificial vagina oriented and maintained by an operator in an underground room which allows him full observation of mating behavior and ejaculatory pattern. (Photos from Hemeida et al. 2001)

Fig. 2.17 Semen collection of male dromedary camel bull using artificial vagina held between the female's hind legs. (Photo reprinted with permission of the authors: Ahmed Tibary and Abdelhak Anouassi, from Tibary and Anouassi 1997b)

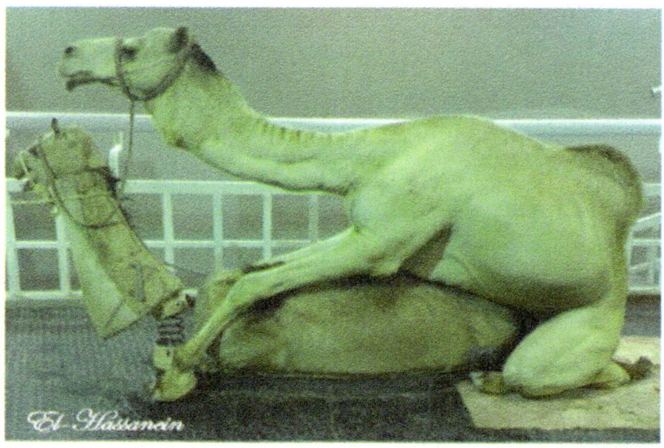

Fig. 2.18 Semen collection of male dromedary camel bull using artificial vagina mounted in a camel dummy. (Photo reprinted with permission of the editor: Daryoush Babazadeh, from El-Hussanein 2017)

male to the other and to carry out rapid evaluation and partial extension of the collected semen (El-Hussanein 2003, 2017; Ziapour et al. 2014; Fig. 2.19).

- Place condoms within the vagina of female teasers which showed acceptable result regarding ejaculated sperm volume and its characteristics (Tibary and Anouassi 2018; Fig. 2.20).
- Training camel bull to semen collection using dummy device and condoms. In this regard, serious behavioral limitations were reported if males are not used exclusively for AI (Tibary and Anouassi 2018; Monaco and Lacalandra 2019).
- To prevent semen contamination, a rope circled around the thorax could be used to attach the tail and the prepuce could be washed and dried before beginning the semen collection session (Monaco and Lacalandra 2019).

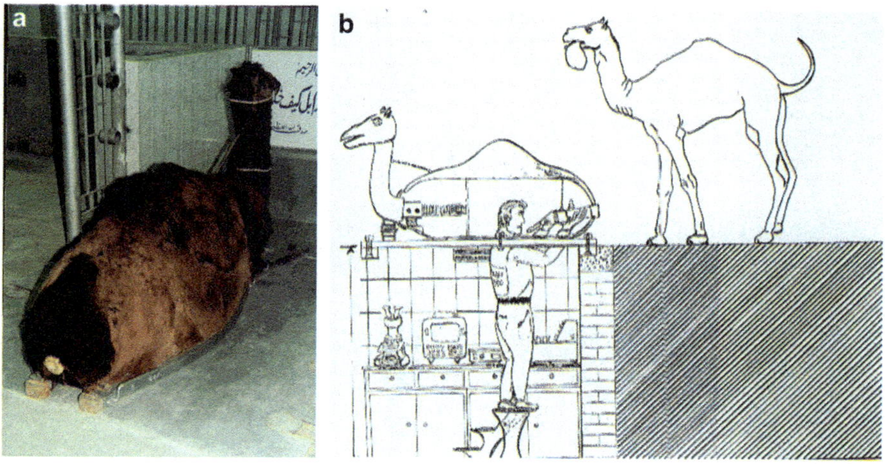

Fig. 2.19 Semen collection of male dromedary camel bull using a dummy and mounted artificial vagina with operator stays in a small laboratory underneath the dummy. (**a**) A hind view for the dummy to indicate its external feature and the mounted AV, (**b**) A diagrammatic drawing for the dummy and the small lab underneath. (Photo and diagram reprinted with permission of IVIS editor from El-Hussanein 2003)

2.6.4 Initial Sperm Examination

The available studies on the ejaculate initial quality of camel sperm showed high variability of the camel sperm characteristics.

2.6.4.1 Macroscopic Examinations

Volume

The sperm volume of the dromedary varies considerably depending on the technique and means of collection as well as the male's libido and the time of stimulation before collection. Using electro-ejaculator the volume varies from 1 ml to 10 ml (Tibary and Anouassi 1997b) even this volume varies from 2 ml to 12.6 ml when semen collection is carried out using AV (Wani et al. 2008; Tibary and Anouassi 1997b). According to Monaco et al. (2018); more than four ejaculates per collection session of about 45 min is possible resulting in total ejaculation volume of 12–16 ml. Even though allowing time for the male to recover after each ejaculation, the concentration of sperm was slightly lower after the third ejaculate.

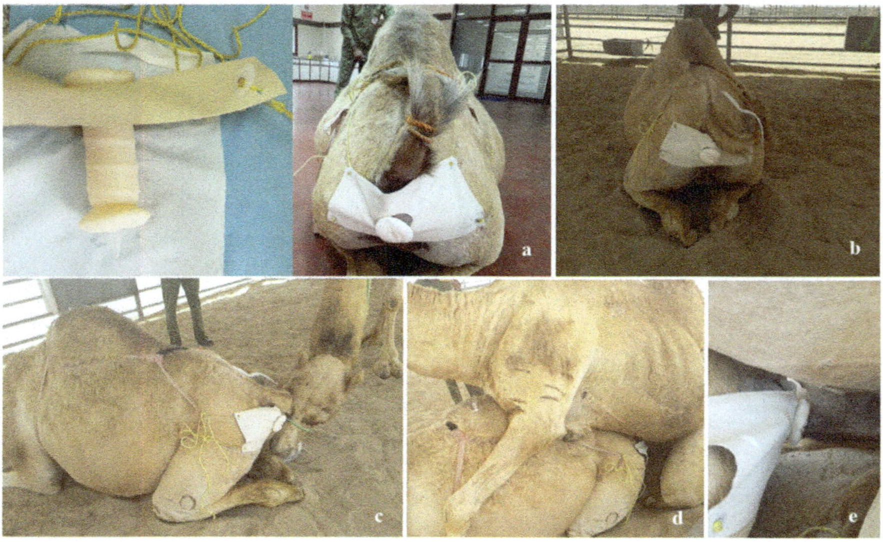

Fig. 2.20 Semen collection of male dromedary camel bull using candom device mounted within the vagina of receptive females and secured with a harness. (**a**) Condom device (left) inserted in the vaginal of an estrous female (right), (**b**) Female with secured intravaginal condom in the breeding pen (**c**) Male teasing the female with intravaginal condom (**d** and **e**) Male breeding within the intravaginal condom. (Photos reprinted with permission of the editor: Ahmed Bamouh, from Tibary and Anouassi 2018)

Appearance and Consistency

Semen of camels is of gray to milky white color (Bravo et al. 2000). Sand contamination can change semen color, for this reason the proposed camel dummy with inserted internal AV prevent semen contamination (El-Hussanein 2017; Skidmore 2019). Camel semen is in gel form and the proportion of seminal fluid varies leading to high variability of viscosity and sperm concentration (Deen et al. 2003). Depending on individual males, semen collection process, stimulation duration and the volume of the gel fraction in the ejaculate, this proportion of the seminal fluid is highly variable. (Tibary and Anouassi 2018; Skidmore et al. 2013).

The main constraint related to camel sperm viscosity is due to the distribution evenly throughout ejaculation of the liquid fraction which induces a difficulty of its separation (Skidmore et al. 2013). It was suggested that the high semen viscosity is resulted of higher seminal plasma levels of glycosaminoglycans or mucin 5B (Kershaw-Young et al. 2013).

It is important that the sperm of the dromedary undergoes a total liquefaction before its microscopic examination and dilution. Tibary and Anouassi (1997b) reported that sperm with little seminal fluid liquefies faster than sperm that contains large volume of seminal fluid. In addition, some studies have shown that liquefaction time of the camel sperm at 25–38 °C is 8–20 min (Tibary and Anouassi 1997b),

15–30 min at 30–37 °C (Al-Qarawi et al. 2002) and 90 min at 37 °C (Wani et al. 2008). However, during these time ranges, change in sperm viscosity also constitutes a partial liquefaction that is subject to a high variance between males and between ejaculates from the same male, ejaculate manipulation technique during incubation, type of extender and quantity of mucin in the ejaculate.(Tibary and Anouassi 2018; Al-Bulushi et al. 2018). The complete liquefaction of dromedary ejaculates took 8 h (Tibary and Anouassi 1997b). In recent study, this time was estimated to 23.9 ± 1.5 h (Mal et al. 2016). In practice, this time would certainly result in loss of quality of semen (Tibary and Anouassi 2018). Fast liquefaction (after 1.5 h at 37 °C) of camel diluted 1: 1 using Tris – lactose egg yolk, Tris – tes egg yolk,, sucrose egg yolk, citrate egg yolk and Tris – fructose egg yolk extenderscompared to semen without an added extender (Wani et al. 2008). Attempts at rapid liquefaction of dromedary sperm using mucolytic and proteolytic compounds such as plasmin, trypsin, chemotrypsin, amylase, collagenase and ficin have given variable results (Tibary and Anouassi 1997b; Bravo et al. 2000; Ghoneim et al. 2010; El-Bahrawy et al. 2015; Monaco et al. 2016; Keshavarz et al. 2016). Results using collagenase (0.5–1%) or papain (1.7 Unit/ml) seems to provide the best semen liquefaction in DC(20–40 min) (Kershaw-Young and Maxwell 2012; Tibary and Anouassi 2018; Monaco et al. 2016). These enzymatic treatments improve post-thaw motility and velocity patterns of spermatozoa but also can be deleterious to spermatozoa (Rateb et al. 2019; Skidmore 2019). Mechanical processes were also investigated to improve the duration of semen liquefaction using prolonged method of dilution (1: 1–1: 5) (Wani et al. 2008; Malo et al. 2017) and application of ultrasound (Rateb 2016; Skidmore 2019). The camel semen treatment by amylase improves viscosity score and mobility. The association of this enzymatic treatment with centrifugation for seminal plasma removal decreases semen viscosity and motility but increases the rate of spermatozoa abnormalities (El-Bahrawy 2017a, b).

2.6.4.2 Microscopic Examination of Semen

After first dilution (1:1) using semen extender and complete semen liquefaction the microscopic evaluation of camel sperm can be performed by measuring concentration, total motility, individual motility, viability and morphological aspects (Table 2.4). After that the second dilution can be applied to final ratio (1: 3).

Table 2.4 Mean characteristics of camel semen collected by the artificial vagina (Bravo et al. 2000)

Volume (ml)	Concentration	Motility (%)	Stained dead sperm (%)	Sperm showing morphological abnormalities (%)	Sperm showing acrosome abnormalities (%)
7.5	325	51	18	28	8.5

Concentration

The concentration expresses the number of spermatozoa per mm^3 (or per ml). It can be determined directly by counting spermatozoa using a hematimetric cell or indirectly by visual comparison of the sperm with standard solutions, by electronic counting or by nephelometry (spectrophotometer or colorimeter). This last method can however be indirectly increased following the presence of debris (sand) in the sperm of the dromedary and thus a variation of the opacity.

Direct counting is done using a hemacytometer. This type of count assumes the prior dilution of the sperm. The dilution rate depends on the apparent concentration of the sperm and the desired final concentration. There are different types of hemacytometer which are characterized in particular by their surface (S) and the depth of their counting chamber (P): example Thoma (S: 1 mm^2, P: 0.1 mm). The hemacytometer consists of a glass slide, hollowed out of a small cuvette whose bottom is lined with a grid. Thoma's cell has a grid of 16 large squares each comprising 16 small squares. The area of the large squares is equal to 1 mm^2. The counting chamber is 0.1 mm high.

After depositing a drop of diluted sperm and covering it with a coverslip, the number of spermatozoa is determined at magnification 400 on a surface corresponding to 4 large squares. Conventionly, only the heads of spermatozoa located inside the two parallel lines delimiting each large square or whose head is on the left and upper lines delimiting a large square are taken into account. The concentration is calculated as follows: Concentration = $N \times 4 \times 10 \times D$ (N is the number of sperm counted in 4 large squares, 4 since the hemacytometer has 16 large squares with a total area equal to 1 mm^2, 10 since the height of the enumeration chamber is equal to 0.1 mm and D is the degree of the initial dilution).

The sperm concentration and the total number per ejaculate of male DC varies among studies ranging from 80 to 1300×10^6/ml (Tibary and Anouassi 1997b; Wani et al. 2008; Al-Bulushi et al. 2018) and from 240 to 2576×10^6 (Tibary and Anouassi 1997b; Morton et al. 2011; Tibary et al. 2014; Tibary and Anouassi 2018). Camel semen concentration is qualified as oligozoospermia when it shows less than 40 millions spermatozoa/ml (Waheed et al. 2014). Specific individual variations, frequency of collection and technique/time needed for semen collection can lead to significant variation of sperm numbers in the dromedary ejaculate. (Tibary and Anouassi 2018; Al-Bulushi et al. 2018). It is also interesting that authors have mentioned that the absence of sperm in many ejaculates is due to incomplete ejaculation (Hassan et al. 1995). Weekly semen collection during the peak breeding season for 3 weeks or twice weekly semen collection over 8 weeks did not affect the semen quality (Al-Bulushi et al. 2018). The pregnancy rate using fresh camel sperm with low concentration (75×10^6 sperm/ml) showed significantly low rate compared to sperms with concentration of 150 and 300×10^6 sperm/ml (et al. 2019b).

Massal Motility

Sperm mobility is evaluated according to two criteria: mass and individual mobility. The mass mobility is evaluated on diluted sperm by placing a drop on a preheated slide and observes it under a phase contrast microscope in low light intensity at 100× magnification. It refers to the intensity of the waves caused by the movement of sperm. A score of 0 to 5 is assigned to the observed sample: 0: no movement of the sperm, 1: slight noticeable movement, no waves, 2: few waves, 3: many waves, 4: fast and intense waves and 5: very fast waves.

The nomenclature of asthenozoospermia was given to camel sperm with decreased spermatozoa motility (Waheed et al. 2014). In Al-Bulushi et al. (2019b) study it was showen that insemination using raw camel sperm with 84–86% total motility result on high and comparable pregnancy rate (81%) to that obtained by natural mating (83%).

Individual Mobility

Individual mobility is assessed on a drop of diluted and sufficiently liquefied sperm placed on a preheated slide under ×400 magnification. The dilution is done in a diluter ("extender") previously heated to a ratio of 1 to 20 (0.25 ml of sperm and 4.75 ml of diluter). These media should be ideally prepared before the examination to avoid any change in pH, detrimental to the motility of the spermatozoa. Three to five fields close to the center of the drop will thus be examined and the average calculated. The individual motility is the proportion (%) of sperm that pass through the field of the microscope relatively quickly with rotational movements of the head (flexing or tracer sperm). Some sperm spinning in small circles or moving backwards is due to the abaxial implantation of their tail. Others move in a curvilinear fashion or more slowly. They are not considered mobile. Computer Assisted Sperm Analysis (CASA) type image analyzers make it possible to quantify more precisely (the nature and speed of movements).

A very good quality sperm must have at least 50% of motile spermatozoa in the camel (Bravo et al. 2000). Numerous reports from the literature have not revealed sperm mobility problems of dromedary sperm, however, we find that the proportion of motile sperm varies from 0 to 80% (Tibary and Anouassi 1997b; Deen et al. 2003) and that is significantly influenced by several factors such as means of collection, technique and conditions of collection, sperm viscosity, nature of the diluter, the synthetic material of the inner lining of the AV and the length of time that the sperm spends inside AV (Tibary and Anouassi 1997b; Deen et al. 2003). Addition of caffeine could improve the individual motility of the camel sperm (Deen et al. 2003).

The examination of individual motility is interesting because it indirectly provides information on the integrity of the sperm membrane and its morphological integrity. Thus, a high percentage of motile sperm combined with a high percentage of dead sperm suggests improper handling of sperm rather than abnormal sperm.

Likewise, low motility is often correlated with a high percentage of abnormal forms or dead sperm.

Morphological Examination

Morphological examination requires staining of the sperm by placing a drop of semen on the end of a slide and spreading it in a thin layer with another blade tilted at 45°. The preparation is dried in air for few minutes. The smear is then fixed by immersion in a solution of methyl alcohol or in a solution of 5% formalin. Free air fixation is also possible such as passing the smear quickly over an alcohol lamp. Any thermal shock will nevertheless be avoided during these preparations. The preliminary dilution of the sperm (2 drops in 0.5 ml of diluent) will facilitate the morphological examination. Some stains are intended to better show the morphology of the spermatozoon (total staining), the others called vital make possible to differentiate between dead and living spermatozoa. Among the total stains, some are said to be simple (India ink, methylene blue, toluidine blue, gentian violet, fuschin, etc.): they provide uniform coloration of the spermatozoa while the second called double (Giemsa, Williams) and better show the structural differences at the level of the head, acrosome or intermediate piece. The principle of vital staining is to use a stain which only crosses the membranes of dead cells (eosin, rose bengal, cresyl green) and a background stain which facilitates reading (methylene blue, nigrosine). Eosin-nigrosin staining is conventionally used with a sperm-stain mixture of between 1/10 and 1/20.

The examination will ideally be carried out with an immersion objective at ×1000 magnification. One hundred spermatozoa will be counted or even more if the number of sperm is greater. Dead sperm will be stained red due to eosin having entered owing to changes in membrane permeability.

Sperm morphological abnormalities can be said to be primary (1) if they originate during the spermatogenesis phase (testis) or secondary (2) if they occur during their maturation phase (epididymis). The majority of sperm lesions are said to be primary. Some can be both primary and secondary such as the presence of droplets, tailless heads. Sperm abnormalities can also be classified as major or minor depending on whether or not they have a negative effect on fertility. Finally, they can affect the various parts of the spermatozoon individually or simultaneously.

The head may show abnormalities in the shape, size, duplication, position or structure of the acrosome. The anomalies of the intermediate piece concerned cytoplasmic rest and angulation. The tail can be too short, too long, absent, duplicated, coiled at the end of the flagellum, or even under the head, and even around it or angulated (angulation exceeding 90°).

A rate of 10% of abnormal spermatozoa has been reported in camel sperm (Al-Bulushi et al. 2018) without significant effect of the frequency of collection on this parameter. The teratozoospermia of camel sperm was given to ejaculates with more than 40% abnormal spermatozoa (Waheed et al. 2014). The proportion of protoplasmic droplet spermatozoa was 1.02 ± 0.2 on average, while 2.7 ± 0.6 and 9.7 ± 2.9% had mid-piece and tail defects, respectively (Wani et al. 2008).

2.6.5 Sperm Dilution and Conservation

2.6.5.1 Fresh Semen Dilution

The purpose of camel sperm dilution is to increase the total mass of the sperm fraction to split the ejaculate into fertilizing doses while adding substances that ensure sperm survival during storage. The latter can last a few hours or from a few months to several years.

In the dromedary, the separation of the liquid fraction from the sperm is difficult due to the distribution evenly throughout ejaculation of the liquid fraction. In addition, the sperm consists of a very variable volume of glandular secretions and that the testicular secretions constitute the essential of the semen.

The dilution medium requires the following characteristics: osmotic pressure which must be isotonic with the sperm and be able to maintain it during the storage period, contains colloidal substances (egg yolk, lipoproteins, lecithins) to ensure spermatozoa protection, contains buffer substance which makes possible to maintain a pH favorable to spermatozoa but its presence is not too important given the low concentration of spermatozoa in the sperm of the dromedary leading to a slight decrease in pH, contains nutrients to support metabolism, vitality and longevity of spermatozoa and contains antimicrobial agents.

The well diluted sperm offer best condition to spermatozoa and allow them to fulfill their 4 functions prior to fertilization: metabolic activity producing energy, mobility to progress in the female genital tract, protective enzymes on the acrosome to facilitate penetration into the oocyte and the presence of proteins on the plasma membrane to ensure their optimal survival in the female genital tract and their attachment to the pellucid of the oocyte.

To use camel sperm few minutes to a few hours after collection, it must be diluted at a ratio of 1:1 to 1:3 in a suitable diluter previously warmed. The dilution depends on the concentration of the ejaculate and has objective to achieve a standard concentration of 100×10^6 sperms/ml (Anouassi et al. 1992). This value can be revised downwards or upwards depending on the quality of collected sperm and insemination method (Bravo et al. 2000; Skidmore and Billah 2006a). Diluted sperm can be stored at room temperature (37 °C) until insemination (Tibary and Anouassi 1997b).

2.6.5.2 Short Term Preservation

If the sperm is stored for up to 24 h, it is mandatory to gently decrease the temperature of the mixture sperm-diluter to 4 °C by placing the tube in a water bath at room temperature and then refrigeration to reach 5 °C after about an hour (Tibary and

Anouassi 1997b). Equitainer of equine semen can be used for the semen transport during 18–36 h. The researches on camelid semen extenders remind scarce. Most extenders used for camel sperm are adapted from bull and stallion commercial sperm extenders as Glucose-EDTA, Dimitropolous 11, INRA-96, Sodium-citrate egg yolk, Skim milk, Kenny's equine extender, lactose egg yolk (Tibary and Anouassi 1997b, 2018) and have showed a low impact on sperm motility and viability. A commercial extender has been available for camel semen since the early 90s (Camel Buffer Green®, IMV, L'aigle, France). Recently, Al-Bulushi et al. (2019a) recorded that INRA is less suitable than Green Buffer to preserve semen at refrigeration for 24/48 h. It is mandatory to remember that for short-term preservation of camel sperm, it is essential to leave the time necessary for the complete liquefaction of the sperm before adding the diluter in order to obtain a homogeneous mixture (Tibary and Anouassi 1997b) and possibly reduce the effects of dilution on sperm mobility. It will be remembered that there are still great variations in the characteristics of the sperm of dromedaries subjected to short-term storage.

2.6.5.3 Long Term Preservation

If semen is to be used beyond 24 h after collection, freezing should be considered. There is no consensus on successful strategies for maintaining the fertilizing lifetime of frozen-thawed camel spermatozoa. The large scale application of AI in dromedary camels is impeded by this constraint. Freezing requires use of cryoprotectant agents. In view of the potential deleterious effects of cryoprotective agents on the sperm, they must be used at an optimal dilution which allows their benefits to be obtained and prevent osmotic and thermal shock from occurring both during dilution, cooling, freezing and thawing processing (Watson 2000).

Semen from DC was frozen successfully in a variety of extenders: Green Buffer® Egg yolk-glycerol, INRA and Egg yolk (Crichton et al. 2015), Tris-egg yolk-glycerol (Deen and Sahani 2006). Two steps are required for the cryopreservation of dromedary spermatozoa: dilution with cooling extender (80 ml Lactose 11%, 20 ml Egg Yolk) and dilution with freezing extender (95.5 m cooling extender, 06.0 ml Glycerol and 1.5 ml Orvus paste –Equex). Immediately after collection, the cooling extender is added to the semen. The freezing extender is characterized by the fact that it contains a 7% of glycerol and an Orvus paste (emulsifying agent) that plays a role in plasma membrane stabilization of the sperm.(Tibary and Anouassi 2018). The first diluter is maintained at 32 °C and the second diluter at 4 °C. The Tris based extender with 7% glycerol and sucrose based extender (73.0 ml Sucrose 12%) showed superiority to preserve post-thaw spermatozoa quality (mobility and acrosome integrity) and fertility of camel bull (Zhao et al. 1994; Tibary and Anouassi 2018; Akbar et al. 2018). Likewise cholesterol-loaded cyclodextrin enhanced cryosurvival, post thaw sperm motility (0 h and 3 h post thaw) at room temperature and post-thaw sperm progressive status (Crichton et al. 2015, 2016).

Packaging sperm in pellets is rarely practiced in dromedary camels. Pellets have the disadvantage of not being able to be identified correctly. More conventionally,

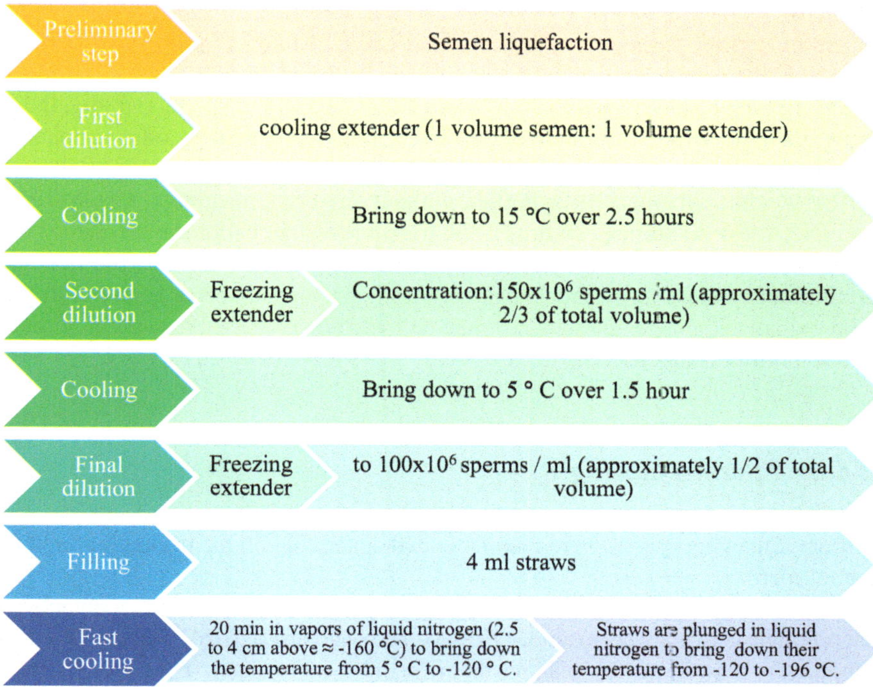

Fig. 2.21 Steps of camel sperm freezing in 4 ml straws

the sperm is packaged in polypropylene or polyvinyl straws with volumes between 0.25 and 4 ml. The required sperm dose has typically been a 4 ml straw that will contain 100×10^6 sperms/ml (Tibary and Anouassi 2018).

After semen liquefaction, freezing process depends mainly on the volume of the straws. When using large straws (4 ml) we proceed as follows (Fig. 2.21):

When we use semen package of 0.25 or 0.5 ml straws, the freezing process starts after total liquefaction according to the following steps (Tibary 2001; Deen et al. 2003; Niasari-Naslaji et al. 2006, 2007; El-Bahrawy 2010; Crichton et al. 2015):

- Dilution and cooling to 5 °C over 1 h.
- Semen package and equilibration at 5 °C for 2 h.
- Fast cooling by straws exposure to liquid nitrogen (4 cm above liquid nitrogen level) for 10 min
- Transfer of the cooled straws directly into liquid nitrogen.

Recent study performed at Camel Reproduction Centre, (Dubai, United Arab Emirates) by Malo et al. (2020), showed that camel sperm quality (total motility, viability and acrosome status) did not change significantly after 24 h cooling at 4 °C using two extenders (INRA 96 and Green Buffer) with or without cryoprotectant (3% glycerol or 3% ethylene glycol). Same authors reported also that shorter equilibration times (0–60 min) did not affect results of post-thaw sperms although

Glycerol 3% and ethylene glycol provided the best motility recovery rates compared to methyl formamide, dimethyl sulfoxide (Malo et al. 2017, 2020). In Addition, the post thawing results using Green Buffer showed highest rates of total and progressive motility and viability (Malo et al. 2020). For sperm viability, INRA with Ethylene Glycol 3% showed the lower rates compared to other extenders-cryoprotectant mixture (Green Buffer with glycerol 3% or Ethylene Glycol 3% and INRA 96with glycerol 3%) or buffers without any cryoprotectants (Green Buffer and INRA 96). Al-Bulushi et al. (2019a, b) recorded that better semen quality was observed using Triladyl (containing glycerol as cryoprotectant). In addition, the addition of catalase to extenders can act during prolonged sperm handling to prevent oxidative damage and thus promote chilled and cryopreserved sperm to retain better motility characteristics (Medan et al. 2008; Malo et al. 2019, 2020).

2.6.6 Sperm Thawing

Different thawing speeds were recommended depending on the packaging used for camel sperm.

The content of pellets is thawed out by dropping them into heated receptacles or incorporate it to thawed extender (Tibary and Anouassi 2018). Semen frozen in ampules is thawed out by placing them for 30 s to 1 min in a water bath set at 45–55 °C (Bravo et al. 2000). Straws (0.25 or 0.5 ml) can be thawed by placing them in water at 40 °C for 8 s or 30–60 s if the water is at 35–37 °C. The 4 ml straws are thawed at 50 °C for 40 s by continuous agitation in the water bath (Tibary and Anouassi 2018; Bravo et al. 2000). Cryovials are thawed out for 2 min by immersion in a water bath at 40 °C (Deen et al. 2003).

In recent study, Malo et al. (2020) reported that sperm extended in Green Buffer maintained better post-thaw vitality than sperm in INRA 96. The analyze of the effect the interaction of extenders with cryoprotectants on sperm vitality showed that INRA96-ethylen glycol 3% exerted most adverse effect on sperm vitality compared to INRA96-Glycerol 3%, Green Buffer-ethylene glycol 3%, Green Buffer-Glycerol 3% and both extenders INRA96 and Green Buffer without cryoprotectants. In other hand, all aforementioned sperm treatments did not showed any post-thaw effect on acrosome integrity (Malo et al. 2020).

Liquefaction of camel semen using Tris lactose supplemented with mucolytic agent (amylase) followed by centrifugation of seminal plasma removal in the presence of antioxidant agent (caffeine) improved significantly post-thaw motility and recovery rate suggesting a particular interest of this refined protocol for camel semen cryopreservation (El-Bahrawy 2017a, b). Similarly, semen extender supplemented with trace elements zinc and selenium nanoparticles (ZnONPs and SeNPs) improved significantly camel sperm vitality, progressive motility, ultrastructural morphology, sperm membrane integrity and decreased apoptosis when frozen and thawed (Shahin et al. 2020). These results meet the previous established conclusion about advantageous effects of nanotechnology in sperm function by incresing antioxidant effect and bioavailability and reducing the undesired liberation of toxic concentrations (Falchi et al. 2018; Shahin et al. 2020).

Damage that occurs during freezing-thawing processes primarily affects cell membranes (plasma and mitochondrial) and, in more serious cases, the nucleus (Ahmed et al. 2017). Evaluating the percentage of motile spermatozoa at various intervals during incubation is a classic method of evaluating the method of freezing/thawing semen. The development of computerized mobility analysis systems helped to reduce the subjectivity part of this assessment. (Tibary and Anouassi 2018) reported other evaluation methods proposed for camel species such as hypoosmotic swelling test (incubation in fructose or sucrose solution of 50–100 mOsm at 37 °C for 45 min), special staining techniques (Isothiocynate-conjugated peanut agglutinin) (Morton et al. 2007, 2008), Chlortetracycline staining for spontaneous capacitation (Crichton et al. 2015). The correlation between all these methods and the pregnancy rate remains to be demonstrated after a wide application of AI in camel herds.

2.6.7 *Insemination Technique*

Compared to other animals such as saw and ewe, camel insemination is simpler since the cervix is shorter and straighter and the uterus is less coiled, so it is easier to pass a catheter through the cervix and to direct it up the uterine horn per rectum (Skidmore et al. 2013). The ovarian function of the female camels must be monitored by series of ultrasonography exams to follow the growth of the dominant follicle which will be ovulated before AI. Ovulation induction is commonly performed out with a single injection of GnRH/hCG given intravenously when there is a 1.3–1.7 cm antral follicle in the ovaries (Skidmore et al. 1996; Medan et al. 2008). AI must be carried out 24 h after the injection of GnRH/hCG. The equipment consists of an insemination gun of 40–45 cm length and 5–6 mm diameter with an outer body and an inner mandrel. It is completed by an external plastic sheath attached to the insemination gun by means of a small washer. The fresh or thawed semen straw is wiped to remove all traces of water (water = spermicide) and the identity of the camel bull is immediately verified. It is then sectioned approximately 1 cm from its closed end and introduced into the insemination gun previously heated by friction to avoid any thermal shock. A plastic sheath (disposable) ensures the sanitary protection and the tightness of the device. It is possible to fit a flexible catheter within the rigid outer catheter preheated to 38 °C to avoid thermal shock for deep uterine insemination (Fig. 2.22) (Tibary and Anouassi 1997c).

The insemination technique is that of cervical catheterization with immobilization of the latter rectally (Fig. 2.22). The right or left hand introduced into the rectum, grasps the cervix and the other hand introduces the catheter into the vulva (previously cleaned), pushing it forward and following the ceiling of the vagina (angle of 45°) to avoid the urinary meatus. Vaginal folds are avoided by pushing the cervix held with the right or left hand forward (this clears the folds).

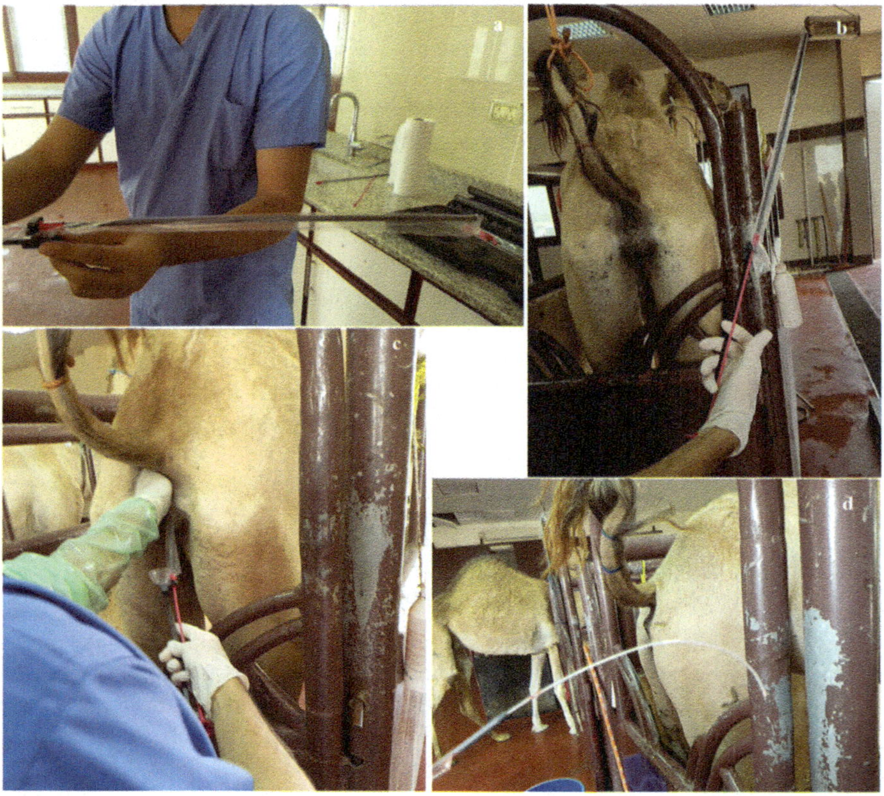

Fig. 2.22 Artificial insemination technique using special catheter for deep uterine deposit of camel semen. (**a**) Sterile deep horn insemination gun (**b**) Female prepared for deep horn insemination (**c**) Insertion of the deep horn insemination gun and transrectal guidance of the flexible catheter (**d**) Deep horn AI gun after insemination. Note the flexible catheter within the rigid outer catheter. (Photos reprinted with permission of the editor: Ahmed Bamouh, from Tibary and Anouassi 2018)

Locating the orifice of the cervix into which the catheter tip is to enter requires skill and must be done without any brutality. The hand that moves the cervix must, by the way a glove is put on, manipulate the cervix so that it comes over the catheter while avoiding the cervical folds until it reaches the cervico-uterine portion. Although it is easy to pass the catheter through the open cervix of the female during oetrus, under no circumstances should the catheter be forcefully pushed into the cervix (risk of injury).

It was therefore suggested that the semen should be deposited at the tip of the ipsilateral uterine horn to the ovary showing the mature follicle, thus closer to the uterine tubule junction rather than the uterine body (Skidmore 2019; Al-Bulushi et al. 2019b). The index finger of the hand inserted into the rectum controls, through the rectal wall, the correct position which allows the semen to be deposited (push

Table 2.5 Pregnancy rate of female camel according to different methodologies of artificial insemination

Author	Ovulation induction	Insemination time	Type of semen		Collection method	Females inseminated	Place of deposition	Pregnancy rate (%)
Deen et al. (2003)	hCG	24 h post hCG	WF	300×10^6	AV	13	UH	40
	hCG	48 h post hCG	EC	300×10^6	AV	10	UH	0
	hCG	24post hCG	FT	300×10^6	AV	10	UH	7
Medan et al. (2008)	hCG	48 h post hCG	WF	100×10^6	AV	13	UB	46
		48 h post hCG	EC (free-catalase)	100×10^6	AV	9	UB	22
		48 h post hCG	EC (with catalase)	100×10^6	AV	8	UB	37
Anouassi et al. (1992)	Vasectomized male		WF	100×10^6	AV		UB	80
McKinnon et al. (1994)	hCG		WF		AV			32
Skidmore and Billah (2006a)	GnRH		E	150×10^6	AV	15	UB	53
	GnRH		E	150×10^6	AV	14	UH	43
	GnRH		E	80×10^6	AV	14	UB	7
	GnRH		E	80×10^6	AV	15	UH	40
	GnRH		E	40×10^6	AV	14	UB	0
	GnRH		E	40×10^6	AV	14	UH	7
Al-Bulushi et al. (2019b)	GnRH	Immediately after ovulation treatment	WF (3–8 ml)	474×10^6	AV	11	UB	81
	GnRH	Immediately after ovulation treatment	WF (1.5–3.5 ml)	442×10^6	AV	10	UB	60
	GnRH	Immediately after ovulation treatment	WF (1 ml)	454×10^6	AV	12	UB	33

(continued)

Table 2.5 (continued)

Author	Ovulation induction	Insemination time	Type of semen		Collection method	Females inseminated	Place of deposition	Pregnancy rate (%)
Al-Bulushi et al. (2019b)	GnRH	Immediately after ovulation treatment	E	300×10^6	AV	17	UB	35
	GnRH	Immediately after ovulation treatment	E	301×10^6	AV	18	UH	72
Al-Bulushi et al. (2019b)	GnRH	Immediately after ovulation treatment	E	302×10^6	AV	19	UH	68
	GnRH	24 h post GnRH	EC	303×10^6	AV	20	UH	70
	GnRH	30 h post GnRH	EC	304×10^6	AV	17	UH	23
Al-Bulushi et al. (2019b)	GnRH	24 h post GnRH	E (3.5 ml)	300×10^6	AV	20	UH	70
	GnRH	24 h post GnRH	E (1.7 ml)	150×10^6	AV	23	UH	65
	GnRH	24 h post GnRH	E (0.9 ml)	75×10^6	AV	22	UH	40
Al-Bulushi et al. (2019b)	GnRH	24 h post GnRH	EC (Triladyl)	300×10^6	AV	18	UH	11
	GnRH	24 h post GnRH	EC (Optixcell)	300×10^6	AV	17	UH	5
	GnRH	24 h post GnRH	EC (Triladyl)	500×10^6	AV	19	UH	21
	GnRH	24 h post GnRH	EC (Optixcell)	500×10^6	AV	16	UH	12
	GnRH	24 h post GnRH	FT	300×10^6	AV	23	UH	0
	GnRH	24 h post GnRH	FT	500×10^6	AV	20	UH	0
Morton et al. (2010)	GnRH	24-48 h post GnRH	FT	150×10^6	AV	11	UB	27
Akbar et al. (2018)	GnRH	30 h post GnRH	FT	100×10^6	AV	21	UB	42[a]

UH uterine horne, *UB* uterine body, [a]pregnancy at 90 days post insemination, *WF* whole fresh, *E* extended, *EC* extended cooled, *FT* frozen thawed, *AV* artificial vagina exerted on the plunger) at just this optimal level. The utero-tubal zone acts as a reservoir for spermatozoa. These are released regularly and continuously in order to ensure fertilization in the upper third of the oviduct without there being polyspermia (Skidmore 2019).

2.6.8 Factors Affecting Fertility After AI

AI with raw undiluted camel ejaculates resulted in wide variability of pregnancy rate: 40–81.8% pregnancy rate (Table 2.5; Anouassi et al. 1992; Deen et al. 2003; Skidmore and Billah 2006a, b; Medan et al. 2008; Morton et al. 2010; Al-Bulushi et al. 2019b). These results are somewhat comparable to results of timed natural mating programs: 67–71.4% (Manjunatha et al. 2015, 2018; Al-Bulushi et al. 2019b). Low volume (1 ml) insemination with undiluted ejaculate resulted in lower pregnancy compared to whole and split undiluted ejaculate. (Al-Bulushi et al. 2019b). However, some studies showed lower or null results of pregnancy rate using chilled and non diluted semen for AI (El-Hussanein 2003; Medan et al. 2008; Morton et al. 2013).

The pregnancy rate after AI using chilled camel semen diluted by different extenders show large variability of pregnancy rate (Table 2.5): 0% using skimmed milk (Skidmore 2000); 5.9–12.5% using Optixcell (Al-Bulushi et al. 2019b); 11.1–21.1% using Triladyl (Al-Bulushi et al. 2019b); 17–34% using INRA 96 (Morton et al. 2010, 2013; Malo et al. 2020); 50% using lactose plus egg yolk (Anouassi et al. 1992), 53% using Laciphos (Skidmore 2000); 10–72.7% using Green buffer plus egg yolk (Al-Bulushi et al. 2019a, b; Skidmore and Billah 2006a; Morton et al. 2010, 2011, 2013; Malo et al. 2020). The wide variability of the pregnancy rate after AI by liquid chilled semen would be difficult to explain especially when considering that in vitro semen quality appears to be high.

The post-thaw frozen semen showed promising consistency, but this is not evidenced by the pregnancy outcomes to date. Single early pregnancy in DC has (1/13 animals inseminated) resulted from AI with cryopreserved spermatozoa (Deen et al. 2003). The use of glycerol as cryoprotectant with Green Buffer extender during chilling did not interfere with cryosurvival, but it may be toxic to the fertility of fresh chilled sperm leading to null pregnancy rate (Malo et al. 2020). After that, Morton et al. (2010) reported a pregnancy rate of about 27% in single ovulated female camels using frozen-thawed semen diluated in Green Buffer and fertilization rates of about 11.1% in multiple ovulating camels. The first successful encouraging results for large scale utilization of frozen semen insemination of dromedary camels was recently reported by Akbar et al. (2018) after a series of experiments on semen processing, cryopreservation and AI giving live birth of camel calves at Camel Breeding Center, Dubai, United Arab Emirates date 2018 (Akbar et al. 2018). In this experiment; 21 inseminated females by frozen -thawed semen with tris based extender and 7% glycerol gave the following results: On the 30th day of insemination, 15 (71.42%) females were found to be pregnant. Between the 18th and 30th days, 6 she-camels faced embryonic mortality (Akbar et al. 2018) and pregnancy was confirmed in the remaining females at 60 and 90 days of insemination (09). At mid gestation, two females aborted and the remaining seven females (33.33%) were carried to full term. More investigations are needed to confirm the fertility rate using frozen thawed semen of eventually large scale application of AI as important step of assisted reproductive techniques in camelids.

The monitoring of the ovarian follicular wave patterns of the female camel by transrectal ultrasound examinations is mandatory for AI success. Ovulation is induced by a single intravenous injection of GnRH (ex: 20 mg buserelin) accompanied by timed artificial insemination once the dominant follicle reaches 1.3–1.8 cm (Skidmore and Billah 2006b). Female camels inseminated 24 h after ovulation induction resulted in higher pregnancy rate (53.7–70.0%) (Skidmore and Billah 2006b; Al-Bulushi et al. 2019a, b) compared to insemination performed at the time of ovulation induction or 30 h avert induction (68.4 and 23.5%, respectively) (Al-Bulushi et al. 2019a, b).

The results of the insemination into the uterine body or the tip of uterine horn ipsilateral to the ovary containing the dominant follicle showed respective increasing conception rate of 0, 7, 53% and 7, 40, 43% according to increased sperm concentration per milliliter of 40, 80 and 150 × 10^6 (Skidmore and Billah 2006a). Deposition of semen in the uterine body resulted in lower pregnancy rates compared to deposition in the tip of the horn (35.3% versus 72.2%) (Al-Bulushi et al. 2019a, b). AI with 75 × 10^6 motile spermatozoa resulted in lower pregnancy rates compared to 150 and 300 × 10^6 motile spermatozoa doses (40.9% versus 65.2 and 70.0%, respectively) and pregnancy rate without significant correlation with semen extenders (Green buffer, Optixcell or Triladyl) (Al-Bulushi et al. 2019a, b). Indeed, the use of insemination with small sperm concentration requires a deep uterine insemination technique. If semen is deposited only in the uterine body rather than at the tip of the horn a substantial loss of spermatozoa occurs due to the semen's backflow through the short and open cervix during the female camel's oestrus.

The short, open cervix that occurs during oestrus of the female camel would lead to considerably loss of spermatozoa, due to backflow of semen through the cervix, when the semen is deposited just into the body of the uterus rather than at the tip of the horn (Skidmore and Billah 2006a; Al-Bulushi et al. 2019a, b). Uterine manipulation during the intrauterine insemination may simulate penile stimulation by male camel or aide sperm transport by a greater release of uterine hormones involved in uterine contractility (Martinez et al. 2002).

Form the given studies; it appears that large scale utilization of AI in camels faces many technical difficulties particularly those relative to oestrus synchronization, preservation of camel sperm and low fertility of females inseminated with frozen semen. In addition, many basic concepts relevant to AI application are still not available, such as infectious disease screening, biosecurity controls, male management, training procedures and hygienic collection and processing of semen (Monaco and Lacalandra 2019).

2.7 Conclusion

Good reproductive performances allow obtaining a high numerical productivity of camel herds and consequently increasing the information available on the siblings and shorten the average generation interval. However, in traditional reproductive management, increasing reproductive efficiency of camel herds faces certain constraints such as long gestation period, short breeding season and traditional reproductive and herd management methods leading to widespread venereal infections and reducing fertility. Oestrus synchronization and AI are simplest and cheapest reproductive technologies to overcome some of these problems. They have a significant influence on the genetic improvement rate. A higher reproduction rate means a lower number of breeding animals for a given population size and, thus, a greater selection strength. A greater number of offspring per breeder also promotes a more accurate estimate of genetic values. Another advantage is the faster diffusion of higher genetic material and international movement of genetic material. To date, it is difficult to accept large scale practice of artificial insemination in DCbecause of the technical and zootechnical difficulties associated with this biotechnology. Investigations on the technical aspects are still in progress; such as semen collection methods, semen liquefaction and cooling, deep freezing and thawing of camel spermatozoa to optimize AI protocols and standardize procedures. The main zootechnical constraint is the nature of the breeding systems of camel herds which are in continuous movement throughout the year in pastoral areas. This makes difficult the ultrasound monitoring of ovarian activity, ovarian activity synchronization followed by timed ovulation induction and AI or blind mating. In addition, there is the risk of inbreeding and the reduction in reproductive capacities by using a reduced number of breeding males over a large number of females during short breeding season.

Acknowledgements The authors would like to express their gratitude to the Dr. Monaco David; Researcher at the Department of Veterinary Medicine (DiMeV), University of Bari "Aldo Moro", Valenzano (Bari), Italy, for his enthusiasm and assistance. We have also to express our appreciation to the scientific journal editors and authors of the book "Theriogenology in Camelidae: Anatomy, Physiology, BSE, Pathology and Artificial Breeding" for allowing as sharing some pictures of reproduction management and AI in camels during the course of this research.

References

Abdoon ASS (2001) Factors affecting follicular population, oocyte yield and quality in camels (*Camelus dromedarius*) ovary with special reference to maturation time in vitro. Anim Reprod Sci 66:71–79

Abo El-Maaty AM, Mohamed RH, Abd El Hameed AR, Hozyen HF, Ali AH (2019) Ovarian hormones and antioxidant biomarkers in dromedary camels synchronized with new and re-used controlled intravaginal drug release (CIDR)/GPG (Ovsynch) program during breeding season. Trop Anim Health Prod 1(6):1619–1625. https://doi.org/10.1007/s11250-019-01850-0

Ahmed H, Andrabi SM, Anwar M, Jahan S (2017) Use of post-thaw semen quality parameters to predict fertility of water buffalo (*Bubalus bubalis*) bull during peak breeding season. Andrologia 49

Ainani H, Achaâban MR, Tibary A, Pévet P, Simonneaux V, El-Allali K (2018) Environmental and neuroendocrine control of breeding activity in the dromedary camel. Rev Mar Sci Agron Vét 6(2):143–157

Akbar SJ, Hassan SM, Ahmad M (2018) Studies on semen processing, cryopreservation and artificial insemination in dromedary camel. Int J Anim Sci 2(2):1018

Akral SN, Khanna ND (1995) Ovarian activity during breeding season in Indian camel. J Anim Sci 65:889–890

Al Abri MA, Faye B (2019) Genetic improvement in dromedary camels: challenges and opportunities. Front Genet 10:167. https://doi.org/10.3389/fgene.2019.00167

Al Eknah MM (2000) Reproduction in Old World camels. Anim Reprod Sci 60–61:583–592

Al-Bulushi S, Manjunatha BM, Bathgate R, Rickard JP, de Graaf SP (2018) Effect of semen collection frequency on the semen characteristics of dromedary camels. Anim Reprod Sci 197:145–53. https://doi.org/10.1016/j.anireprosci.2018.08.022

Al-Bulushi S, Manjunatha BM, Bathgate R, Rickard JP, de Graaf SP (2019a) Liquid storage of dromedary camel semen in different extenders. Anim Reprod Sci 207:95–106

Al-Bulushi S, Manjunatha BM, Bathgate R, Rickard JP, de Graaf SP (2019b) Artificial insemination with fresh, liquid stored and frozen thawed semen in dromedary camels. PLoS One 14(11):e0224992. https://doi.org/10.1371/journal.pone.0224992

Al-Eknah M, Homeida N, Al-Haider A (2001) A new method for semen collection by artificial vagina from the dromedary camel. J Camel Prac Res 8:127–130

Al-Fatlawi AK, Al-Hamedawi TM (2017) Induction of fertile estrus in Iraqi camel (*Camelus dromedarius*) during seasonal anoestrus. Kufa J Vet Med Sci 8(1):58–63

Ali A, Al-Sobayil FA, Tharwat M, Alhawas A, Ahmed AF (2010) Causes of infertility in female camels (*Camelus dromedarius*) in middle of Saudi Arabia. J Agric Vet Sci Qassim Univ 2(2):59–66

Almutairi SE, Boujenane I, Musaad A, Awad-Acharari F (2010) Non-genetic factors influencing reproductive traits and calving weight in Saudi camels. Trop Anim Health Prod 42:1087–1092. https://doi.org/10.1007/s11250-010-9529-y

Al-Qarawi AA, Abdel-Rahman HA, El-Mougy SA, El-Belely MS (2002) Use of a new computerized system for evaluation of spermatozoal motility and velocity characteristics in relation to fertility levels in dromedary bulls. Anim Reprod Sci 74:1–9

Al-Sobayil KA (2003) Hormonal treatments for inducing fertile estrus in young dromedary females at the end of their seasonal anestrus in Qassim region, Saudi Arabia. J Camel Prac Res 5:179–181

Al-Sobayil KA (2008) The use of estrus synchronization and timed artificial insemination in dromedary she-camels in Saudi Arabia. J Agric Vet Sci 1:3–9

Al-Sobayil KA, Khalil MH, Al-Jobeile HS, Mohamed KM, Salal SA (2006) Quantitive genetic analyses and evaluation for early growth performance in Saudi camels. Int Sci Conf Camel 4:1759–1771

Anouassi A, Adnani M, Raed EL (1992) Artificial insemination in the camel requires induction of ovulation to achieve pregnancy. In: Allen WR, Higgins AJ, Mayhew IG, Snow DH, Wade JF (eds) Proceedings of the 1st international camel conference. R & W Publications (Newmarket) Ltd., London, pp 179–182

Bahbahani H, Musa HH, Wragg D, Shuiep ES, Almathen F, Hanotte O (2019) Genome diversity and signatures of selection for production and performance traits in dromedary camels. Front Genet. https://doi.org/10.3389/fgene.2019.00893

Boeta M, Porras MSA, Zarco LA, Aguirre-Hernández R (2006) Ovarian activity of the mare during winter and spring at a latitude of 19°21′ North. J Equine Vet Sci 26:55–58. https://doi.org/10.1016/j.jevs.2005.12.003

Bogle OA, Ratto MH, Adams GP (2012) Ovulation-inducing factor (OIF) induces LH secretion from pituitary cells. Anim Reprod Sci 133:117–122

Bravo PW, Skidmore JA, Zhao XX (2000) Reproductive aspects and storage of semen in Camelidae. Anim Reprod Sci 62:173e93

Brigitte AK (2005) Reproductive performance of camels (*Camelus dromedarius*) under pastoral management and its influence on herd development. Live Pro Sci 92:17–29

Campbell AJ, Pearson LK, Spencer TE, Tibary A (2015) Double ovulation and occurrence of twinning in alpacas (Vicugna pacos). Theriogenology 84:421e4

Chen BX, Yuen ZX, Pan GW (1985) Semen-induced ovulation in the bactrian camel (*Camelus bacterianus*). J Reprod Fertil 73:335–339

Chhaibi H, Campbell A, Pasha K, Tibary A (2016) Pharmacokinetics of a long-acting progesterone formulation in female camels. In: Camelid reproduction satellite symposium. Tours France, pp 23–26

Cooper MJ, Skidmore J, Ali M, Billah A, Wensvoort S, Billah M, Allen WR (1990) An attempt to induce and synchronize ovulation and superovulation in dromedary camels for embryo transfer. Proceedings of the Unite de Coordination pour L. Elevage Camel, (UCDEC'90), Paris, France, pp 313–326

Cooper MJ, Skidmore J, Allen WR, Wensvoort S, Billah M, Chaudhry M, Billah AM (1992) Attempts to stimulate and synchronize ovulation and superovulation in dromedary camels for embryo transfer. Proceedings of the 1st International Camel Conference, February 2–6, 1992, Dubai, pp 187–191

Crichton EG, Pukazhenthi BS, Billah M, Skidmore JA (2015) Cholesterol addition aids the cryopreservation of dromedary camel (*Camelus dromedarius*) spermatozoa. Theriogenology 83(2):168–174. https://doi.org/10.1016/j.theriogenology.2014.09.005

Crichton EG, Malo C, Pukazhenthi BS, Nagy P, Skidmore JA (2016) Evaluation of cholesterol-treated dromedary camel sperm function by heterologous IVF and AI. Anim Reprod Sci 174:20–28. https://doi.org/10.1016/j.anireprosci.2016.08.013

Deen A, Sahani MS (2006) Cryopreservation of camel semen. J Camel Pract Res 13:1–6

Deen A, Vyas S, Sahani MS (2003) Semen collection, cryopreservation and artificial insemination in the dromedary camel. Anim Reprod Sci 77:223–233

Derar R, Ali A, Al-Sobayil FA (2014) The postpartum period in dromedary camels: uterine involution, ovarian activity, hormonal changes, and response to GnRH treatment. Anim Reprod Sci 151:186–193. https://doi.org/10.1016/j.anireprosci.2014.10.024

Dholpuria S, Vyas S, Purohit GN, Pathak KML (2012) J Ultrasound 15:135–141

Dogan I, Toker MB, Alcay S, Udum Kucuksen D (2020) Comparison and assessment of ovarian follicular dynamics during the breeding and non-breeding season in Saanen goats. Vet Med Austria 107:63–71

El-Agamy ESI (2006) Camel milk. Handbook of milk of non-bovine mammals 2006

El-Allali K, Achaaban MR, Vivien-Roels B, Bothorel B, Tligui NS, Pevet P (2005) Seasonal variations in the nycthemeral rhythm of plasma melatonin in the camel (*Camelus dromedarius*). J Pineal Res 39:121e8

El-Allali K, Achaâban MR, Ouassat M (2017) Anatomy of the dromedary head skeleton revisited. J Morphol Sci 34:73–88. https://doi.org/10.4322/jms.100916

El-Allali K, Sghiri A, Bouâouda H, Achâaban MR, Ouzir M, Bothorel B, El Mzibri M, El Abbadi N, Moutaouakkil A, Tibary A, Pévet P (2018) Effect of Melatonin implants during the non-breeding season on the onset of ovarian activity and the plasma prolactin in dromedary camel. Front Vet Sci 44:1–9. https://doi.org/10.3389/fvets.2018.00044

El-Bahrawy KA (2010) Cryopreservation of dromedary camel semen supplemented with alpha-amylase enzyme. J Camel Pract Res 17:211–216

El-Bahrawy KA (2017a) The influence of caffeine supplementation and concerted utilization of enzymatic and mechanical semen liquefaction on freezability of dromedary camel spermatozoa. Int J Vet Sci Med 5:121–127. https://doi.org/10.1016/j.ijvsm.2017.09.005

El-Bahrawy KA (2017b) Influence of enzymatic and mechanical liquefaction of seminal plasma on freezability of dromedary camel semen. World Vet J 7(3):108–116. https://doi.org/10.5455/wvj.201708100
El-Bahrawy KA, Khalifa MA, Rateb SA (2015) Recent advances in dromedary camel reproduction: an Egyptian field experience. Emir J Food Agric 27:350–354
El-Hussanein EE (2003) An intervention for easy semen collection from dromedary camels, El-hussanein camel dummy. In: Skidmore JA, Adams GP (eds) Recent advances in camelid reproduction, vol 2003. International Veterinary Information Service, Ithaca
El-Hussanein EE (2017) Prospects of improving semen collection and preservation from elite dromedary camel breeds. World Vet J 7(2):47–64
Elias E, Bedrak E, Yagil R (1984) Estradiol concentration in the serum of the onehumped camel during the various reproductive stages. Gen Comp Endocrinol 56(2):258–264. https://doi.org/10.1016/0016-6480(84)90039-x
Elliot FI (1961) Artificial insemination of Bactrian camel (*Camelus Bactrianus*). Int Zoo Yearb 3:94
Emami Mibody MA, Iranmanesh M, Motamedi-Mojdehi R, Ghasemi Meymandi M (2016) Quantitative genetic analysis of dromedary camel (*Camelus dromedarius*) under pastoral management in central desert of Iran. Int J Adv Biol Biomed Res 4(1):10–14
Falchi L, Khalil WA, Hassan M, Marei WFA (2018) Perspectives of nanotechnology in male fertility and sperm function. Int J Vet Sci Med 6:265–269. https://doi.org/10.1016/j.ijvsm.2018.09.001
Faye B (1997) Guide de l'élevage du dromadaire. Éditions Sanofi, Libourne, 126 p
Faye B (2018) The improvement of camel reproduction performances: just a technical question? Rev Mar Sci Agron Vét 6(2):265–269. https://www.agrimaroc.org/index.php/Actes_IAVH2/article/view/607/618
Gherissi DE, Afri-Bouzebda F, Bouzebda Z (2017) Gestational age estimation of Sahraoui dromedary camel based on fetomatemal measures and phenotypic characteristics. J Anim Vet Adv 16(2):32–39
Gherissi DE, Afri-Bouzebda F, Bouzebda Z, Bonnet X (2018) Are female camels capital breeder? Influence of age, seasons, and body condition on reproduction in an extremely arid region. Mamm Biol 93:124–134
Gherissi DE, Chacha F, Lamraoui R, Gherissi A, Miloudi AL, Afri-Bouzebda F, Bouzebda Z (2019) Cross sectional survey of congenital and acquired genital disorders in Sahraoui female camels (*Camelus dromedarius*) at El Oued Abattoir, Southeastern Algeria. Anim Res Int 16(3):693–1733
Gherissi DE, Monaco D, Bouzebda Z, Afri Bouzebda F, Gaouar SBS, Ciani E (2020a) Camel herds' reproductive performance in Algeria: objectives and thresholds in extreme arid conditions. J Saudi Soc Agric Sci 19(7):482–491. https://doi.org/10.1016/j.jssas.2020.09.002
Gherissi DE, Bouzebda Z, Bouzebda-Afri F, Chacha F, Lamraoui R, Gherissi A, Miloud AL (2020b) Ecophysiology of camel ovarian functioning under extremely arid conditions in Algeria Euro-Mediterranean. J Environ Integr 5:56. https://doi.org/10.1007/s41207-020-00194-8
Ghoneim IM, Al-Eknah MM, Waheed MM, Alhaider AK, Homeida AM (2010) Effect of some extenders and enzymes on semen viscosity and sperm viability in dromedary camels (*Camelus dromedarius*). J Camel Pract Res 17:85–89
Hafez ES, Hafez B (2001) Reproductive parameters of male dromedary and bactrian camels. Arch Androl 46(2):85–98
Hammadi M, Khorchani T, Khaldi G, Majdoub A, Abdouli H, Slimane N, Portetelle D, Renaville R (2001) Effect of diet supplementation on growth and reproduction in camels under arid range conditions. Biotechnol Agron Soc Environ 5:69–72
Hanzen C (2015) Sémiologie: Propédeutique de l'appareil génital mâle des ruminants. Documents pédagogiques: Notes de cours et syllabus. Université de Liège. Available via https://orbi.uliege.be/bitstream/2268/70541/1/R03_Proped_genitale_femelle_2016.pdf
Hassan MM, Saeed M, Rizwan-ul-Muqtadir M (1995) Semen collection by artificial vagina and cryopreservation of camel (*Camelus dromedarius*) spermatozoa. Pak Vet J 15:105–108

Helmy MM (1991) A pharmacological method for synchronization of oestrous in the she-camel (*Camelus dromedarius*). In: Egyptian Society of Animal Reproduction and Fertility, 3rd Ann. Cong., Cairo, 12–14 January

Hemeida N, Al-Eknah M, Ismail ST, Al-haider A (2001) A new technique for collection of semen from dromedary camels. Emir J Agric Sci 13:18–22

Hussein FM, Metwelly Mona KK, Mahmoud A, Ragab MH (2015) Effect of CIDR application duration (7-10-14 days) on circulating estrogen and progesterone during breeding and non-breeding season in she-camels. Alexandria J Vet Sci 44:125e9

Kadim Isam T, Al-Amri Issa S, Al Kindi Abdulaziz Y, Msafiri M (2018) Camel meat production and quality: a review. J Camel Pract Res 25(1):9–23. https://doi.org/10.5958/2277-8934.2018.00002.4

Kershaw-Young CM, Maxwell WMC (2012) Seminal plasma components in camelids and comparisons with other species. Reprod Domest Anim 47:369–375

Kershaw-Young CM, Evans G, Maxwell WM (2013) Glycosaminoglycans in the accessory sex glands, testes and seminal plasma of alpaca and ram. Reprod Fertil Dev 24:368–369

Keshavarz M, Niasari Naslaji A, Zare H, Ziapour S, Mirtavoosi M, Cmidi M, Kalantari A, Moosavi-Movahedi AA (2016) Effect of Ficin enzyme on semen viscosity in dromedary camel. J Camel Pract Res 23:219–222

Konuspayeva G, Faye B, Loiseau G (2009) The composition of camel milk: a meta-analysis of the literature data. J Food Compos Anal 22:95–101

Mal G, Vyas S, Srinivasan A, Patil NVR, Pathak KML (2016) Studies on liquefaction time and proteins involved in the improvement of seminal characteristics in dromedary camels (Camelus dromedarius). Scientifica:4659358. https://doi.org/10.1155/2016/4659358

Malo C, Crichton EG, Skidmore JA (2017) Optimization of the cryopreservation of dromedary camel semen: cryoprotectants and their concentration and equilibration times. Cryobiology 74:141–147

Malo C, Grundin J, Morrell JM, Skidmore JA (2019) Individual male dependent improvement in post-thaw dromedary camel sperm quality after addition of catalase. Anim Reprod Sci 209:106168. https://doi.org/10.1016/j.anireprosci.2019.106168

Malo C, Crichton EG, Skidmore JA (2020) Preservation of the spermatozoa of the dromedary camel (Camelus dromedarius) by chilling and freezing: the effects of cooling time, extender composition and catalase supplementation. Theriogenology 153:9–18. https://doi.org/10.1016/j.theriogenology.2020.04.043

Manjunatha B, Al-Bulushi S, Pratap N (2015) Synchronisation of the follicular wave with GnRH and PGF 2α analogue for a timed breeding programme in dromedary camels (*Camelus dromedarius*). Anim Reprod Sci 160:23–29. https://doi.org/10.1016/j.anireprosci.2015.06.023

Manjunatha BM, Al-Hosni A, Al-Bulushi S (2018) Resynchronization of synchronized follicular wave in dromedary camels of unknown pregnancy status (Camelusdromedarius). Theriogenology 119:208–213. https://doi.org/10.1016/j.theriogenology.2018.07.001

Marie M, Anouassi A (1986) Mating-induced luteinizing hormone surge and ovulation in the female camel (*Camelus dromedarius*). Biol Reprod 35:792–798

Marie M, Anouassi A (1987) Induction of luteal activity and progesterone secretion in the non-pregnant one-humped camel (*Camelus dromedarius*). J Reprod Fertil 80:183–192

Martinez EA, Vazquez JM, Roca J, Lucas X, Gil MA, Parrilla I, Vazquez JL, Day BN (2002) Minimum number of spermatozoa required for normal fertility after deep intrauterine insemination in non-sedated sows. Reproduction 123:163–170

McKinnon AO, Tinson AH, Nation G (1994) Embryo transfer in dromedary camels. Theriogenology 41:145–150

Medan MS, Absy G, Zeidan AE, Khalil MH, Khalifa HH, Abdel-Salaam AM, Abdel-Khalek TM (2008) Survival and fertility rate of cooled dromedary camel spermatozoa supplemented with catalase enzyme. J Reprod Dev 54(1):84–89. https://doi.org/10.1262/jrd.19124

Minoia P, Moslah M, Lacalandr GM, Khorchan T, Zarrilli A (1992) Induction of oestrous and management of reprodction in the female dromedary. In: Allen WR, Higgins AJ, Mayhew IG, Snow DH, Wade JF (eds) Proceedings of the first international camel coference. R&W Publications, Ž. Newmarket, pp 119–123

Monaco D, Lacalandra GM (2019) Considerations for the development of a dromedary camel (*Camelus dromedarius*) semen collection centre. Anim Reprod Sci 212:106239. https://doi.org/10.1016/j.anireprosci.2019.106239

Monaco D, Lacalandra GM, El-Bahrawy KA (2013) Ovarian monitoring and effects of Controlled Intravaginal Drug Releaser (CIDR) on vaginal environment and follicular activity in dromedary camels, during non-breeding season in Egypt. Emir J Food Agric 25:296–300

Monaco D, Fatnassi M, Padalino B, Hammadi M, Khorchani T, Lacalandra GM (2016) Effect of α–amylase, papain, and spermfluid® treatments on viscosity and semen parameters of dromedary camel ejaculates. Res Vet Sci 105:5–9

Monaco D, Fatnassi M, Doghbri L, Padalino P, Seddik MM, Khorchani T, Hammadi M, Lacalandra GM (2018) Mating behaviour and semen parameters in dromedary camel bulls (*Camelus dromedarius*): a comparison between two types of artificial vagina. Emir J Food Agric 30(4):326–334. https://doi.org/10.9755/ejfa.2018.v30.i4.1669

Morton KM, Bathgate R, Evans G, Maxwell WMC (2007) Cryopreservation of epididymal alpaca (Vicugna pacos) sperm: a comparison of citrate, tris and lactose based diluents, and pellets and straws. Reprod Fertil Dev 19:792–796

Morton KM, Vaughan JL, Maxwell WMC (2008) The continued development of artificial insemination technology in Alpacas. Rural Industries Research and Development Corporation, Kingston

Morton KM, Billah M, Skidmore JA (2010) Artificial insemination of dromedary camels with fresh and chilled semen: effect of diluent and sperm dose, preliminary results. Reprod Domest Rum VII Abs

Morton K, Billah M, Skidmore JA (2011) Effect of green buffer storage on the fertility of fresh camel semen after artificial insemination. Reprod Domest Anim 46(3):554–557. https://doi.org/10.1111/j.1439-0531.2010.01678.x

Morton K, Billah M, Skidmore JA (2013) Effect of sperm diluent and dose on the pregnancy rate in dromedary camels after artificial insemination with fresh and liquid-stored semen. J Camel Sci 6:49–62

Mostafa TH, Abd El-Salaam AM, Abdoon ASS, Abdel-Khalek AE (2016a) Improving fertility of maghrabian she-camels treated with different hormonal treatments during breeding season in Egypt. IOSR-JAVS 9(10):20–28. https://doi.org/10.9790/2380-10012028

Mostafa TH, Abd El-Salaam AM, Farag ME, Refae MM (2016b) Effect of feeding treatments on productive and reproductive performance of she-camels. Glob Adv Res J Agric Sci 5(7):263–276

Musa BE, Merkt H, Sieme H, Hago B, Hoppen H (1990) The female camel (*Camelus dromedarius*) and the artificial insemination. Proc. Unite de Coordination pour L'Elevage Camelin. In: Is it possible to improve the reproductive performance of the camel? Proceeding UCDEC workshop, Paris, pp 257–261

Niasari-Naslaji A, Mosaferi S, Bahmani N, Gharahdaghi AA, Abarghani A, Ghanbari A, Gerami A (2006) Effectiveness of a tris-based extender (SHOTOR diluent) for the preservation of Bactrian camel (Camelus bactrianus) semen. Cryobiology 53:12–21

Niasari-Naslaji A, Mosaferi S, Bahmani N, Gerami A, Gharahdaghi AA, Abarghani A, Ghanbari A (2007) Semen cryopreservation in Bactrian camel (*Camelus bactrianus*) using SHOTOR diluent: effects of cooling rates and glycerol concentrations. Theriogenology 68:618–625

Quzy I, Anwar S, Purohit GN (2013) Hormonal management of ovarian activity in breeding camels two months ahead of the natural breeding season. Camel Int J Vet Sci 1:37–49. Retrieved from: https://ndpublisher.in/admin/issues/camelv1n1d.pdf

Raiymbek G, Kadim I, Konuspayeva G, Mahgoub O, Serikbayeva A, Faye B (2015) Discriminant amino-acid components of Bactrian (*Camelus bactrianus*) and dromedary (*Camelus dromedarius*) meat. J Food Compos Anal 41:194–200

Rateb SA (2016) Ultrasound-assisted liquefaction of dromedary camel semen. Small Ruminant Res 141:48–55. https://doi.org/10.1016/j.smallrumres.2016.07.005

Rateb SA, Khalifa MA, Abd El-Hamid IS, Shedeed HA (2019) Enhancing liquid-chilled storage and cryopreservation capacities of ram spermatozoa by supplementing the diluent with different additives. Asian-Australas J Anim Sci 33(7):1068–1076. https://doi.org/10.5713/ajas.19.0338

Shahin MA, Khalil WA, Saadeldin IM, Swelum AA-A, El-Harairy MA (2020) Comparison between the effects of adding vitamins, trace elements, and nanoparticles to shotor extender on the cryopreservation of dromedary camel epididymal spermatozoa. Animals 10(1):78

Sanjay K, Sharma VK, Sudhuman S, Hariprasad GR, Gorakh M, Alagiri S et al (2012) Proteomic identification of camel seminal plasma: purification of b-nerve growth factor. Anim Reprod Sci 136(4):289–295. https://doi.org/10.1016/j.anireprosci.2012.11.001

Sghiri A, Driancourt MA (1999) Seasonal effects on fertility and ovarian follicular growth and maturation in camels (*Camelus dromedarius*). Anim Reprod Sci 55:223e37

Silva M, Ulloa-Leal C, Norambuena C, Fernandez A, Adams GP, Ratto MH (2014) Ovulation-inducing factor (OIF/NGF) from seminal plasma origin enhances Corpus Luteum function in llamas regardless the preovulatory follicle diameter. Anim Reprod Sci 148:221e7

Skidmore JA (2000) Reproductive physiology in male and female camels. Recent advances in camelid reproduction 2000. International Veterinary Information Service, Ithaca

Skidmore JA (2019) The use of some assisted reproductive technologies in old world camelids. Anim Reprod Sci 207:138–145

Skidmore JA, Billah M (2006a) Comparison of pregnancy rates in dromedary camels (*Camelus dromedarius*) after deep intra-uterine versus cervical insemination. Theriogenology 66:292–296. https://doi.org/10.1016/j.theriogenology.2005.11.013

Skidmore JA, Billah M (2006b) Investigation of the most appropriate time for insemination of female camels (*camelus dromedarius*) after GnRH injection and comparison of pregnancy rates after deep intra-uterine versus cervical insemination. In: Proceedings of first conference of International Society of Camelids Research and Development (ISOCARD), p 54

Skidmore JA, Billah M, Allen WR (1994) The ovarian follicular wave pattern in the mated and non-mated dromedary camel (*Camelus dromedarius*). J Reprod Fertil 49:545–548. https://doi.org/10.1530/jrf.0.1060185

Skidmore JA, Billah M, Allenm WR (1996) The ovarian follicular wave pattern and induction of ovulation in the mated and non-mated one humped camel (Camelus dromedarius). J Reprod Fertil 106:185–192

Skidmore JA, Starbuck GR, Lamming GE, Allen WR (1998) Control of luteolysis in the one-humped camel (*Camelus dromedarius*). J Reprod Fertil 114:201–209

Skidmore JA, Adams GP, Billah M (2009) Synchronisation of ovarian follicular waves in the dromedary camel (Camelus dromedarius). Anim Reprod Sci 114:249–255. https://doi.org/10.1016/j.anireprosci.2008.08.024

Skidmore JA, Morton KM, Billah M (2013) Artificial insemination in dromedary camels. Anim Reprod Sci 136:178–186

Swelum AA, Alowaimer AN (2015) The efficacy of controlled internal drug release CIDR in synchronizing the follicular wave in dromedary camels (*Camelus dromedarius*) during the breeding season. Theriogenology 84:1542–1548

Swelum AA, Saadeldin IM, Ba-Awadh H, Alowaimer AN (2018a) Shortened daily photoperiod during the non-breeding season can improve the reproductive performance of camel bulls (*Camelus dromedarius*). Anim Reprod Sci 195:334–344. https://doi.org/10.1016/j.anireprosci.2018.06.014

Swelum AA, Saadeldin IM, Ba-Awadh H, Alowaimer AN (2018b) Effects of melatonin implants on the reproductive performance and endocrine function of camel (Camelus dromedarius) bulls during the non-breeding and subsequent breeding seasons. Theriogenology 119:18–27

Swelum AA, Saadeldin I, Moumen A, Ba-Awadh H, Alowaimer AN (2018c) Efficient follicular wave synchronization using a progesterone-releasing intravaginal device (PRIDΔ) in Camelus dromedaries. Theriogenology 118:203–211

Tibary A (2001) Semen preservation in Camelids. In: Proceedings of the annual conference of the Society for Theriogenology, September 12–15, 2001, Vancouver, BC, Canada, pp 369–378
Tibary A (2018) Monitoring and controlling follicular activity in camelids. Theriogenology 109:22–30. https://doi.org/10.1016/j.theriogenology.2017.12.011
Tibary A, Anouassi A (1997a) Reproductive physiology in the female Camelidae. In: Tibary A, Anouassi A (eds) Theriogenology in Camelidae: anatomy, physiology, BSE, pathology and artificial breeding. Actes Editions Rabat- Maroc, Printed in UAE, p 489
Tibary A, Anouassi A (1997b) Male breeding soundness examination. In: Tibary A, Anouassi A (eds) Theriogenology in Camelidae: anatomy, physiology, BSE, pathology and artificial breeding. Actes Editions Rabat- Maroc, Printed in UAE, p 79
Tibary A, Anouassi A (1997c) Artificial breeding and manipulation of reproduction in camelidae. In: Tibary A, Anouassi A (eds) Theriogenology in Camelidae: anatomy, physiology, BSE, pathology and artificial breeding. Actes Editions Rabat- Maroc, Printed in UAE, p 413
Tibary A, Anouassi A (2018) Challenges in the development of artificial insemination in the dromedary camel. Rev Mar Sci Agron Vét 6(2):178–188
Tibary A, Anouassi A, Sghiri A (2005) Factors affecting reproductive performance of camels of the herd and individual level. In: Faye B, Esenov P (eds) Desertification combat and safety food. The added P.N. 97
Tibary A, Pearson LK, Anouassi A (2014) Applied andrology in camelids. CABI, Wallingford, pp 418–449
Tingari MD, Manna MM, Rahim AT, Ahmed AK, Hamad MH (1986) Studies on camel semen. I. Electroejaculation and some aspects of semen characteristics. Anim Reprod Sci 12:213–222
Van Arendonk JAM, Bijma P (2003) Factors affecting commercial application of embryo technologies in dairy cattle in Europe – a modeling approach. Theriogenology 59:635–649
Vyas S, Rai AK, Sahani MS, Khanna ND (2004) Use of real-time ultrasonography for control of follicular activity and pregnancy diagnosis in the one humped camel (*Camelus dromedarius*) during the non-breeding season. Anim Reprod Sci 84:229–233
Vyas S, Singh R, Purohit GN, Pareek PK, Sahani MS (2008) Ultrasound evaluation of ovarian response to photoperiodic control measures in *Camelus dromedarius*. Vet Arch 78:39–48
Waheed MM, Ghoneim IM, Hassieb MM, Alsumait AA (2014) Evaluation of the breeding soundness of male camels (*Camelus dromedarius*) via clinical examination, semen analysis, ultrasonography and testicular biopsy: a summary of 80 clinical cases. Reprod Domest Anim 49:790–796. https://doi.org/10.1111/rda.12370
Wani NA, Billah M, Skidmore JA (2008) Studies on liquefaction and storage of ejaculated dromedary camel (*Camelus dromedarius*) semen. Anim Reprod Sci 109(1–4):309–318
Watson PF (2000) The causes of reduced fertility with cryopreserved semen. Anim Reprod Sci 60:481–492
Zarrouk A, Souilem O, Beckers JF (2003) Actualités sur la reproduction chez la femelle dromadaire (*Camelus dromedarius*). Revue Elev Med Vét Pays Trop 56(1–2):95–102
Zhao XX, Huang YM, Chen BX (1994) Artificial insemination and pregnancy diagnosis in the Bactrian camel (*Camelus bactrianus*). J Arid Environ 26:61–65
Ziapour S, Niasari-Naslaji A, Mirtavousi M, Keshavarz M, Kalantari A, Adel H (2014) Semen collection using phantom in dromedary camel. Anim Reprod Sci 151:15–21

Chapter 3
Biotechnological Advancements in Livestock Production

Bhaskar Sharma, Dixita Chettri, and Anil Kumar Verma

Abstract Animal-based agriculture is being used to surmount the food demand for the ever-increasing human population under limited land and water resources. The meat and milk consumption has severely grown over the past few decades with increasing population and the demand is expected to grow by double in the next three decades. The fulfillment of milk, meat, and related products demand has turned into a greater challenge for poor and developing countries which is leading to malnutrition. The traditional methods for livestock production are facing challenges due to limited offspring production, unstable genetic improvement, tedious methods, high cost, and disease susceptibility. Here we review the modern technological developments offering interesting solutions to the issues associated with conventional methods for animal production. The advanced biotechnological interventions such as *in vitro* fertilization, somatic nuclear transfer, embryonic stem cell technology, multiple ovulations, and embryo transfer could improve the nutritional composition and animal health. The capabilities of precise and desired interventions at the cellular and molecular levels renders unprecedented developments in animal reproduction. Molecular marker-assisted selection and transfer of useful and superior traits to the recipient line ensures improvements such as better growth, quality, and disease-resistance for several generations. The chapter covers biotechnological methodologies adopted in modern animal breeding, including marker-assisted selection methods for stable and improved livestock production.

Keywords Marker assisted selection · Somatic cell nuclear transfer · Artificial insemination · Embryo transfer · Stem cell technology

---

B. Sharma
School of Life and Environmental Sciences, Faculty of Science, Engineering, and Built Environment, Deakin University, Geelong, VIC, Australia

Department of Biotechnology, TERI School of Advanced Studies, New Delhi, India
e-mail: sharmabha@deakin.edu.au

D. Chettri · A. K. Verma (✉)
Department of Microbiology, Sikkim University, Gangtok, Sikkim, India
e-mail: dchettri02@sikkimuniversity.ac.in; akverma@cus.ac.in

V. K. Yata et al. (eds.), *Sustainable Agriculture Reviews 54*, Sustainable Agriculture Reviews 54, https://doi.org/10.1007/978-3-030-76529-3_3

Abbreviations

AFLPs	Amplified fragment length polymorphisms
AI	Artificial insemination (AI)
BCWD	Bacterial cold-water disease
bFGF	Basic fibroblast growth factor
BLUP	Best linear unbiased prediction
BOEC	Bovine oviduct epithelial cells
CRISPR Cas9	Cluster regularly interspaced short palindromic repeats
DNA	De-oxy ribonucleic acid
eCG	Equine chorionic gonadotropins
ESC	Embryonic stem cell technology (ESC)
ET	Embryo transfer (ET)
FAO	Food and Agriculture Organization
FSH	Follicle-stimulating hormones
GAS	Gene-assisted selection
IETS	International Embryo Technology Society
IVF	In vitro fertilization (IVF)
JAK/STAT	Janus kinase and signal transducer and activator of transcription proteins
LD	Linkage disequilibrium
LE	Linkage equilibrium
LH	Luteinizing hormone
LIF	Leukemia inhibitory factor
MAI	Molecular marker-assisted introgression
MAS	Marker Assisted Selection
MOET	Multiple ovulation and embryo transfer (MOET)
NGS	Next-generation sequencing
PCC	Premature chromosome condensation
QTL	Quantitative Trait Locus
RAPD	Random amplified polymorphic DNA
RFLPs	Restriction fragment length polymorphisms
SCNT	Somatic cell nuclear transfer (SCNT)
SCR	Sire conception rate
SNP	Single nucleotide polymorphism
SSR	Simple sequence repeats
TALEN	Transcription activator-like effector nuclease
TGF-beta	Transforming growth factor beta
TGS	Third-generation sequencing

3.1 Introduction

Livestock production began with the domestication of sheep, chicken, pigs, goats, and other cattle about 8000–10,000 years ago (Porter 2020). Eventually, the need for the genetic improvement of livestock turned up to produce better characteristics through breeding. Later, with the rediscovery of Mendel's law (Mendel 1866) and the establishment of modern genetics, the selection and crossbreeding of livestock revolutionized genetic improvement and production. The growth rate of the commercial production of the pig, chicken, and other poultry was substantially enhanced in unprecedently lesser time (Muir and Aggrey 2003). Therefore, the development of genetic improvement tools led the increased livestock production in developed and developing countries. The global livestock production attributes to around 40% of total agricultural production. Livestock production contributes more than half in developed countries and approximates one-third in developing countries. Livestock demand is increasing in developing countries due to the larger population size and dietary habits. On the other side, the agricultural food consumption was decreased to 1.1% from 2.2% in the last 40 years. The annual per capita meat consumption was increased by 100% in developing countries between 1980 and 2002. The FAO (Food and Agriculture Organization) projected an increase of 60% in the global animal food production demand by 2050, considering the huge population growth (Alexandratos and Bruinsma 2012). There is an urgent requirement for the Livestock revolution to meet the population food demand while sustaining limited natural resources. The emerging trends in the livestock sector are influenced by increasing competition for grazing and water resources, enhanced usage of pigs and poultry, livestock production in hot, moist, and disease-prone habitat, and health risks due to industrialization and urbanization (Harrison et al. 2002). There is a need to introduce effective public policies for natural resource management, food safety, environmental pollution, and public health considering large transformation in the livestock sector (Harrison et al. 2002).

Recently, the demand for animal-based foods such as eggs, milk, and meat has increased with the growing population in developing countries. The share of total calories from animal-based food has risen in the past few decades, while the total calories from traditional staples and root-based products such as cassava, potatoes, and plantains have decreased by approximately 4%, making animal-based products the second major source for the calories after cereals in developing countries. The consumption of animal-based products is still low in developing countries compared to developed countries but continuously rising over time. The most organized structural organizations for livestock production and maintenance have been intensive establishments in urban areas for poultry and meat production in East Asia and Latin America (Seré C 1996). The production systems for the livestock have dramatically evolved because of a surge in the demand and increasing pressure on natural resources. The use of the grazing system contributes to a large proportion of total livestock production. However, the grazing systems are severely affected by climate change, land degradation, land enforcement, and increasing population. Similarly, using limited land with crop-livestock production, where ruminant livestock is grown with crops in a mixed system, is a popular choice in developing countries (Seré C 1996). Crops are used to feed the animals and livestock that produce manure and food that generate diversified income for rural farmars. Such livestock production systems are critical for household food security and immediate income generation with limited resources. Another type of production system is an intensive industrial livestock production system that involves heavyweight machinery, limited land area, improved production, and low-cost production. It is particularly beneficial for meeting the demand of the growing population but could reduce the value of the livestock (Seré C 1996).

Biotechnology can be defined as a technological application using a biological system to produce or modify any method or product for human benefits. The animal, plants, and microorganisms are modified using a set of techniques to produce desired products for human use. The application of the modern knowledge of the techniques and methods results in the higher qualitative and quantitative improvement in the desired traits of the selected organisms at a lower cost. The discovery of the genetic code led to the origin of the new modern method known as 'biotechnology'. The first-ever product developed using biotechnology in the field of animal sciences was bovine somatotropin metabolite, commercially produced at a large scale and administered to more than half cattle in the United States of America. Such technologies are necessary for ensuring nutritive food for the growing population. The major concern with the use of biotechnology has been observed with regulatory policies where potential risks of releasing genetically modified materials in the environment have attracted the attention of the wider scientific community. Biotechnology largely deals with the modification of the genetic material that poses a potential threat to human health and the environment. Sometimes, the quality of the product is compromised when producing in large quantities. Given the safety concerns, industrial manufacturing must be regulated in a robust and organized manner to establish a solid livestock production system. Overall, biotechnological interventions have revolutionized animal-based production, making it a stable industry for feeding future generations.

3.2 Biological Methods in Animal Breeding

3.2.1 Somatic Cell Nuclear Transfer (SCNT)

The cellular ability to generate all cell types is termed as totipotency (Lu and Zhang 2015). In mammalian cells, the totipotency is observed in the blastomere of preimplantation embryos and eventually lost during embryo development to produce inner cell mass. The totipotency is a vital characteristic of somatic cell nuclear transfer (SCNT) that can be used to clone animals such as sheep (Wilmut et al. 1997a), cows (Cibelli et al. 1998), pigs (Onishi et al. 2000), and rabbits (Chesné et al. 2002) from differentiated somatic cells (Gurdon 1962). The somatic cell nuclear transfer is a method where DNA from the somatic cell of the donor is transferred to the enucleated metaphase II oocyte of the acceptor for the production of genetically identical cells to the donor. The discovery of Dolly through SCNT led the researchers to think about cellular differentiation reprogramming. The differentiated nucleus are reversed to attain totipotency using nuclear reprogramming. The reprogramming of the nucleus through SCNT involves enucleation from recipient cells followed by injecting the nucleus from donor cells and finally activation of the reconstructed embryos. The nuclear membrane breaks down during an enucleation process to make a premature chromosome condensation (PCC) promoted by M-phase promoting factors (Campbell et al. 1996). PCC is necessary for initiating the reprogramming and in which the chromatin-bound proteins dissociate from the genome. After fertilization, the phospholipase from sperm triggers the oocyte activation through calcium signaling (Saunders et al. 2002). In the case of the absence of phospholipases, the activation is induced by either treatment of the strontium chloride or electro pulse treatment (Liu et al. 2014). Later, nuclear expansion occurs in the fertilized zygote to form a paternal and maternal pronucleus followed by DNA replication. The zygote further activates the genome by restoring transcription in the new genome that is known as a zygotic genome activation. Further, the SCNT process extends to chromatin, transcriptional, and epigenetic reprogramming that leads to the development of the clone. SCNT is an excellent tool for the study of genetic disease, growth regulation, gene therapy (Capecchi 2000), and genetic imprinting (Solter 2004). SCNT can be used to produce pharmaceutically important organisms, therapeutic proteins, and the preservative of endangered species. Despite the advancements, developmental abnormalities and low transformation efficiencies were observed that limit the widespread usage of the SCNT (Hill et al. 2000b; Kubota et al. 2000). The low transformation efficiencies varied from 0–10% and were predicted due to errors in the epigenetic signal reprogramming. The abnormalities could result in immunodeficiency, early death, respiratory defects, and obesity (Loi et al. 2016). Several factors can affect nuclear reprogramming, such as the origin and age of the donor and recipient cells (Wells et al. 2003; Hill et al. 2000b), epigenetic signals (Enright et al. 2003), tissue, and cell culture (Zakhartchenko et al. 1999), and drugs (Kubota et al. 2000). There were instances when the SCNT efficiency was enhanced through optimized experimental handling of the SCNT (Wilmut et al. 1997b). Overall, the SCNT method is highly suitable for cattle and other animals for genetic improvements. The method can undergo a wide variety of genetic improvements and produce quality transgenic lines without affecting animal biological properties. The somatic cells are very easy to collect, store, and transport that increases the success rate and stability of the method. SCNT is very useful in the biomedical applications and the therapeutic treatments of the animals. However, high cost and tedious process is limiting its extensive application. Moreover, poor oocyte recovery has been observed in few cases that need to address before successful implementation of the method (Ogura et al. 2002). There are several legal and ethical restriction imposed on the SCNT products in many countries (Kishigami et al. 2006). The SCNT needs significant improvement to be easily accessible to most of the part of the world.

3.2.2 Artificial Insemination (AI)

Artificial insemination is a method in which the semen from a male donor is placed into the reproductive tract of the female recipient. The method does not use any natural method for mating or reproduction (Salisbury et al. 1978). Artificial insemination is a part of a group of technologies known as assisted reproduction technologies. The methodology allows the transfer of the conception products into females and the collection of semen, the development of reliable methods for inserting semen into the female reproductive organ is a prerequisite. Usually, the semen is collected into a sterilized insulated vessel using an artificial vagina consisting of suitable temperature and pressure for stimulation and ejaculationl (Salisbury et al. 1978). Few other methods may include electroejaculation apparatus for collecting semen from animals with abnormalities or as an alternative for extreme conditions. Semen is sensitive to enzymatic and chemical exposure (Wani et al. 2008). The semen exists in liquid form and consists of various energy molecules such as protein, magnesium, calcium, zinc, and fructose. The semen plasma helps spermatozoa reach the fertilization site and survival for successful fertilization. The collected semen is used fresh or can be stored at 4 °C. The spermatozoa are sensitive to a higher temperature whereas the lower temperature minimizes metabolic activities and bacterial infection that increase the lifespan of spermatozoa. Cryopreservation can be opted for long-term preservation at a much lower temperature (Brinsko and Varner 1992). The semen sample is mixed with the sugar, lipoprotein, and cryopreserving glycerol for longer time preservation. Natural detection of estrus and ovulation is a critical task for timely insemination (Bó and Baruselli 2014). The ovulation can be induced using vasectomized boys or the administration of the using hormones. Artificial insemination has been improvised and upgraded continuously with the latest technological developments.

Artificial insemination is the most common method for improving domestic livestock such as dairy cattle, pigs, horses, and sheep (Table 3.1). The usage of artificial insemination has been extended to human fertility treatments. The physical contact between livestock was the major reason for the various disease transmission, which resulted in a shorter life span and loss in productivity. Artificial insemination avoids contact between livestock and retains the quality of the livestock for a longer time. The strength of artificial insemination is the genetic improvement and higher production yield using the semen of superior quality male for best female (Morrell 2011). The methodology is highly useful for the conservation of genetic material of the endangered species and the revival of the genetic merits of the animals. The breeding became possible for animals under physical challenges or abnormalities, and different relocation. Despite the advantages of technology, few limitations were observed for the artificial reproduction of the animals. The viral and bacterial

Table 3.1 Major advantages and limitations of biological methods in animal breeding

Sr. No.	Technology/ method	Advantages	Constraints/limitations	References
1	Somatic cell nuclear transfer (SCNT)	1. Introduction of selective genetic enhancements 2. Capable of producing efficient transgenic offspring and transgenic animal lines 3. Wide range of genetic alteration to *in vitro* cultured cells 4. No major biological properties and nutritional quality alteration in cloned animals compared to non-cloned animals 5. Easy to obtain, culture, and cryopreserve somatic cells 6. Therapeutic and biomedical applications	1. Poor quality oocyte recovery 2. Abnormalities in generated clones such as defected phenotypes 3. Resource intensive and costly procedure 4. Several legal and ethical issues related to the use of SCNT products	Pickering et al. (2005), Hill et al. (2000a), Ogura et al. (2002), and Kishigami et al. (2006)
2	Artificial insemination (AI)	1. Limits the spread of disease and sterility 2. Early detection of semen defects ensures better breeding efficiency 3. Semen transportation, storage, and extensive uses are possible 4. Helpful to inseminate physically different sized animals and increased rate of conception 5. Reduce the number of sires maintained for insemination	1. Time-consuming, and tedious operations 2. Well trained staff needed, and special instruments required 3. Advanced hygiene and maintenance of the instruments 4. The spread of genetic defects is possible 5. Reduction of the effective population of breed and inbreeding can spread	Noakes et al. (2018)
3	Embryo transfer (ET)	1. Reduced risk of disease transmission due to limited transportation of live animals 2. Easy to exchange genetic material and promote international trading 3. Increased number of offspring per female 4. Storage and expansion of rare and endangered species genetic stock 5. Desired mating and faster genetic process	1. High cost involved in the method 2. Advanced expertise needed to conduct the embryo transfer 3. The success rate can be low sometimes 4. Spread of the disease through recipient is possible	Bó and Mapletoft (2018)

(continued)

Table 3.1 (continued)

Sr. No.	Technology/ method	Advantages	Constraints/limitations	References
4	Embryonic stem cell technology (ESC)	1. Lesser animal testing, diagnosis, and research needed 2. Efficient preservation and revival of rare and endangered species 3. Targeted treatment and care for animals 4. Advanced genetic improvement is possible	1. Require skilled experts for the successful implementation of the technology 2. The costly method requires *in vitro* maintenance 3. Several legal and ethical issues associated with the use of ESCs	Deckha and Xie (2008)
5	Multiple ovulation and embryo transfer (MOET)	1. Improve the reproductive potential of the animals 2. Production of numerous offspring is possible 3. Easy collection, storage, and transportation of the embryos 4. No surgical methods are mandatory for the successful MOET 5. High yield and growth in a short period 6. MOET can be undertaken at farm area which reduces handling damages	1. Costly method and require skilled experts 2. Sometimes a longer waiting period for animals to become productive 3. Hormone dosage optimization to reduce the risk of abnormalities	Bergstein-Galan et al. (2019)
6	In vitro fertilisation (IVF)	1. Preservation of valuable genetic stocks for a longer period 2. Lesser hormonal treatments to animals 3. A shorter generation interval is needed 4. No impact of physiological abnormality on offspring 5. Semen cost is highly reduced 6. A range of donors can be used	1. IVF is a costly method and requires specialized instruments 2. Skilled and experienced staff needed 3. Special hygiene and maintenance required in every step 4. The success rate can decline if animals are facing oocyte inactivation	del Collado et al. (2017) and Abdalla et al. (2009)

infection in the semen is a major threat to the successful implementations of artificial reproduction (Morrell 2011). The widespread use of artificial reproduction methods compromised the fertility of cattle and animals and eventual loss of genetic variation (Morrell 2011). Artificial insemination can adopt the latest biotechnological advancement and integrate new tools to improve the process of large-scale livestock production. Artificial insemination can be beneficial for insemination of different size animals, semen transportation, early detection of the disease, and limiting the disease transmission through animals (Noakes et al. 2018). However, the AI method cannot overcome with any possible genetic defect in donors or recipients. The AI has been used traditionally to breed animals at a larger scale and can be readily improved with new technological advancements.

3.2.3 Embryo Transfer (ET)

The embryo transfer has emerged as a complementary method of artificial insemination. The embryo transfer is an artificial breeding method where the embryos are retrieved from an elite donor female before implantation and transferred to a recipient female that serves as surrogate mother (Wilmut et al. 1985; Betteridge 1981). The embryos are recovered through ovulation by hormonal treatment. The embryo transfer methods spread the desired genetic traits from superior female animals (Table 3.1). Embryo transfer can be achieved surgically or non-surgically. In the surgical method, the uterus of the donor animal is accessed through an incision on the midline or the flank (Tan et al. 1990). The uterine horns are flushed by penetrating the uterine wall using a tube from one end and with a hypodermic needle from another end. The medium is collected in a dish usually forced by a syringe. The embryos are roughly found to be 0.2 mm in size in the 32-cell stage of development. The embryos are examined under a microscope and picked up with a small glass pipette attached to a micro-syringe. These embryos are directly transferred to the recipient female uterus by puncturing the uterine wall with a needle. Further, the glass pipette with an embryo is inserted into the reproductive tract of the animal and embryo is released into the uterine lumen with slight pressure. The embryos are transferred within 4–5 h of recovery to minimize the risk associated with temperature and pH. Additionally, it is necessary to note that the estrous cycle of the donor and recipient animals should be in synchronization. Moreover, many chemical agents can be used to induce estrous-synchronization or prostaglandins.

The non-surgical method is based on the use of the foley catheter for collecting the embryos (Wright 1981). The genital area of the animal is washed properly, and local anesthetic injection is given. Later, the catheter is inserted into the vagina and goes through the cervix followed by the pressure release of the catheter. The flushed fluid is collected, and embryos are filtered from the collection of the medium. The procedure of embryo transfer involves the selection and management of the donor and recipient female. The estrus cycle detection should be normal for the superovulation to happen on the time. The superovulation and embryo production occur eventually followed by the embryo transfer event. The embryo transfer reduces disease transmission, circumvent infertility, induce genetic testing, rapid exchange of genetic material, and increase the offspring per female (Smith 1988; Betteridge 1981). The embryo transfer technique has been applied to most domestic animals. Overall, embryo transfer is another major technique developed to enhance the impact of superior genotype in the population. The latest developments are introduced to embryo transfer to establish an easy and less destructive method. The ET ensure faster genetic improvements in major animal species and comparatively higher offspring generation (Bó and Mapletoft 2018). The ET reduces the risk of disease transmission and allows better storage and transportation of the embryo which makes international trade easy.

3.2.4 Embryonic Stem Cell Technology (ESC)

Embryonic stem cells (ESCs) are self-renewable and pluripotent cells that can differentiate into any cell type of primary germ layer. They originate from the inner cell mass of the pre-implanted blastocyst. These characteristics make them ideal candidates for genetic modification, gene therapy, and cell regeneration. ESCs are categorized into naive and primed (Nichols and Smith 2009). The mouse ESCs are naïve pluripotency while human ESCs are developed into an advanced primed pluripotent state (Hanna et al. 2010). The naïve pluripotent stage in the mouse is characterized as domed colonies, depends on JAK/STAT signaling, and contains reactivated X-chromosome. These cells can contribute to the chimera formation into the blastocyst and exhibit higher pluripotent homogeneity. The primed ESCs are however, the naïve ESCs have failed to further differentiate into functional cell types in the various cell culture medium, and protocol optimization and establishment of the stable embryonic stem cell line are necessary before attempting the differentiation (Talbot et al. 1995; Wang et al. 2005). The derivation of embryonic stem cells in domestic cattle has not yet been established. Several studies including different feeder layers (Cong et al. 2014), micro-drop culture (Kim et al. 2012), and 2i conditions (Furusawa et al. 2013; Maruotti et al. 2012), combinations of different culture medium LIF and bFGF (Gong et al. 2010; Maruotti et al. 2012) showed poor pluripotency and limited cell proliferation. The derivation of the stable bovine pluripotent has been challenging and does not qualify the in vivo teratoma assay, in vitro embryoid body formation, and chimera formation (Blomberg and Telugu 2012; Ezashi et al. 2016). The bovine ESCs often show poor proliferation and derivation capabilities, and the disappearance of pluripotency markers (Soto and Ross 2016; Wang et al. 2005; Talbot et al. 1995). Recently, a study demonstrated pluripotency marker gene expression, stable morphology, similar karyotype, and transcriptome profiles, and population doubling time. The cell-culturing conditions used were custom base medium lacking TGF-beta supplemented with bFGF and Antagonist I (Wu et al. 2015). Antagonist I is an inhibitor of the WNK-β catenin signaling pathway that decides the cell fate and stem cell pluripotency. The conditions were the same as 'region-selective" primed type pluripotent stem cells sharing molecular functions with gastrula-stage-epiblast in human ESCs. Apart from the latest developments in cattle, similar trials are being attempted to establish stable ESCs in sheep and pigs. The establishment of the bovine embryonic stem cells will facilitate early development and biotechnological applications such as genome editing for the genetic improvement of the livestock. The full potential of ESC technology has not been experienced yet but this can improve stable genetic integration and treatments in the animals. ESCs can differentiate into any cell of the animal body which will help increase integration of desired genetic changes (Deckha and Xie 2008). However, legal, and ethical restrictions have made the technology inaccessible to large part of the world and will take longer time to be adoptable. Genome editing and livestock genetic improvement using embryonic stem cell technology have greater potential to revolutionize the livestock breeding.

3.2.5 Multiple Ovulation and Embryo Transfer (MOET)

Artificial insemination and embryo transfer methods for breeding have been popularly used for the improvement of animal characteristics. These methods are not well established in cattle and are usually expensive. The Multiple ovulation and embryo transfer (MOET) method is an advancement of the embryo transfer method where the fertilized eggs are removed from the super-ovulated female donors and transferred in the genetically different multiple surrogate mothers (Bergstein-Galan et al. 2019). The MOET can be used in the sheep, camel, goat, buffalo, and other cattle except for horses which are difficult to super-ovulate. This method produces significantly higher offspring in cattle with improved characteristics with lower cost compared to traditional methods. Thus, genetically superior donors can generate numerous progeny than it could earlier with other traditional methods such as artificial insemination and embryo transfer (Vettical et al. 2019). The screening of the health and breeding capabilities of the suitable donors and recipients is a crucial part of the process of embryo transfer (Seidel Jr. and Seidel 1991). Animals without any disease or infection and reproductive abnormalities such as ovarian-bursal adhesion should be selected for a successful embryo transfer and super-ovulation (Tibary and Anouassi 1997). The super-ovulation is achieved through various doses of follicle-stimulating hormones (FSH), progesterone, equine chorionic gonadotropins (eCG) for a few days (Anouassi and Tibary 2013). The optimization of hormone dosage and administration time is necessary for better stimulation (McKinnon et al. 1994). The synchronization of the embryo stages in the recipient is achieved through stimulating hormones based on their follicular size. Even a few days difference in synchronization often lead to decreased pregnancy (Skidmore et al. 2002). Finally, the embryo transfer is achieved using either surgical or non-surgical methods. The non-surgical methods commonly include insertion of catheter and embryo flushing media for the careful recovery of the embryos (Anouassi and Tibary 2013). The morphological and biochemical investigations of the recovered embryos are performed to categorize the transferable and non-transferable embryos. Healthy and suitable embryos can be cryopreserved for future use. The method is versatile and frequently used for a better reproductive rate. MOET can be used to identify or screen the genetic defects by cross mating the parental and super-ovulated progenies and the transmission of the disease is largely eliminated from parental generation.

3.2.6 In Vitro Fertilisation (IVF)

In vitro fertilization is a method for embryo culture and manipulation without in-vivo maternal signaling. The embryo propagation is vulnerable to environmental factors such as temperature, oxygen availability, light, nutrition, and embryo-maternal signaling. In-vitro fertilization is a technique to generate an embryo in a suitable synthetic environment or external environment through the fertilization of oocytes and sperm cells. The process of in vitro fertilization involves the recovery of the oocytes, maturation, fertilization, and propagation of the embryo. The oocytes are removed using the castration method from ovaries of the domestic animals followed by in vivo maturation (de Collado et al. 2017). The immature oocytes are usually aspirated in vivo from unstimulated or mild stimulated ovaries and cultured in the standard maturation medium (Rose and Bavister 1992). During aspiration of the follicles from the ovaries, a granular, multilayered, tight, and rough cumulus is identified as cumulus-oocyte complexes (Yang and Lu 1990). The propagation media allow resuming of oocyte first meiotic division. The luteinizing hormone (LH) and follicle-stimulating hormones (FSH) are commonly added with the maturation medium. The growth of the oocyte can be affected by any alteration in the standard maturation conditions. The qualitative losses can be assessed during maturation through blastula and morula yield estimation (Khosla et al. 2001; Luciano and Sirard 2017). During maturation, the culture conditions affect developmental competence and epigenetic growth of the embryo. Several cytoplasmic changes undergo before nuclear maturation during the maturation process (Hyttel et al. 1997). In vitro maturation process involves the messenger RNA and protein synthesis, mobilization, and storage which is vital for the early development of the embryo (Motlik and Fulka 1986; Luciano and Sirard 2017). After maturation of oocytes, the medium is changed to induce sperm capacitation and fertilization which can last up to one day. The ejaculated semen is preferably used for fertilization that gives rise to more eggs than epididymal spermatozoa. The fresh semen is incubated with heparin or any high ionic strength medium for maximum oocyte penetration (Pavlok et al. 1988). A variable concentration of heparin, depending on the species, was effective for better fertility (Leibfried-Rutledge et al. 1989). The serum albumin, epinephrine, heparin, caffeine, calcium, and hypo-taurinechemicals can be used for sperm capacitation (Parrish 2014; Polisca et al. 2013). The co-culture of the oocyte and sperm cells and the fertilization time strictly depend on the type of animal species and conditions used for the early embryo development and maturation. The fertilized embryo is propagated in the enriched culture medium such as bovine oviduct epithelial cells (BOEC) or supplemented with serum, growth regulators, proteins, and hormones (Fukui and Ono 1989; Li et al. 2018). The chemical composition and fluid properties are adjusted to mimic the in vivo environment of embryo development. The methods, protocols, and guidelines for embryo evaluation, classification, transfer, and maintenance are adapted according to the International Embryo Technology Society (IETS) (Bo and Mapletoft 2018; Stringfellow and Seidel 1998). The morula or early blastocyst stages of the embryo are suitable for transferring into the recipient and later blastocyst stages can be considered with special hygiene and following guidelines of IETS for the safe and disease-free transfer into the recipient (Stringfellow and Seidel 1998). The IVF methods in cattle face difficulties due to oocyte inactivation after injection of the spermatozoon (Abdalla et al. 2009). The inactivation could be a result of the difference in morphological characteristics of the oocytes or physiological deficiencies. IVF has great potential in animal reproduction and recently the integration of IVF with advanced genomic methods can provide precise and comprehensive information of the desired characteristics in animals. IVF can be used in the modern genome editing methods such as CRISPR Cas9 and TALEN for efficient genome editing to achieve desired genetic selection or modification in the animals.

3.3 Marker Assisted Selection (MAS) for Livestock Production

With technological progress in the identification of different genetic markers, research is being done for the development of breeding techniques using genetic markers. This technology is used to identify the genomic segments that control the variation seen in the quantitative trait of a species and their application in breed selection (Dekkers and Van der Werf 2007). Information about the identified Quantitative Trait Locus (QTL) is used for screening traits at the DNA level using an associated marker in the newly emerged technique of Marker Assisted Breeding (MAB) or Marker Assisted Selection (MAS). It is an indirect method of trait selection based on genotype composition, where the presence of a marker linked to a desired trait in the polymorphic loci is used as the criteria for selection and not the trait itself. The main objective of MAS is to collect the marker and QTL-related information and to complement it together with the phenotypic information to improve the selection process (Wakchaure et al. 2015). Since quantitative traits are targeted for various breeding programs and classical selection methods depend on phenotypic characteristics, the development of genetic markers has helped make a step forward in the development of strain improvement programs (Moniruzzaman et al. 2014). Integrated use of the classical method with this new genetically based approach will facilitate the identification and selection process in the early developmental phase.

The MAS has its utility for those traits which are expressed, for example, in the later stages of life, depends on the sex, have a low level of inheritance, or are regulated by a few genes. The traits with expensive measurement procedures can also be selected using this technique. An increase in the sensitivity of the selection response from −0.7% to 64% for a single trait is estimated when MAS is used along with

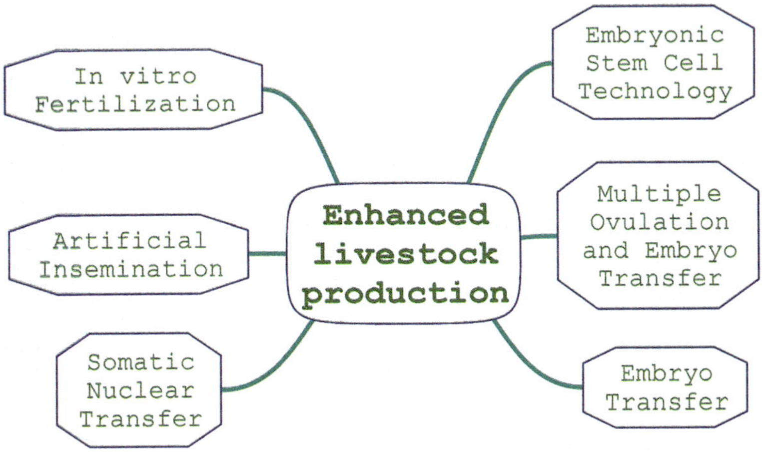

Fig. 3.1 Biotechnological strategies for enhanced livestock production

BLUP (a linear mixed model for predicting the random effect) than the BLUP alone. The trait studied using MAS is the amount and quality of carcass, longevity, milk production, and inheritance-related diseases (Ashraf et al. 2019) (Fig. 3.1).

3.3.1 Molecular Markers for MAS

Molecular markers are the point of reference in the genome. They do not necessarily play an important role but help identify specific DNA segments in the genome. Several genetic markers have been used to study the trait for their application in MAS, such as single nucleotide polymorphism (SNP), restriction fragment length polymorphisms (RFLPs), random amplified polymorphic DNA (RAPD), amplified fragment length polymorphisms (AFLPs), and simple sequence repeats (SSR) or microsatellites. Each of these markers has its limitations and disadvantages, however, the relationship between the marker and the target gene influences the success rate of MAS (Dekkers and Van der Werf 2007).

3.3.2 Properties of an Ideal Marker

The molecular markers are required to exhibit some unique characteristics such as codominance for easy identification of allelic nature of a diploid organism i.e. whether heterozygous or homozygous. Moreover, the molecular marker should be highly polymorphic, frequently appearing in the genomic compositions of the organism and no external factor should influence the selective nature of the marker. Other important features include high availability and duplicability, simple and rapid assays with ease of exchange of the data (Yadav et al. 2017).

3.3.3 Marker Types

Three types of markers are used based on the linkage between the marker and genes for the trait of economic value:

(i) Direct marker: The marker can be positioned within the target gene sequence. This relationship is highly favorable, and the MAS technique for such genes is termed gene-assisted selection (GAS). However, the difficulty in the identification of such markers makes their use less frequent.
(ii) LD- marker: The marker may be in close association with the gene of interest so that they are always inherited together showing a high value of linkage disequilibrium (LD).
(iii) LE markers: The maker may be in linkage equilibrium (LE) with the gene of interest. The random combination of these genes will have the same value as that of their independent expression showing zero linkage disequilibrium values.

Based on the techniques utilized for their identification, the DNA molecular markers can further be classified into three types (Yadav et al. 2017):

(i) Hybridization-based DNA Markers: Radio-labelled or enzyme-conjugated DNA segments or oligonucleotides are used in this method for DNA hybridization that helps easy detection of the marker molecules. Restriction Fragment Length Polymorphisms (RFLPs) and Oligonucleotide fingerprinting are the two common approaches that use this technology.
(ii) PCR-based DNA Markers: In this technique, the presence of the DNA molecule is determined by its PCR amplification using specific primers. The Random Amplified Length Polymorphic DNAs (RAPDs), Simple Sequence Repeats, or microsatellites (SSRs) and Amplified Fragment Length Polymorphisms (AFLPs) are the examples of this technique.
(iii) DNA Chip and Sequencing-based DNA Markers: Development of rapid genome sequencing methods like Next-generation sequencing (NGS) and third-generation sequencing (TGS) has led to the discovery of single nucleotide polymorphism (SNP) based markers. These markers have a high frequency of occurrence with ease of access and other desirable features for their use in MAS (Singh and Singh 2015).

3.3.4 Designing the MAS Program

The MAS has been applied in various livestock breeding programs, from improvement of the traits to the traceability of products. The prerequisite for design of a successful MAS breeding program is a strong and close linkage between the targeted gene and a marker gene. The success of the MAS program also depends on the progress made in other techniques of genetics, namely (i) gene mapping: it is

Table 3.2 The technologies adopted for genetic improvement in a few livestock

Sr. No.	Animal/ ivestock	Methodologies adopted for genetic improvement	References
1.	Cow/ buffalo	Somatic cell nuclear transfer, Artificial insemination, embryo transfer, embryonic stem cell technology, IVF, MOET	Cibelli et al. (1998), Chebel et al. (2004), Drost et al. (1986), Gugjoo et al. (2019), Thalkar, and Wei et al. (2010)
2.	Sheep/ goat	Somatic cell nuclear transfer, Artificial insemination, embryo transfer, embryonic stem cell technology, IVF, MOET	Wilmut et al. (1985), Evans and Maxwell (1987), Cognie (1999), Notarianni et al. (1991), Bari et al. (2003), and Catalá et al. (2012)
3.	Pig	Somatic cell nuclear transfer, Artificial insemination, embryo transfer, embryonic stem cell technology, IVF, MOET	Gupta et al. (2013), Knox (2016), Brüssow et al. (2000), Romagnuolo et al. (2019), and Hicks et al. (2020)
4.	Chicken	Somatic cell nuclear transfer, Artificial insemination, embryo transfer, embryonic stem cell technology, IVF	Gupta et al. (2013), Burrows and Quinn (1939), Jones et al. (2006), Lavial and Pain (2010), and Davidson (2004)
5.	Rabbit	Somatic cell nuclear transfer, Artificial insemination, embryo transfer, embryonic stem cell technology, IVF, MOET	Chesné et al. (2002), Morrell (1995), Song et al. (2017), Wang et al. (2007), Arias-Álvarez et al. (2017), and Garcia-Dominguez et al. (2020)

necessary for the identification and mapping of the target gene and the associated marker, (ii) marker genotyping: it is used to study the genetic differences in large numbers of individuals based on a large number of markers available. It must be made cost-effective for processing many samples for detecting QTL for the application of MAS. (iii) QTL detection: for the identification of the gene and the associated gene marker of the economically valuable traits. The degree of association must also be estimated. (iv) Genetic evaluation: for the statistical analysis of genotypic and phenotypic data to estimate the breeding value of individuals and (v) MAS for the efficient use of information at the DNA level to develop strategies for successful breeding programmes (Moniruzzaman et al. 2014).

The design of MAS includes the selection of the parent as a critical initial step. The selected parent breeds should have contrasting characteristics so that the genome of the individual parent and the heritability of its segment in the different recombination formed in the next generation can be easily identified. To achieve this, the germplasm of the parent breed is screened. Once the parents have been selected, a breeding population is developed by crossbreeding between the parents. Finally, the DNA from successive generation is used to select traits based on MAS (Table 3.2).

3.3.5 Application of MAS

The MAS finds its application in various breeding programs for selecting different traits.

(i) Molecular marker-assisted introgression (MAI): Molecular markers can also be used as a guide to identify the introgressive genes expressing individuals. While traditional breeding techniques require a series of backcrosses to identify individuals with introgressive genes, the use of MAI reduces the workload, time, and cycles of backcrossing. Growth, meat or milk quality, etc. are the traits targeted for breeding programmes (Fertilization). The most common application of MAI is its use in the introgression of disease resistance genes (Prajapati et al. 2017). The MAS has been studied for its accuracy in the identification of the Bacterial cold-water disease (BCWD) resistance-related genes in rainbow trout (*Oncorhynchus mykiss*). SNP-based marker was applied for the study and it was observed that MAS provided accuracy of about 0.5 which is higher than the classical approach (Liu et al. 2018).

(ii) Progeny testing and multiple ovulation and embryo transfer (MOET) programs: The different molecular markers are linked to the gene of interest for QTL in the livestock industry and used as GAS or LD-MAS for trait preselection and breeding. For the accuracy of selection, progeny testing of the traits before widespread use is required for assuring accuracy of selection. However, the high interval between the generations reduces the genetic gain. The integration of MAS reduces the interval but maintains selection accuracy to increase the genetic gain. A significant gain of 65% in the traits of the carcass has been reported. Similarly, the response of milk production increased to 15% compared to the previous report of 6% when the MAS and MOET were used together (Wani et al. 2008).

(iii) Product traceability: To ensure the quality of the various animal-based products and their food safety, molecular markers have been used as a trustworthy tool. The molecular markers have been used to identify the genotype and parentage of the various animal products other than their use in the selection process (Moniruzzaman et al. 2014).

(iv) Assessment of undesirable genes: Like the selection of desirable traits, the MAS can be used for screening of the presence of undesirable genes, physical or genetic defects in breeding livestock. The MAS help identify the genes responsible for the various defects in the individual and their removal from breeding programmes (Fertilization).

(v) Fertility study: The MAS can also be used for fertility studies in different breeding programs whereby the breeding soundness of the cattle involved can be evaluated. Since a large number of insemination dosages i.e. approximately three lakh doses are formulated from a single bull, it is essential to assure its fertility for successful conception and economic realization of artificial insemination (Raina et al. 2020). Conventionally the fertility of the bull is determined through phenotypic analysis which includes the study of sperm cell concentration, morphology, motility along with the absence of any physical or genital disease in the bull and Sire Conception Rate (SCR) (Morrell et al. 2017). However, all these parameters could not give a clear representation with reports indicating a continuous decline in the fertility rate of the bulls. Since the fertility trait of the bull has high variability and low heritability, the phenotypic selection is not competent enough for the identification of bulls with high fertility. Identification of molecular markers linked with the fertility traits like the SCR and quality of semen have been carried out for their application in MAS for differentiation and selection of bulls with high fertility for the process of artificial insemination (Raina et al. 2020).

3.3.6 Advantages and Limitation of MAS

The MAS has a major advantage of faster selection in the early stages of life cycle, as the phenotype can be calculated based on the genotype information. This helps reduce the generation gap for selection and expression of the desired trait. Since the genotype is considered, the traits that need to be transferred to the successive generation can be identified without taking environmental influences into account. Moreover, when the phenotypic selection is difficult or unfeasible to examine for the different features such as disease resistance, protein content of the product to be derived, the MAS has the lead. The traits, with low heritability, are sex-linked (milk or egg production) or can only be estimated after the slaughter of the animals (meat quality), can also be selected using MAS giving an advantage over the conventional methods (Wakchaure et al. 2015).

The major setback of MAS is the high cost of genotyping. Sample collection and gathering of genotypic information from the entire population is unfeasible for developing cattle breeding program. Missing markers or incomplete information on the genotype, lack of consistency of QTL data over some time, or varied population does not make the QTL a reliable source of information. Furthermore, the validity of the markers used must be continuously confirmed for each breeding population. Although the molecular marker may not be affected by environmental factors, its associated trait may get influenced and show different phenotypes. These factors must also be considered, and the marker analysis needs to be done in a varied environment making the process cumbersome. Finally, the availability of experts with the necessary technical knowledge is another obstacle to the application of MAS in developing countries (Moniruzzaman et al. 2014) (Table 3.3).

3.4 Conclusion

The current scenario of the world population and climate change indicate a higher food demand in the next two decades. Livestock production and consumption in the world has increased over time. Livestock will be one of the major sources of food for developing counties. The genetic improvements of this livestock through the latest biotechnological interventions and molecular marker-assisted selection of the beneficial traits and transfer into the recipient line can enhance the food quantity while maintaining the quality in a relatively shorter time. The latest technologies can provide increased accuracy and intensity of the improved trait selection for the

Table 3.3 Industrial application and advantage of MAS

Sr. No.	MAS applications	Advantages	References
1	Dairy cattle breeding schemes	Increased milk production	Meuwissen and Van Arendonk (1992)
2	Closed nucleus breeding schemes	Gain in carcasses	Goddard (1996)
3	Assessment of genetic defect	Removal of carriers of genes related to bovine leukocyte adhesion deficiency	Shuster et al. (1992) and Dekkers (2004)
4	Introgression of genes under NARI-Suwarna scheme	Introgression of the *FecB* mutation in sheeps for increased ovulation rate	Nimbkar
5	QTL mapping and introgression of genes	Increased meat quality of Qinchuan cattle breed	Abd El-Hack et al. (2018)
6	SNP marker assisted BLUP	Increase breeding value of Turbot	Lyu et al. (2019)
7	Marker linked sex identification	Selective production of faster growing (XX) females for mass production and (YY) super males for basic research in the Southern catfish (*Silurus meridionalis*)	Zheng et al. (2020)

livestock. The inception of the embryonic stem cell technology for livestock breeding has opened new avenues of the reprogramming genetic material for the in-vitro selection of the desired characteristics. The livestock sector requires aggressive research for the successful implementation of the existing technologies.

Acknowledgments The authors would like to thank the Department of Microbiology, Sikkim University, and School of Life and Environmental Sciences, Deakin University, Geelong, Australia for providing the computational infrastructure and central library facilities for procuring references and plagiarism analysis (Turnitin: Plagiarism Detection Software).

Conflict of Interest Statement The authors declare that the study was conducted in the absence of any commercial or financial relationships that could be construed as a potential conflict of interest.

References

Abd El-Hack ME, Abdelnour SA, Swelum AA, Arif M (2018) The application of gene marker-assisted selection and proteomics for the best meat quality criteria and body measurements in Qinchuan cattle breed. J Mol Biol Rep 45:1445–1456

Abdalla H, Shimoda M, Hirabayashi M, Hochi S (2009) A combined treatment of ionomycin with ethanol improves blastocyst development of bovine oocytes harvested from stored ovaries and microinjected with spermatozoa. Theriogenology 72:453–460

Alexandratos N, Bruinsma J (2012) World agriculture towards 2030/2050: the 2012 revision

Anouassi A, Tibary A (2013) Development of a large commercial camel embryo transfer program: 20 years of scientific research. Anim Reprod Sci 136:211–221

Arias-Álvarez M, García-García RM, Lopez-Tello J, Rebollar PG, Gutiérrez-Adán A, Lorenzo PL (2017) In vivo and in vitro maturation of rabbit oocytes differently affects the gene expression profile, mitochondrial distribution, apoptosis and early embryo development. Reprod Fertil Dev 29:1667–1679

Ashraf WM, Giuma EA, Magdy AM, Abdelatefm AA, Abdulraouf AM, Ghada EA, Ismail HM, Katarzyna J-W (2019) Genomic technologies in improving the efficiency of animal health and husbandry. Int J Prog Sci Technol 17:265–278

Bari F, Khalid M, William H, Murray A, Merrell B (2003) Factors affecting the survival of sheep embryos after transfer within a MOET program. Theriogenology 59:1265–1275

Bergstein-Galan TG, Weiss RR, Kozicki LE (2019) Effect of semen and donor factors on multiple ovulation and embryo transfer (MOET) in sheep. Reprod Domest Anim 54:401–407

Betteridge KJ (1981) An historical look at embryo transfer. Reproduction 62:1–13

Blomberg LA, Telugu BPVL (2012) Twenty years of embryonic stem cell research in farm animals. Reprod Domest Anim 47:80–85

Bó GA, Baruselli PS (2014) Synchronization of ovulation and fixed-time artificial insemination in beef cattle. Animal 8:144–150

Bo GA, Mapletoft RJ (2018) Evaluation and classification of bovine embryos. Anim Reprod (AR) 10:344–348

Bó GA, Mapletoft RJ (2018) Embryo transfer technology in cattle. In: Animal biotechnology 1. Springer, Cham

Brinsko SP, Varner DD (1992) Artificial insemination and preservation of semen. Vet Clin N Am Equine Pract 8:205–218

Brüssow K-P, Torner H, Kanitz W, Rátky J (2000) In vitro technologies related to pig embryo transfer. Reprod Nutr Dev 40:469–480

Burrows WH, Quinn JP (1939) *Artificial insemination of chickens and turkeys*. US Department of Agriculture, Washington, DC

Campbell KH, Loi P, Otaegui PJ, Wilmut I (1996) Cell cycle co-ordination in embryo cloning by nuclear transfer. Rev Reprod 1:40–46

Capecchi MR (2000) How close are we to implementing gene targeting in animals other than the mouse? Proc Natl Acad Sci 97:956–957

Catalá MG, Izquierdo D, Rodríguez-Prado M, Hammami S, Paramio M-T (2012) Effect of oocyte quality on blastocyst development after in vitro fertilization (IVF) and intracytoplasmic sperm injection (ICSI) in a sheep model. Fertil Steril 97:1004–1008

Chebel RC, Santos JEP, Reynolds JP, Cerri RLA, Juchem SO, Overton M (2004) Factors affecting conception rate after artificial insemination and pregnancy loss in lactating dairy cows. Anim Reprod Sci 84:239–255

Chesné P, Adenot PG, Viglietta C, Baratte M, Boulanger L, Renard J-P (2002) Cloned rabbits produced by nuclear transfer from adult somatic cells. Nat Biotechnol 20:366–369

Cibelli JB, Stice SL, Golueke PJ, Kane JJ, Jerry J, Blackwell C, de Leon FAP, Robl JM (1998) Cloned transgenic calves produced from nonquiescent fetal fibroblasts. Science 280:1256–1258

Cognie Y (1999) State of the art in sheep-goat embryo transfer. Theriogenology 51:105–116

Cong S, Cao G, Liu D (2014) Effects of different feeder layers on culture of bovine embryonic stem cell-like cells in vitro. Cytotechnology 66:995–1005

Davidson TR (2004) The chick embryo amnion as an in vitro culture system for IVF and NT embryos. Louisiana State University, Baton Rouge

Deckha M, Xie Y (2008) The stem cell debate: why should it matter to animal advocates. Stanford J Anim Law Policy 1:69

Dekkers JCM (2004) Commercial application of marker-and gene-assisted selection in livestock: strategies and lessons. J Anim Sci 82:E313–EE28

Dekkers JCM, Van der Werf J (2007) Strategies, limitations and opportunities for marker-assisted selection in livestock. In: Marker-assisted selection-current status and future perspectives in crops, livestock, forestry and fish. FAO, Rome, pp 167–184

del Collado M, da Silveira JC, Oliveira MLF, Alves BMSM, Simas RC, Godoy AT, Coelho MB, Marques LA, Carriero MM, Nogueira MFG, Eberlin MN, Silva LA, Meirelles FV, Perecin F (2017) In vitro maturation impacts cumulus–oocyte complex metabolism and stress in cattle. Reproduction 154:881

Drost M, Wright JM, Elsden RP (1986) Intergeneric embryo transfer between water buffalo and domestic cattle. Theriogenology 25:13–23

Enright BP, Kubota C, Yang X, Tian XC (2003) Epigenetic characteristics and development of embryos cloned from donor cells treated by trichostatin a or 5-aza-2′-deoxycytidine. Biol Reprod 69:896–901

Evans G, Maxwell WMC (1987) Salamons' artificial insemination of sheep and goats. Butterworths, Sydney

Ezashi T, Yuan Y, Michael Roberts R (2016) Pluripotent stem cells from domesticated mammals. Annu Rev Anim Biosci 4:223–253

Fertilization, In-vitro. Biotechnology for the livestock industry

Fukui Y, Ono H (1989) Effects of sera, hormones and granulosa cells added to culture medium for in-vitro maturation, fertilization, cleavage and development of bovine oocytes. Reproduction 86:501–506

Furusawa T, Ohkoshi K, Kimura K, Matsuyama S, Akagi S, Kaneda M, Ikeda M, Hosoe M, Kizaki K, Tokunaga T (2013) Characteristics of bovine inner cell mass-derived cell lines and their fate in chimeric conceptuses. Biol Reprod 89(28):1–12

Garcia-Dominguez X, Juarez JD, Vicente JS, Marco-Jiménez F (2020) Impact of embryo technologies on secondary sex ratio in rabbit. Cryobiology

Goddard ME (1996) The use of marker haplotypes in animal breeding schemes. Genet Sel Evol 28:1–16

Gong G, Roach ML, Le Jiang XY, Tian XC (2010) Culture conditions and enzymatic passaging of bovine ESC-like cells. Cell Reprogram (Formerly "Cloning and Stem Cells") 12:151–160

Gugjoo MB, Fazili MR, Shah RA, Sharma GT (2019) Mesenchymal stem cell: basic research and potential applications in cattle and buffalo. J Cell Physiol 234:8618–8635

Gupta MK, Das ZC, Heo YT, Joo JY, Chung H-J, Song H, Kim J-H, Kim N-H, Lee HT, Ko DH (2013) Transgenic chicken, mice, cattle, and pig embryos by somatic cell nuclear transfer into pig oocytes. Cell Reprogram (Formerly "Cloning and Stem Cells") 15:322–328

Gurdon JB (1962) The developmental capacity of nuclei taken from intestinal epithelium cells of feeding tadpoles. J Embryol Exp Morphol 10:622–640

Hanna J, Cheng AW, Saha K, Kim J, Lengner CJ, Soldner F, Cassady JP, Muffat J, Carey BW, Jaenisch R (2010) Human embryonic stem cells with biological and epigenetic characteristics similar to those of mouse ESCs. Proc Natl Acad Sci 107:9222–9227

Harrison P, Bruinsma J, de Haen H, Alexandratos N, Schmidhuber J, Bödeker G, Ottaviani MG (2002) World agriculture: towards 2015/2030. Online. http://www.fao.org/documents

Hicks E, Mentler M, Arena HA, Current JZ, Whitaker BD (2020) Cyanidin improves oocyte maturation and the in vitro production of pig embryos. Vitro Cell Dev Biol Anim 56:577–584

Hill JR, Burghardt RC, Jones K, Long CR, Looney CR, Shin T, Spencer TE, Thompson JA, Winger QA, Westhusin ME (2000a) Evidence for placental abnormality as the major cause of mortality in first-trimester somatic cell cloned bovine fetuses. Biol Reprod 63:1787–1794

Hill JR, Winger QA, Long CR, Looney CR, Thompson JA, Westhusin ME (2000b) Development rates of male bovine nuclear transfer embryos derived from adult and fetal cells. Biol Reprod 62:1135–1140

Hyttel PFTCH, Fair T, Callesen H, Greve T (1997) Oocyte growth, capacitation and final maturation in cattle. Theriogenology 47:23–32

Jones TA, Jones SM, Paggett KC (2006) Emergence of hearing in the chicken embryo. J Neurophysiol 96:128–141

Khosla S, Dean W, Reik W, Feil R (2001) Culture of preimplantation embryos and its long-term effects on gene expression and phenotype. Hum Reprod Update 7:419–427

Kim EY, Noh EJ, Park HY, Park MJ, Noh EH, Lee JB, Jeong CJ, Lee DS, Riu KZ, Park SP (2012) Establishment of bovine embryonic stem cell lines using a minimized feeder cell drop. Cell Reprogram (Formerly "Cloning and Stem Cells") 14:520–529

Kishigami S, Mizutani E, Ohta H, Hikichi T, Van Thuan N, Wakayama S, Bui H-T, Wakayama T (2006) Significant improvement of mouse cloning technique by treatment with trichostatin A after somatic nuclear transfer. Biochem Biophys Res Commun 340:183–189

Knox RV (2016) Artificial insemination in pigs today. Theriogenology 85:83–93

Kubota C, Yamakuchi H, Todoroki J, Mizoshita K, Tabara N, Barber M, Yang X (2000) Six cloned calves produced from adult fibroblast cells after long-term culture. Proc Natl Acad Sci 97:990–995

Lavial F, Pain B (2010) Chicken embryonic stem cells as a non-mammalian embryonic stem cell model. Develop Growth Differ 52:101–114

Leibfried-Rutledge ML, Critser ES, Parrish JJ, First NL (1989) In vitro maturation and fertilization of bovine oocytes. Theriogenology 31:61–74

Li X-X, Cao P-H, Han W-X, Xu Y-K, Wu H, Yu X-L, Chen J-Y, Zhang F, Li Y-H (2018) Non-invasive metabolomic profiling of culture media of ICSI-and IVF-derived early developmental cattle embryos via Raman spectroscopy. Anim Reprod Sci 196:99–110

Liu Y, Han XJ, Liu MH, Wang SY, Jia CW, Yu L, Ren G, Wang L, Li W (2014) Three-day-old human unfertilized oocytes after in vitro fertilization/intracytoplasmic sperm injection can be activated by calcium ionophore a23187 or strontium chloride and develop to blastocysts. Cell Reprogram 16:276–280

Liu S, Vallejo RL, Evenhuis JP, Martin KE, Hamilton A, Gao G, Leeds TD, Wiens GD, Palti Y (2018) Retrospective evaluation of marker-assisted selection for resistance to bacterial cold water disease in three generations of a commercial rainbow trout breeding population. Front Genet 9:286

Loi P, Iuso D, Czernik M, Ogura A (2016) A new, dynamic era for somatic cell nuclear transfer? Trends Biotechnol 34:791–797

Lu F, Zhang Y (2015) Cell totipotency: molecular features, induction, and maintenance. Natl Sci Rev 2:217–225

Luciano AM, Sirard M-A (2017) Successful in vitro maturation of oocytes: a matter of follicular differentiation. Biol Reprod 98:162–169

Lyu D, Hu Y, Wu H, Sun S, Wang W (2019) Estimating breeding values for juvenile body weight using trait associated SNP assisted BLUP in turbot (Scophthalmus maximus). Aquaculture 508:46–51

Maruotti J, Muñoz M, Degrelle SA, Gómez E, Louet C, Monforte CD, De Longchamp PH, Brochard V, Hue I, Caamaño JN (2012) Efficient derivation of bovine embryonic stem cells needs more than active core pluripotency factors. Mol Reprod Dev 79:461–477

McKinnon AO, Tinson AH, Nation G (1994) Embryo transfer in dromedary camels. Theriogenology 41:145–150

Mendel G (1866) Versuche uber pflanzen-hybriden. Verhandlungen des naturforschenden Vereins in Brunn fur 4:3–47

Meuwissen THE, Van Arendonk JAM (1992) Potential improvements in rate of genetic gain from marker-assisted selection in dairy cattle breeding schemes. J Dairy Sci 75:1651–1659

Moniruzzaman M, Khatun R, Mintoo AA (2014) Application of marker assisted selection for livestock improvement in Bangladesh. Bangladesh Vet 31:1–11

Morrell JM (1995) Artificial insemination in rabbits. Br Vet J 151:477–488

Morrell JM (2011) Artificial insemination: current and future trends. Artif Inseminn Farm Anim 1:1–14

Morrell JM, Nongbua T, Valeanu S, Verde IL, Lundstedt-Enkel K, Edman A, Johannisson A (2017) Sperm quality variables as indicators of bull fertility may be breed dependent. Anim Reprod Sci 185:42–52

Motlik J, Fulka J (1986) Factors affecting meiotic competence in pig oocytes. Theriogenology 25:87–96

Muir WM, Aggrey SE (2003) Poultry genetics, breeding, and biotechnology. CABI, Wallingford

Nichols J, Smith A (2009) Naive and primed pluripotent states. Cell Stem Cell 4:487–492

Nimbkar C. Indian experiences in implementing marker assisted breeding in livestock

Noakes DE, Parkinson TJ, England GCW (2018) Arthur's veterinary reproduction and obstetrics-e-book. Elsevier Health Sciences

Notarianni E, Gallit C, Lauriet S, Moor RM, Evans MJ (1991) Derivation of pluripotent, embryoniccell lines from the pig and sheep. Methods 10:250

Ogura A, Inoue K, Ogonuki N, Lee J, Kohda T, Ishino F (2002) Phenotypic effects of somatic cell cloning in the mouse. Cloning Stem Cells 4:397–405

Onishi A, Iwamoto M, Akita T, Mikawa S, Takeda K, Awata T, Hanada H, Perry ACF (2000) Pig cloning by microinjection of fetal fibroblast nuclei. Science 289:1188–1190

Parrish JJ (2014) Bovine in vitro fertilization: in vitro oocyte maturation and sperm capacitation with heparin. Theriogenology 81:67–73

Pavlok A, Torner H, Motlik J, Fulka J, Kauffold P, Duschinski U (1988) Fertilization of bovine oocytes in vitro: effect of different sources of gametes on fertilization rate and frequency of fertilization anomalies. Anim Reprod Sci 16:207–213

Pickering SJ, Minger SL, Patel M, Taylor H, Black C, Burns CJ, Ekonomou A, Braude PR (2005) Generation of a human embryonic stem cell line encoding the cystic fibrosis mutation ΔF508, using preimplantation genetic diagnosis. Reprod Biomed Online 10:390–397

Polisca A, Orlandi R, Troisi A, Brecchia G, Zerani M, Boiti C, Zelli R (2013) Clinical efficacy of the Gn RH agonist (deslorelin) in dogs affected by benign prostatic hyperplasia and evaluation of prostatic blood flow by Doppler ultrasound. Reprod Domest Anim 48:673–680

Porter V (2020) Mason's world dictionary of livestock breeds, types and varieties. CABI, Wallingford

Prajapati BM, Gupta JP, Pandey DP, Parmar GA, Chaudhari JD (2017) Molecular markers for resistance against infectious diseases of economic importance. Vet World 10:112

Raina VS, Kour A, Chakravarty AK, Vohra V (2020) Marker-assisted selection vis-à-vis bull fertility: coming full circle – a review. Mol Biol Rep 47(11):9123–9133

Romagnuolo R, Masoudpour H, Porta-Sánchez A, Qiang B, Barry J, Laskary A, Qi X, Massé S, Magtibay K, Kawajiri H (2019) Human embryonic stem cell-derived cardiomyocytes regenerate the infarcted pig heart but induce ventricular tachyarrhythmias. Stem cell Rep 12:967–981

Rose TA, Bavister BD (1992) Effect of oocyte maturation medium on in vitro development of in vitro fertilized bovine embryos. Mol Reprod Dev 31:72–77

Salisbury GW, VanDemark NL, Lodge JR (1978) Physiology of reproduction and artificial insemination of cattle. WH Freeman and Company

Saunders CM, Larman MG, Parrington J, Cox LJ, Royse J, Blayney LM, Swann K, Lai FA (2002) PLC zeta: a sperm-specific trigger of Ca(2+) oscillations in eggs and embryo development. Development 129:3533–3544

Seidel GE Jr, Seidel SM (1991) Training manual for embryo transfer in cattle. FAO, Rome

Seré C, Steinfeld H, Groenewold J. 1996. World livestock production systems

Shuster DE, Kehrli ME, Ackermann MR, Gilbert RO (1992) Identification and prevalence of a genetic defect that causes leukocyte adhesion deficiency in Holstein cattle. Proc Natl Acad Sci 89:9225–9229

Singh BD, Singh AK (2015) Sequence-based markers. In: Marker-assisted plant breeding: principles and practices. Springer India, New Delhi

Skidmore JA, Billah M, Allen WR (2002) Investigation of factors affecting pregnancy rate after embryo transfer in the dromedary camel. Reprod Fertil Dev 14:109–116

Smith C (1988) Applications of embryo transfer in animal breeding. Theriogenology 29:203–212

Solter D (2004) Imprinting. Int J Dev Biol 42:951–954

Song J, Yang D, Ruan J, Zhang J, Chen YE, Jie X (2017) Production of immunodeficient rabbits by multiplex embryo transfer and multiplex gene targeting. Sci Rep 7:1–9

Soto DA, Ross PJ (2016) Pluripotent stem cells and livestock genetic engineering. Transgenic Res 25:289–306

Stringfellow DA, Seidel SM (1998) Manual of the International Embryo Transfer Society. The Society, Champaign

Talbot NC, Powell AM, Rexroad CE Jr (1995) In vitro pluripotency of epiblasts derived from bovine blastocysts. Mol Reprod Dev 42:35–52

Tan SL, Bennett S, Parsons J (1990) Surgical techniques of oocyte collection and embryo transfer. Br Med Bull 46:628–642

Thalkar MG. Embryo transfer technology in cattle

Tibary A, Anouassi A (1997) Artificial breeding and manipulation of reproduction in camelidae. In: Theriogenology in camelidae: anatomy, physiology, pathology and artificial breeding. Actes Editions, Rabat, pp 413–414

Vettical BS, Hong SB, Umer MA, Wani NA (2019) Comparison of pregnancy rates with transfer of in vivo produced embryos derived using multiple ovulation and embryo transfer (MOET) with in vitro produced embryos by somatic cell nuclear transfer (SCNT) in the dromedary camel (Camelus dromedaries). Anim Reprod Sci 209:106132

Wakchaure R, Ganguly S, Praveen PK, Kumar A, Sharma S, Mahajan T (2015) Marker assisted selection (MAS) in animal breeding: a review. J Drug Metab Toxicol 6:e127

Wang L, Duan E, Sung L-y, Jeong B-S, Yang X, Cindy Tian X (2005) Generation and characterization of pluripotent stem cells from cloned bovine embryos. Biol Reprod 73:149–155

Wang S, Tang X, Niu Y, Chen H, Li B, Li T, Zhang X, Hu Z, Zhou Q, Ji W (2007) Generation and characterization of rabbit embryonic stem cells. Stem Cells 25:481–489

Wani NA, Billah M, Skidmore JA (2008) Studies on liquefaction and storage of ejaculated dromedary camel (Camelus dromedarius) semen. Anim Reprod Sci 109:309–318

Wei JZ, Jiang QY, Jiang HS, Wei JW, Wei ZJ, Yu Z, Zhang SX, Wei YM (2010) A study on multiple ovulation and embryo transfer (MOET) technology in Australian Holstein cow. Guangxi Agric Sci 41:835–837

Wells DN, Laible G, Tucker FC, Miller AL, Oliver JE, Xiang T, Forsyth JT, Berg MC, Cockrem K, L'Huillier PJ (2003) Coordination between donor cell type and cell cycle stage improves nuclear cloning efficiency in cattle. Theriogenology 59:45–59

Wilmut I, Sales DI, Ashworth CJ (1985) The influence of variation in embryo stage and maternal hormone profiles on embryo survival in farm animals. Theriogenology 23:107–119

Wilmut I, Schnieke AE, McWhir J, Kind AJ, Campbell KH (1997a) Viable offspring derived from fetal and adult mammalian cells. Nature 385:810–813

Wilmut I, Schnieke AE, McWhir J, Kind AJ, Campbell KHS (1997b) Viable offspring derived from fetal and adult mammalian cells. Nature 385:810–813

Wright JM (1981) Non-surgical embryo transfer in cattle embryo-recipient interactions. Theriogenology 15:43–56

Wu J, Okamura D, Li M, Suzuki K, Luo C, Ma L, He Y, Li Z, Benner C, Tamura I (2015) An alternative pluripotent state confers interspecies chimaeric competency. Nature 521:316–321

Yadav AK, Tomar SS, Jha AK, Singh J (2017) Importance of molecular markers in livestock improvement: a review. Int J Agric Innov Res 5:614–622

Yang YB, Lu KH (1990) The influence of bovine oocyte type on in vitro-fertilization and subsequent development in vitro. Theriogenology 33:355

Zakhartchenko V, Alberio R, Stojkovic M, Prelle K, Schernthaner W, Stojkovic P, Wenigerkind H, Wanke R, Düchler M, Steinborn R (1999) Adult cloning in cattle: potential of nuclei from a permanent cell line and from primary cultures. Mol Reprod Dev Inc Gamete Res 54:264–272

Zheng S, Wang X, Zhang S, Long J, Tao W, Li M, Wang D (2020) Screening and characterization of sex-linked DNA markers and marker-assisted selection in the Southern catfish (Silurus meridionalis). Aquaculture 517:734783

Chapter 4
Applications of Stem cells Technology in Livestock Production

Vinay Bhaskar, Satish Kumar, and Dhruba Malakar

Abstract Urbanized population and a gigantic demand for livestock products are expected to rise dramatically in the coming decades. Initiatives have been geared up to intensify the livestock production. Decipher of large animal genomic sequence boost the genetic enhancement. Embryonic stem cells are efficient platforms for genome editing to produce transgenic animals. Moreover, stem cells opens a new avenue to insight into developmental processes. Stem cells increase the production efficiency by maintaining the health and welfare of the livestock. Conservation of elite species also adds advantage to use reprogrammed induced pluripotent stem cells in farm animals. However, uses of embryonic stem cells are bound with ethical concerns adult stem cells promises alternative approach. Based on the ease of availability and isolation, mesenchymal stem cells are extensively studied and widely used stem cells for therapeutic purposes. Multipotency, immunomodulation, homing properties of mesenchymal stem cells plays a vital role to alleviate diseases and increase the production. Stem cell technology used in large animals can also serve as a model for various human diseases. Hence conventional approaches to enhance the productivity and welfare of animals were reformed by the advent of stem cell technology.

Keywords Stem cells · Embryonic stem cells · Mesenchymal stem cells · Livestock production · Therapeutic applications

Abbreviations

ART	Assisted reproductive technology
BMP	Bone morphogenic protein
CD	Cluster of differentiation
CRISPR	Cluster of regularly interspaced short palindromic repeats
CXCR	Chemokine receptor type

V. Bhaskar · S. Kumar · D. Malakar (✉)
Animal Biotechnology Centre, National Dairy Research Institute, Karnal, Haryana, India

V. K. Yata et al. (eds.), *Sustainable Agriculture Reviews 54*, Sustainable Agriculture Reviews 54, https://doi.org/10.1007/978-3-030-76529-3_4

EC	Embryonic carcinoma cells
ES	Embryonic stem cells
FAO	Food and agriculture organization
GCTM	Germ cell tumor marker
HGF	Hepatocyte growth factor
HSC	Hematopoietic stem cells
IBD	Inflammatory bowel disease
IL	Interleukin
INF-γ	Interferon gamma
iPSC	Induced pluripotent stem cells
ISCT	International society for cellular therapy
LIF	Leukemia inhibitory factor
MHC	Major histo compatibility complex
MSC	Mesenchymal stem cells
NO	Nitric oxide
OCT 4	Octamer binding transcription factor 4
PGE2	Prostaglandin E2
QTL	Quantitative trait locus
SDF	Stromal derived factor
SSEA	Stage specific embryonic antigen
TALEN	Transcription activator like effector nuclease
TGDF	Teratocarcinoma derived growth factor
TGF	Transforming growth factor
TNF	Tumor necrosis factor
TRA-1	T cell receptor alpha locus
VCAM	Vascular cell adhesion molecule
ZFN	Zinc finger nuclease

4.1 Introduction

In this new millennium where we are exploring biological solutions for biological problems, We are on the brink of a paradigm change, emerging from basic mechanical care services to considering biological approaches for health promotion, evaluation of risk assessment, treatment, and even prognosis. Slow but steady attempts have made to understand regenerative capacities and underlying mechanisms. The stem cells set to transform the entire health care delivery. Past decade witnessed a significant flux of stem cells technology in human and veterinary research. The properties of stem cells fascinates the researchers to explore its wide range of applications. Stem cells are unspecialized cells and was first reported in bone marrow as self renewing cells in the year 1963. Stem cells have clonogenic capacity which can give rise to same kind of genetically identical cells. Self renewing capacity (Asal and Güven 2020) and the ability to differentiate into wide variety of cell types even to germ cell type (Malik et al. 2020; Ota 2008) were some of the remarkable properties of stem cells. Embryonic stem cells (ESC), Adult stem cells, Induced pluripotent stem cells (iPSC) are some of the extensively explored stem cell types. These cells are primarily undifferentiated cells found in the body throughout the lifespan starting from early embryonic development. Embryonic stem cells first isolated from mouse blastocyst in 1981 then recently in 2006 induced pluripotent stem cells were produced for the first time by reprogramming skin cells to early pluripotent state. After their first discovery they are reported in many livestock species (Table 4.1). Based on their differentiation ability stem cells are categorized into totipotent- Stem Cells from early Embryo (1–3 days) are totipotent as each cell have the ability to develop into whole individual. Pluripotent- Some cells of blastocyst (5–9 days) are pluripotent they can give rise to all adult type of cells into body not placenta. Multipotent-Adult stem cells are multipotent which have capacity to differentiate into many tissues like hematopoietic cells. Oligopotent- Stem cells which have capacity to differentiate only few tissues like lymphoid or myeloid stem cells. Unipotent-Stem cells are unipotent which can give rise to only same kind of cells like skin cells are a good example of unipotent stem cells.

Table 4.1 Stem cells source, types and markers in livestock

Species	Source	Type	Marker	References
Cattle	Adipose, endometrium, bone marrow, mammary tissue, Wharton's jelly	MSC, ES, iPSC	Oct-4,Sox-2,Nanog,TRA-1-81, TRA-1-60, CD73, CD 90, CD 105	Hill et al. (2019)
Buffalo	Mammary tissue, adipose, placenta, endometrium, Wharton's jelly	MSC,ES, iPSC	CD73, CD 90, CD 105, Oct-4,Sox-2,Nanog	Gugjoo et al. (2019a, b)
Sheep	Adipose, umbilical cord Wharton's jelly	ES, iPSC	Oct-4,Sox-2,Nanog	Gugjoo and Amarpal (2018)
Goat	Adipose,umbilical cord	ES, iPSC	Oct-4,Sox-2,Nanog	Gugjoo et al. (2020)
Pig	Bone marrow, adipose, umbilical, endometrium, Wharton's jelly	MSC, iPSC	Oct-4,Sox-2,Nanog, CD73, CD 90, CD 105	Bodek et al. (2015), Xu et al. (2019), and Vassiliev et al. (2010)
Horse	Bone marrow, adipose, umbilical cord, endometrium, Wharton's jelly	MSC, ES	Oct-4,Sox-2,Nanog,TRA-1-81, TRA-1-60, CD73, CD 90, CD 105	Burk et al. (2013) and MacDonald and Barrett (2020)
Dog	Adipose tissue, umbilical cord.	MSC, iPSC	Oct-4,Sox-2,Nanog,CD 73, CD 90, CD105	Kriston-Pál et al. (2017) and Baird et al. (2015)

4.2 Stem cell Types and Their Identification

Embryonic stem cells are pluripotent stem cells derived from ICM of blastocyst. They have ability to differentiate into all three germ layers. ESC derived from Humans, murines and primates shows peculiar cell cycle pattern they spends most of the time in S phase as it lacks G1 check point (White and Dalton 2005; Bárta et al. 2010). Different methods were adopted to isolate the embryonic stem cells like immunosurgery, mechanical dissection, Laser dissection, enzymatic method and mostly cultured on mitotically inactivated fibroblast feeder layers (Yu and Thomson 2014). Intrinsic and extrinsic pathways influence the self-renewing and pluripotency of stem cells. Both self-renewing and pluripotency processes are coordinated processes altered by different signaling pathways like LIF/stat3, wnt/b-catenin, FGF/ERK, TGF/SMAD and PKC signaling (Huang et al. 2015). Induced pluripotent stem cells are another type stem cells which have pluripotency properties. Due to ethical concerns of embryonic stem cells iPSC gain its popularity because of its technical simplicity (Liang and Zhang 2013). iPSCs are transcription factors mediated reprogrammed cells. Yamanaka described Oct4,Sox2,Nanog and c-myc are key transcription factors which helps in reprogramming of cells to pluripotent state (Takahashi and Yamanaka 2006). Oct4, Sox2, Nanog are important factors that maintain pluripotency in both types. Oct-4 is the master regulator of maintaining pluripotency Embryo restricts the development of ICM in the absence of Oct 4. Most of the cases pluripotent identity is maintained by Oct4 in associated with Sox2 form heterodimer and regulate Nanog expression (Wu and Schöler 2014).

Adult stem cells were found in bone marrow, dental pulp, retina, pancreas, spinal cord, cornea, heart, liver, blood etc. almost in all the body parts. They are clonogenic and maintain homeostasis (Nakada et al. 2011). They promote healing by replacing the deceased cells (Kanji and Das 2017). Adult stem cell differentiate into specialized cell type suitable to microenvironment (Tang et al. 2006). As per the standards set by ISCT cells positive for CD90, CD73, CD105 and negative for CD45,CD 34, CD79 are considered as mesenchymal stem cells (Dominici et al. 2006). Hematopoietic stem cells are positive for CD34 and lack in marker in mesenchymal stem cells. Expression of markers varies with species for example CD73 cell surface marker in dog showed negative while it is an established marker in humans. Stem cell surface markers in canines shows positive for CD105, CD90, CD44, STRO-1 (Amanda B.T. Hill et al. 2018). In equine CD105, CD90, CD44 were shown positive but umbilical derived mesenchymal stem cells expresses low levels of CD-90 (Barberini et al. 2014). Identified markers of stem cells so far were detailed in Table 4.2. Bovine mesenchymal stem cells shows positive for CD29, CD166, CD106, CD73, CD44, CD90 (Sampaio et al. 2015; Gao et al. 2014) (Fig. 4.1). Stem cells have great potential in terms of cells for therapeutic applications under different conditions, including metabolic, degenerative and inflammatory diseases, restoring and regenerating damaged or destroyed tissues, as well as a cancer treatment. Immunomodulatory, antimicrobial activity and migration of adult stem cells are discussed further in this chapter.

Table 4.2 Reported markers for stem cells

Markers	Type of stem cells	References
TRA-1-81	EC, ES	Andrews et al. (2005)
TRA-1-60	EC, ES	Zhao et al. (2012)
Telomerase	ES, EC	Hiyama and Hiyama (2007)
c-Kit	MSC,HSC,ES,EC	Lennartsson and Rönnstrand (2012)
SSEA-4	EC, ES	Zhao et al. (2012)
SSEA-3	EC, ES	Andrews et al. (2005)
OCT-4	EC, ES	Pan et al. (2002)
Germ cell nuclear factor	EC, ES	H. Wang et al. (2016)
GCTM-2	EC, ES	Andrews et al. (2005)
TDGF	ES	Baldassarre et al. (1997)
CD133	HSC, neural stem cells	Zhao et al. (2012)
CD30	EC, ES	Zhao et al. (2012)
Alkaline phosphatase	EC, ES	Štefková et al. (2015)
Neurosphere	ES	Campos (2004)
Stro-1	MSC, HSC	Lin et al. (2011)
Lin	MSC, HSC	Seita and Weissman (2010)
Colony forming unit	MSC, HSC	Kaufman et al. (2001)
CD 44	MSC	Jiang et al. (2019)
BMPR	MSC	Mira et al. (2010)
CD 73	MSC	Maleki et al. (2014)
CD 34	MSC, HSC	AbuSamra et al. (2017))
Thy-1	HSC, MSC	Saalbach and Anderegg (2019)

4.3 Application of Stem cells in Livestock

Conventional livestock production should be optimized to reach the demand of daily needs of fast growing population. According to FAO, 60% of animal and crop output has to be increased to meet the requirement The requirement for meat and milk will relatively increased by 70% by 2050 (Herrero et al. 2015). Advancement in genomic technologies during early twenty-first century led to the identification of hundreds of QTLs. This discovery opens a wide opportunity to improve the genomic selection in livestock across the globe. Advancements in Stem cell technology integrated with genomic selection offers a wide range of applications in livestock production.

Stem cells especially embryonic stem cell type increases the efficiency of cloning somatic cell nuclear transfer. It was reported that embryonic stem cells has 10–20 times higher cloning efficiency in comparison with cumulus cells (Singh 2019; van Thuan et al. 2010). Chimeric and transgenic animals can be produced, Desired trait carrying transgenic animal production is ramp by using embryonic stem cells as it was reported that higher genetic modifications is possible in ES than somatic cells. Homologous recombinations were higher in case of ES in comparison with somatic cells (Choi et al. 2020). Manipulating the live stock species

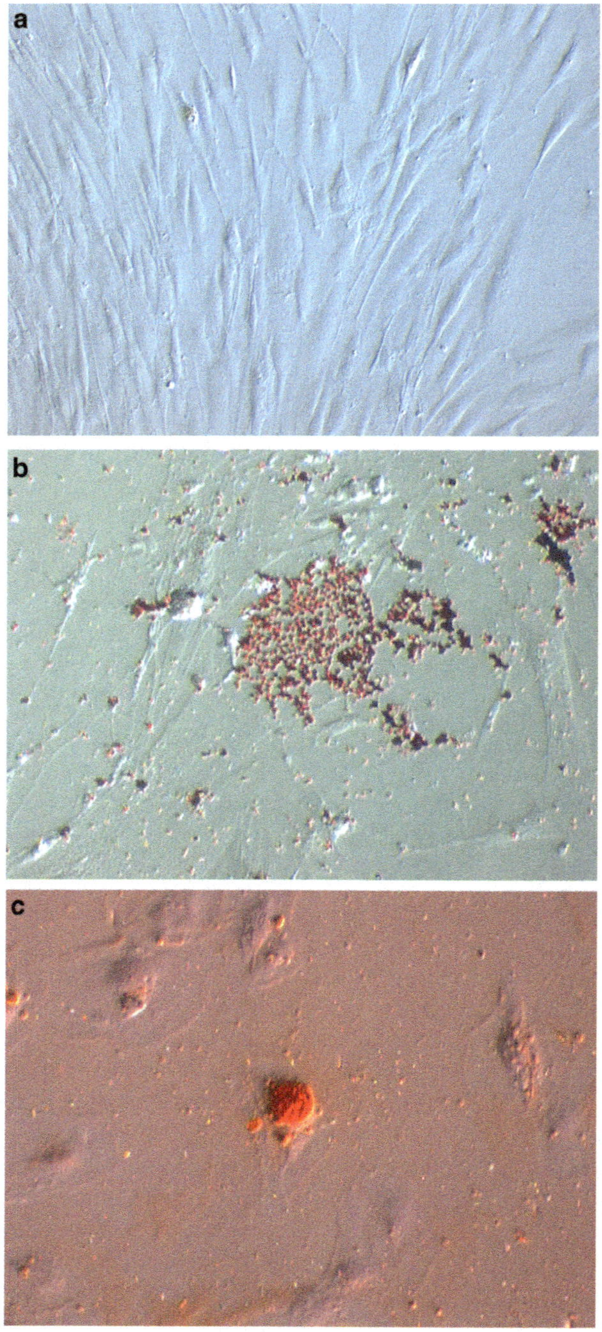

Fig. 4.1 Buffalo stem cells and their differentiation into bilineage (**a**) Fully confluent Buffalo adipose derived mesenchymal stem cells (**b**) Adipogenic differentiated stained with Oil red o staining. (**c**) Osteogenic differentiated stained with alizarin red s

genome could efficiently increase the animal production with desirable traits. Enhancing welfare of animal, increasing food production are the main objectives of gene editing. They also used in resolving the environmental issues. Embryonic stem cells were used as donar cells for gene editing further used in cloning. ZFN, TALEN, CRISPR Genome editing technologies enable production of disease resistance and high yielding milk livestock (Menchaca et al. 2020). Economically developed countries dairy industry facing huge economic loss due to mastitis. These cutting edge technologies can produce the mastitis resilient trait. Introgression of lysostaphin gene in cow (Wall et al. 2005) and lysozyme in goat (Maga et al. 2006) are reported to have resilient to mastitis. Likewise influenza resistant pig, PRRSV (porcine reproductive and respiratory syndrome virus) resilient porcine trait (Lu et al. 2017) and environmental friendly porcine trait which digest the phytate (Golovan et al. 2001) were produced using these technologies. Over the past few years around 300 edited cattle, sheep, pigs goats were produced (Tan et al. 2016)

ES also able to generate germ cells which has its application in alleviation of infertility in livestock. Although ARTs reduced the effectiveness of this problem, ethical issues, invasive, expertise handling and costly treatment hinder the progress of solving the issue. Stem cells offer a new approach of treatment. Oocyte generation, ovarian regeneration and dysfunctional sperm were possible and treated using stem cells. so that elite gene pool can be conserved (Wang et al. 2019). Moreover ES opens a new way for drug development. A vibrant framework ideal for defining innovative molecular targets and improving specific drugs is defined by stem cells. Pathological context of disease, testing of drugs and assess the effectiveness of drugs and its safety can be evaluated using stem cells. Desirable cells were obtained by iPSC then effect of drugs and chemicals were evaluated (Rubin and Haston 2011).

Due to its complexity developmental biology is less understood. Embryonic stem cells and induced pluripotent stem cells are versatile biological cells that open insight into developmental process. Embryonic stem cells have the capacity to differentiate into almost any kind of cells. The pathways involved in embryonic development, fate of cells and regulation of transcription factors can be revealed by exploring the stem cells (Zhu and Huangfu 2013). The list of endangered species increasing alarmingly. Conservation strategies are not sufficient to protect them from becoming extinct. A recent potential technology like iPSC attracts scientists to revive the lost ecosystem. Extinct animals can also be resurrected by using this technology. Still lots to explore its application in conserving the endangered species recent advances are encouraging (Stanton et al. 2019).

Ethical limitations of embryonic stem cells are overcome by adult stem cells. Mesenchymal stem cells are extensively studied stem cells. As they are easy to isolate and expand the popularity gained after the recognition of immune modulation, anti proliferative, migration abilities. Clinical use of stem cells in veterinary medicine was well reviewed by Voga et al. (2020).

4.4 Mesenchymal Stem Cells

MSCs are adult stem cells which have the ability to differentiate into multiple lineages. MSCs considered immunomodulatory since they lack type II MHC antigen and thus do not cause immune reactions.MSCs have recently proven to regulate endogenous tissue and immune cells, initially open the way as a regenerative therapy to reconstruct skeletal tissue. Preclinical action mechanism studies indicate that the potential effects of MSC transplantation are short-lived, and linked to active, paracrine interactions between MSCs and host cells. The shreds of evidence indicate that MSC's primary mechanism of action based on paracrine signalling that results in functional changes in the immune cells, such as natural killer cells, B-cells, T-cells, dendritic cells, and monocytes/macrophages. There have been studies of many factors contributing to MSC's immunomodulatory effects such as transforming growth factorβ, prostaglandin E2 (PGE2), interleukin 10 (IL 10) indolamine 2,3 dioxygenase (IDO) and stimulated gene-6 (TSG-6) tumour necrosis factor- (TNF) (Voga et al. 2020). Moreover, MSCs secretes a variety of soluble factors and Antimicrobial peptides like cathelicidin, lipocalin,β-Defensins,hepcidin etc. (Marrazzo et al. 2019).

4.4.1 MSC Immune Modulation and Antimicrobial Properties

MSCs have anti-inflammatory properties and reduce scar formation. MSCs strongly inhibit T lymphocytes and b lymphocytes, by activating programmed death pathway and proliferation and subsequent cytokine signaling (Augello et al. 2005). MSCs suppress excess production of autoantibodies. However, B cell-MSC interactions are not well known. MSCs interact with other immune cells and inhibit B-cells, and at the same time, it induces bregs (Fan et al. 2016). MSCs interacts with B-cells and inhibit the proliferation of B-cells and other co-stimulatory factors (Rosado et al. 2014). MSCs do not produce any immunological reaction as it lacks MHC 1 (Krampera et al. 2003), MHC 2 (Nauta and Fibbe 2007), CD80 and CD86 molecules on its surface (Le Blanc et al. 2003). The proliferation of NK cells is associated with IL-2. MSCs inhibit IL-2 and NK cells. The activated NK receptors NKp30, NKG2D is responsible for the lysis of MSCs (Spaggiari et al. 2006). VCAM1 and galectin-1 (Ren et al. 2010; Gieseke et al. 2010) are also involved in MSC-cell mediated immunoregulation. Eventually, MSCs strongly upregulates the expression of galectin-9 upon activation INF-γ and tend to be an important proliferative mediator, thus act as an immunoregulatory marker (Ungerer et al. 2014). CD8+ induce apoptosis in MSCs, which offers clinical benefits as they completely lack the immune rejection. This shows the clinical efficiency of MSCs. Apoptosis is important for MSC's anti-inflammatory and regenerative events. In agreement, apoptosis stimulated before treatment of MSC increased its effectiveness (Galleu et al. 2017).

MSCs also secrete some soluble factors which are necessary for immune modulation. The key mediators of MSC's immunosuppressive activity in the presence of pro-inflammatory cytokines are, PGE2, cyclooxygenase 2 (COX-2) and indoleamine-pyrrole 2,3-dioxygenase (IDO) (Krampera et al. 2003). PGE2 has been associated directly in macrophage production of IL-10 (Tam et al. 2009) and in preventing the monocytes differentiation into DCs. (Spaggiari et al. 2009). IFN-γ regulates the immunomodulation of MSCs and TNF-α and produces PGE2 (Ren et al. 2008), IL-6 (Ghannam et al. 2010), TGF-β1 (Nemeth et al. 2010), NO, LIF and HGF (Najar et al. 2010) are other crucial MSC immune regulatory molecules found till date. These factors interact with both innate and adaptive immunity and initiate the immune suppression. T- cells and NK cells suppression is mediated by IDO. Allograft kidney acceptance is mediated by the MSC secreted IDO via treg generation (Ge et al. 2010). IL-10 is secreted by interaction of MSCs with the Antigen presenting cells thus regulate the immune suppression irrespective of allogenic or naïve or pre activated T- cells (Yang et al. 2009). Critical role of nitric oxide secreted by mesenchymal stem cells is reported by Sato et al. 2007. Proliferation of T- cells is repressed by the phosphorylation of stat 5 in the presence of NO and MSC. Inhibition of Nitric oxide synthase of MSC tends to the proliferation of T-cells (Fig. 4.2).

The use of mesenchymal stem cells (MSC) in the treatment of bacterial infections has gained growing attention in recent years. MSCs have direct bactericidal activity. Thus, MSC administration could replace traditional antibiotic therapy (Kadam et al. 2019; Marrazzo et al. 2019) MSC also produces antimicrobial peptides (AMPs), usually found in neutrophils or epithelial cells. AMPs directly destroy bacteria by destroying the integrity of the microbial membrane (Krasnodembskaya et al. 2010) or by triggering the production of pro-inflammatory cytokines. Human MSC produces multiple AMPs, including cathelicidin peptide LL-37 (Krasnodembskaya et al. 2010), hepcidin (Alcayaga-Miranda et al. 2017), β-defensin 2, and lipocalin 2 (Gupta et al. 2012). MSC-produced AMPs which considered to be one of the critical components in the removal of bacterial infections, as investigated in multiple animal models (Johnson et al. 2017).

4.4.2 MSC Migration and Homing

MSC Homing is of two types based on the administration route (1) Systemic homing (2) Non-systemic homing. Systemic homing happened when MSCs administered intravenously or intraarterial. MSC enters the bloodstream and undergoes a multistep process to exit and blood stream and migrates to the targeted region. Non-systemic homing MSCs injected at target tissue MSCs travels to the damaged tissue. Some of the crucial stages of homing are the migration of MSCs in the bloodstream and trans endothelial immigration. MSCs thought to use the same mechanism as the movement of leukocytes to pass into the bloodstream. Adhesion molecules are essential for the trafficking of mesenchymal stem cells (Butcher and Picker 1996). These adhesion molecules include integrins, chemokine receptors, and selectins.

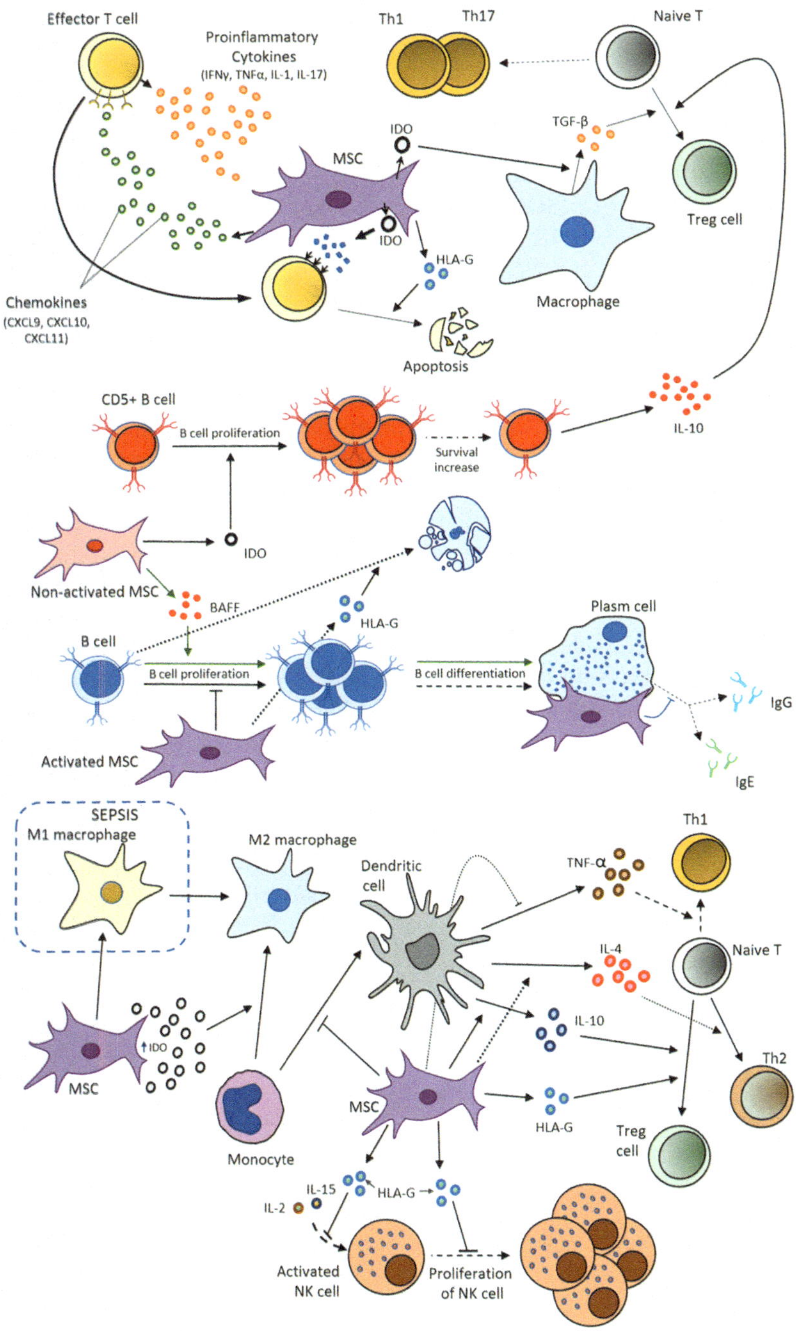

Fig. 4.2 Overview of mesenchymal stem cells immunomodulation. MSC interaction with adaptive and innate immune cells by releasing various soluble factors. (This figure is adapted from Jimenez-Puerta et al. 2020)

Primarily MSCs associate with endothelium by gathering and rolling. In the second, G-protein - coupled receptors activate cells, accompanied in the third step by integrin-mediated, activation-dependent arrest. Eventually, the cells transmigrate in the 4th phase via the endothelium (Ullah et al. 2019). MSCs will move to inflamed tissues Reacting to chemokine and chemokine receptor signals Induced according to inflammatory conditions (Kia et al. 2011). MSCs express a wide variety of receptors of chemokine and chemokine which helps in MSC homing. Some studies have shown chemokine receptor expression; namely, CCR1, CCR2, CCR4, CCR6, CCR7, CCR8, CCR9, CCR10, CXCR1, CXCR2, CXCR3, CXCR4, CXCR5, CXCR6 are present on MSC (Ghaffari-Nazari 2018). Reports have shown that the SDF-1/CXCR4 fulfil the essential function of MSC migration to the bone marrow. CXCR4 takes dose-dependent responses to SDF-1 (Kitaori et al. 2009) (Fig. 4.3).

4.5 Transplantation Modes and Theraputic Applications of Stem Cells

Based on source of stem cells for transplantation possible treatment methods are (1) Autogenic transplantation, (2) Allogenic transplantation and (3) xenogenic transplantation. Autogenic transplant refers to the stem cells that are isolated from the animals own tissues like adipose tissue, cord blood etc. In allogenic transplantation stem cells were isolated from another donar of same species. Stem cells were isolated from different species for transplantation in xenogenic transplation which involves the risk of rejection. Stem cells are introduced into animals body intravenouly or locallly injected based on the condition and severity of disease. Stem cells are reported to be more tolerent to oxidative stress as animals undergo stress inorder to increase the productivity.

4.5.1 Bone, Cartilage and Ligament Injuries

Equines and canines are more susceptible to different injuries. Stem cells has the ability to differentiate into osteocytes and myocytes. Reports increasing on the reparative effect of mesenchymal stem cells in different species and recent studies quoted the effectiveness of MSC treatment in tendon, cartilage and bone injuries (Renzi et al. 2013; Canapp et al. 2016; Gibson et al. 2017). Adequate supply of blood and absence of nerve and lymph to cartilage bounded with reparative limitation favorable results were reported for cartilage repair by stem cells in rabbit and equine (Agung et al. 2006). Although Bones have regeneration capacity, deformity caused by the devastating events can't be compensated in some critical cases. Reproductive inefficiency, loss of meat and loss in production of milk were some results of bone fracture in livestock which increases the proportion to lose the elite

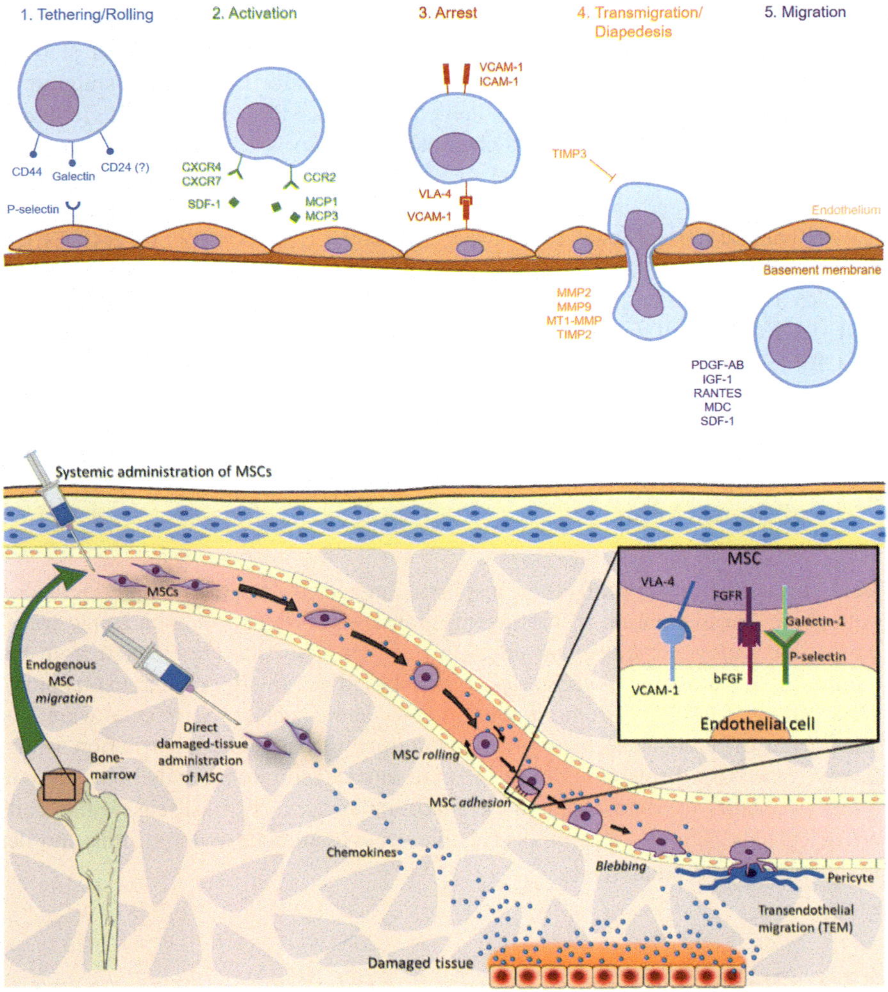

Fig. 4.3 (**a**) Step wise mechanism of homing of Mesenchymal stem cells. (This figure is reproduced with the permission of Ullah et al. 2019) (**b**) On administration migration of MSC to site of infection. (This figure is adapted from Jimenez-Puerta et al. 2020)

species (St. Jean and Anderson 2014). Changed trends in the society to conserve the economically productive animals reformed the treatment methods. Stem cells enhance the regeneration of bone (Iaquinta et al. 2019). Stem cells not only recuperate the animal's structure but also stabilize its functionality (Moritz and Ambrosio 2017).

4.5.2 Degenerative Diseases

Stem cells can differentiate into osteocytes and chondrocytes and possess immune modulatory effect to treat certain degenerative diseases. Degenerative diseases like Osteoarthritis effects the quality of semen (D. F. Wolfe 2018). Promising results of stem cell injection to osteoarthritis condition significantly improves the local cartilage and synovial fluid viscosity in equines (Broeckx et al. 2019; Brehm et al. 2014 & Frisbie et al. 2009). Reports in canines showed infusion of stem cells in cartilage regeneration also infer a reliable approach (Shah et al. 2018; Sasaki et al. 2019). Most common progressive neurodegenerative disease in dogs is degenerative myelopathy (DM). Recent Success in mice model promises to use stem cells in canine suffering with DM (Hoffman and Dow 2016).

4.5.3 Inflammatory Diseases

Inflammatory diseases like Mastitis, laminitis, IBD, metritis, dermatitis in bovine, ovine, canine and equine species cause huge economical loss and has detrimental effect on animals health. Mastitis cause huge loss to the dairy industry. Resistance of bacteria to antibiotics makes its treatment difficult. Stem cells offers an alternative therapy for mastitis becoming popular now a days. Recent reports quote the safety of mesenchymal stem cells for the treatment of mastitis. Intra mammary infusion of mesenchymal stem cells decreases the bacterial population in the milk (Peralta et al. 2020). Rise of inflammatory bowel disorder cases over past decade and failure of present therapies to treat this condition attracts stem cell in order to maintain homeostasis in the gut. IBD is a gastrointestinal inflammatory disease and its etiology is unknown till date. Sub-mucosal and intravenous infusion of stem cells showed the significant improvement in reconstruction of disruptive mucosal tissue after the infusion of stem cells (Shi et al. 2019). Lamella tissue inflammation cause laminitis frequently found in cattle and horse. Use of MSC shown promisable treatment for laminitis in equine and cattle (Gugjoo et al. 2019a, b) Metritis is a reproductive disorder characterized by inflammation of endometrial tissue followed by bacterial infection. Antibiotics are the most common method to treat these diseases. However positive selection of bacteria with antibiotic treatment making microbes drug resistance and yielding huge loss to bovine and equine industry (Lara et al. 2018). Endometrial stem cells holds promise able treatment for this condition. Mambelli et al. (2013) observed the homing of stem cells to injured endometrial tissue. Falomo et al. (2015) observed the increase in the expression of IL10 and decrease in the expression of TNFA indicating the down regulation of chronic inflammation of disease in equine. Canine atopic dermatitis is the most common multifunctional disorder occurs in dogs in which less antimicrobial peptides were produced exposing animal to bacteria. Immunomodulatory and antimicrobial activities of MSCs holds promising treatment for this disease (Dias et al. 2019).

4.5.4 *Animal Models for Human Diseases*

To reduce the gap between clinical practice and translational research animal models are required to alleviate the human diseases. As small animal models failed to reproduce the results for human use large animal models can be alternative and more appropriate models to explore the use of stem cell technology for human welfare. Pigs, Dogs, cattle, goat, rabbit, and buffalo can be alternative sources to explore the stem cells full spectrum of action. Less data availability on safety and efficacy of the stem cells has to be addressed by proving the stem cells ability in large animal models (Harding et al. 2013). The mystery riddled around development process in human can also be untangled with the anatomical similar large animal models. Xenotransplantation of organs from swine to human, transgenic pigs to understand the pathophysiology of cystic fibrosis are some of examples of the stem cell large animal models for the welfare of mankind (Wolfe 2009; Klymiuk et al. 2010). Human Xenogenic implantation of MSC showed decrease in pulmonary odema and enhance oxygenation in acute respiratory distress syndrome model of ovine (Asmussen et al. 2014). Stem cell therapeutic animal model for osteoarthritis, inflammatory bowel disease, crohns disease, ulcerative colitis, diabetes to understand the effect and safety of infusion. Regime of biologics, administration route, and therapy results can be quickly transferred to human use.

4.6 Conclusion

Tremendous properties of stem cells hold a reliable source to solve extensive problems now our community is facing. The impact of stem cells on production and welfare of the livestock is revolutionary. On administration migrating to the site of infection, restoring the damaged tissue and boosting its immunity made stem cells a robust technology to use in farm animals. No optimized protocol to isolate the stem cells from different tissues across different species, effect of age on stemness, contradictory results on therapeutic efficacy and its safety needs extensive research.

References

AbuSamra DB, Aleisa FA, Al-Amoodi AS, Ahmed HMJ, Chin CJ, Abuelela AF, Bergam P, Sougrat R, Merzaban JS (2017) Not just a marker: CD34 on human hematopoietic stem/progenitor cells dominates vascular selectin binding along with CD44. Blood Adv 1(27):2799–2816. https://doi.org/10.1182/bloodadvances.2017004317

Agung M, Ochi M, Yanada S, Adachi N, Izuta Y, Yamasaki T, Toda K (2006) Mobilization of bone marrow-derived mesenchymal stem cells into the injured tissues after intraarticular injection and their contribution to tissue regeneration. Knee Surg Sports Traumatol Arthrosc 14(12):1307–1314. https://doi.org/10.1007/s00167-006-0124-8

Alcayaga-Miranda F, Cuenca J, Khoury M (2017) Antimicrobial activity of mesenchymal stem cells: current status and new perspectives of antimicrobial peptide-based therapies. Front Immunol 8(MAR):1–15. https://doi.org/10.3389/fimmu.2017.00339

Andrews PW, Matin MM, Bahrami AR, Damjanov I, Gokhale P, Draper JS (2005) Embryonic stem (ES) cells and embryonal carcinoma (EC) cells: opposite sides of the same coin. Biochem Soc Trans 33(6):1526–1530. https://doi.org/10.1042/BST20051526

Asal M, Güven S (2020) Stem cells: sources, properties, and cell types. In: Biomaterials for organ and tissue regeneration. Woodhead Publishing, Oxford, pp 177–196. https://doi.org/10.1016/b978-0-08-102906-0.00007-6

Asmussen S, Ito H, Traber DL, Lee JW, Cox RA, Hawkins HK, McAuley DF, McKenna DH, Traber LD, Zhuo H, Wilson J, Herndon DN, Prough DS, Liu KD, Matthay MA, Enkhbaatar P (2014) Human mesenchymal stem cells reduce the severity of acute lung injury in a sheep model of bacterial pneumonia. Thorax 69(9):819–825. https://doi.org/10.1136/thoraxjnl-2013-204980

Augello A, Tasso R, Negrini SM, Amateis A, Indiveri F, Cancedda R, Pennesi G (2005) Bone marrow mesenchymal progenitor cells inhibit lymphocyte proliferation by activation of the programmed death 1 pathway. Eur J Immunol 35(5):1482–1490. https://doi.org/10.1002/eji.200425405

Baird AEG, Barsby T, Guest DJ (2015) Derivation of canine induced pluripotent stem cells. Reprod Domest Anim 50(4):669–676. https://doi.org/10.1111/rda.12562

Baldassarre G, Romano A, Armenante F, Rambaldi M, Paoletti I, Sandomenico C, Pepe S, Staibano S, Salvatore G, De Rosa G, Persico MG, Viglietto G (1997) Expression of teratocarcinoma-derived growth factor-1 (TDGF-1) in testis germ cell tumors and its effects on growth and differentiation of embryonal carcinoma cell line NTERA2/D1. Oncogene 15(8):927–936. https://doi.org/10.1038/sj.onc.1201260

Barberini DJ, Freitas NPP, Magnoni MS, Maia L, Listoni AJ, Heckler MC, Sudano MJ, Golim MA, Da Cruz Landim-Alvarenga F, Amorim RM (2014) Equine mesenchymal stem cells from bone marrow, adipose tissue and umbilical cord: immunophenotypic characterization and differentiation potential. Stem Cell Res Ther 5(1):1–11. https://doi.org/10.1186/scrt414

Bárta T, Vinarský V, Holubcová Z, Doležalová D, Verner J, Pospíšilová Š, Dvořák P, Hampl A (2010) Human embryonic stem cells are capable of executing G1/S checkpoint activation. Stem Cells 28(7):1143–1152. https://doi.org/10.1002/stem.451

Bodek G, Bukowska J, Wisniewska J, Ziecik AJ (2015) Evidence for the presence of stem/progenitor cells in porcine endometrium. Mol Reprod Dev 82(3):182–190. https://doi.org/10.1002/mrd.22459

Brehm W, Burk J, Delling U (2014) Application of stem cells for the treatment of joint disease in horses. Methods Mol Biol 1213:215–228. https://doi.org/10.1007/978-1-4939-1453-1_18

Broeckx SY, Martens AM, Bertone AL, Van Brantegem L, Duchateau L, Van Hecke L, Dumoulin M, Oosterlinck M, Chiers K, Hussein H, Pille F, Spaas JH (2019) The use of equine chondrogenic-induced mesenchymal stem cells as a treatment for osteoarthritis: A randomised, double-blinded, placebo-controlled proof-of-concept study. Equine Vet J 51(6):787–794. https://doi.org/10.1111/evj.13089

Burk J, Ribitsch I, Gittel C, Juelke H, Kasper C, Staszyk C, Brehm W (2013) Growth and differentiation characteristics of equine mesenchymal stromal cells derived from different sources. Vet J 195(1):98–106. https://doi.org/10.1016/j.tvjl.2012.06.004

Butcher EC, Picker LJ (1996) Lymphocyte homing and homeostasis. Science 272(5258):60–66. https://doi.org/10.1126/science.272.5258.60

Campos LS (2004) Neurospheres: insights into neural stem cell biology. J Neurosci Res 78(6):761–769. https://doi.org/10.1002/jnr.20333

Canapp SO, Leasure CS, Cox C, Ibrahim V, Carr BJ (2016) Partial cranial cruciate ligament tears treated with stem cell and platelet-rich plasma combination therapy in 36 dogs: a retrospective study. Front Vet Sci 3(DEC):1–9. https://doi.org/10.3389/fvets.2016.00112

Choi EH, Yoon S, Koh YE, Seo YJ, Kim KP (2020) Maintenance of genome integrity and active homologous recombination in embryonic stem cells. Exp Mol Med 52(8):1220–1229. https://doi.org/10.1038/s12276-020-0481-2

Dias IE, Pinto PO, Barros LC, Viegas CA, Dias IR, Carvalho PP (2019) Mesenchymal stem cells therapy in companion animals: useful for immune-mediated diseases? BMC Vet Res 15(1):1–14. https://doi.org/10.1186/s12917-019-2087-2

Dominici M, Le Blanc K, Mueller I, Slaper-Cortenbach I, Marini FC, Krause DS, Deans RJ, Keating A, Prockop DJ, Horwitz EM (2006) Minimal criteria for defining multipotent mesenchymal stromal cells. The International Society for Cellular Therapy position statement. Cytotherapy 8(4):315–317. https://doi.org/10.1080/14653240600855905

Falomo ME, Ferroni L, Tocco I, Gardin C, Zavan B (2015) Immunomodulatory role of adipose-derived stem cells on equine endometriosis. Biomed Res Int 2015:6–11. https://doi.org/10.1155/2015/141485

Fan L, Hu C, Chen J, Cen P, Wang J, Li L (2016) Interaction between mesenchymal stem cells and B-cells. Int J Mol Sci 17(5). https://doi.org/10.3390/ijms17050650

Frisbie DD, Kisiday JD, Kawcak CE, Werpy NM, McIlwraith CW (2009) Evaluation of adipose-derived stromal vascular fraction or bone marrow-derived mesenchymal stem cells for treatment of osteoarthritis. J Orthop Res 27(12):1675–1680. https://doi.org/10.1002/jor.20933

Galleu A, Riffo-Vasquez Y, Trento C, Lomas C, Dolcetti L, Cheung TS, Von Bonin M, Barbieri L, Halai K, Ward S, Weng L, Chakraverty R, Lombardi G, Watt FM, Orchard K, Marks DI, Apperley J, Bornhauser M, Walczak H et al (2017) Apoptosis in mesenchymal stromal cells induces in vivo recipient-mediated immunomodulation. Sci Transl Med 9(416):1–12. https://doi.org/10.1126/scitranslmed.aam7828

Gao Y, Zhu Z, Zhao Y, Hua J, Ma Y, Guan W (2014) Multilineage potential research of bovine amniotic fluid mesenchymal stem cells. Int J Mol Sci 15(3):3698–3710. https://doi.org/10.3390/ijms15033698

Ge W, Jiang J, Arp J, Liu W, Garcia B, Wang H (2010) Regulatory T-cell generation and kidney allograft tolerance induced by mesenchymal stem cells associated with indoleamine 2,3-dioxygenase expression. Transplantation 90(12):1312–1320. https://doi.org/10.1097/TP.0b013e3181fed001

Ghaffari-Nazari H (2018) The known molecules involved in MSC homing and migration. J Stem Cell Res Med 3(1):1–4. https://doi.org/10.15761/jscrm.1000127

Ghannam S, Bouffi C, Djouad F, Jorgensen C, Noël D (2010) Immunosuppression by mesenchymal stem cells: mechanisms and clinical applications. Stem Cell Res Ther 1(1):1–7. http://www.embase.com/search/results?subaction=viewrecord&from=export&id=L361605013%5Cn; https://doi.org/10.1186/scrt2%5Cn; http://elvis.ubvu.vu.nl:9003/vulink?sid=EMBASE&issn=17576512&id=doi:10.1186%2Fscrt2&atitle=Immunosuppression+by+mesenchymal+stem+cell

Gibson MA, Brown SG, Brown NO (2017) Semitendinosus myopathy and treatment with adipose-derived stem cells in working German shepherd police dogs. Can Vet J 58(3):241–246

Gieseke F, Böhringer J, Bussolari R, Dominici M, Handgretinger R, Müller I (2010) Human multipotent mesenchymal stromal cells use galectin-1 to inhibit immune effector cells. Blood 116(19):3770–3779. https://doi.org/10.1182/blood-2010-02-270777

Golovan SP, Meidinger RG, Ajakaiye A, Cottrill M, Wiederkehr MZ, Barney DJ, Plante C, Pollard JW, Fan MZ, Hayes MA, Laursen J, Hjorth JP, Hacker RR, Phillips JP, Forsberg CW (2001) Pigs expressing salivary phytase produce low-phosphorus manure. Nat Biotechnol 19(8):741–745. https://doi.org/10.1038/90788

Gugjoo MB, Amarpal (2018) Mesenchymal stem cell research in sheep: current status and future prospects. Small Rumin Res 169:46–56. https://doi.org/10.1016/j.smallrumres.2018.08.002

Gugjoo MB, Amarpal, Fazili MR, Shah RA, Sharma GT (2019a) Mesenchymal stem cell: basic research and potential applications in cattle and buffalo. J Cell Physiol 234(6):8618–8635. https://doi.org/10.1002/jcp.27846

Gugjoo MB, Amarpal, Makhdoomi DM, Sharma GT (2019b) Equine mesenchymal stem cells: properties, sources, characterization, and potential therapeutic applications. J Equine Vet 72:16–27. https://doi.org/10.1016/j.jevs.2018.10.007

Gugjoo MB, Amarpal, Fazili M u R, Shah RA, Saleem Mir M, Sharma GT (2020) Goat mesenchymal stem cell basic research and potential applications. Small Rumin Res 183:106045. https://doi.org/10.1016/j.smallrumres.2019.106045

Gupta N, Krasnodembskaya A, Kapetanaki M, Mouded M, Tan X, Serikov V, Matthay MA (2012) Mesenchymal stem cells enhance survival and bacterial clearance in murine Escherichia coli pneumonia. Thorax 67(6):533–539. https://doi.org/10.1136/thoraxjnl-2011-201176

Harding J, Roberts RM, Mirochnitchenko O (2013) Large animal models for stem cell therapy. Stem Cell Res Ther 4(2):1–9. https://doi.org/10.1186/scrt171

Herrero M, Wirsenius S, Henderson B, Rigolot C, Thornton P, Havlík P, De Boer I, Gerber P (2015) Livestock and the environment: what have we learned in the past decade? Annu Rev Environ Resour 40:177–202. https://doi.org/10.1146/annurev-environ-031113-093503

Hill ABT, Hill JEBT, Bressan FF, Miglino MA, Garcia JM (2018) Derivation and differentiation of canine ovarian mesenchymal stem cells. J Vis Exp 2018(142):2–7. https://doi.org/10.3791/58163

Hill ABT, Bressan FF, Murphy BD, Garcia JM (2019) Applications of mesenchymal stem cell technology in bovine species. Stem Cell Res Ther 10(1):1–13. https://doi.org/10.1186/s13287-019-1145-9

Hiyama E, Hiyama K (2007) Telomere and telomerase in stem cells. Br J Cancer 96(7):1020–1024. https://doi.org/10.1038/sj.bjc.6603671

Hoffman AM, Dow SW (2016) Concise review: stem cell trials using companion animal disease models. Stem Cells 34(7):1709–1729. https://doi.org/10.1002/stem.2377

Huang G, Ye S, Zhou X, Liu D, Ying QL (2015) Molecular basis of embryonic stem cell self-renewal: from signaling pathways to pluripotency network. Cell Mol Life Sci 72(9):1741–1757. https://doi.org/10.1007/s00018-015-1833-2

Iaquinta MR, Mazzoni E, Bononi I, Rotondo JC, Mazziotta C, Montesi M, Sprio S, Tampieri A, Tognon M, Martini F (2019) Adult stem cells for bone regeneration and repair. Front Cell Dev Biol 7. https://doi.org/10.3389/fcell.2019.00268

Jiang B, Yan L, Wang X, Li E, Murphy K, Vaccaro K, Li Y, Xu RH (2019) Concise review: mesenchymal stem cells derived from human pluripotent cells, an unlimited and quality-controllable source for therapeutic applications. Stem Cells 37(5):572–581. https://doi.org/10.1002/stem.2964

Jimenez-Puerta GJ, Marchal JA, López-Ruiz E, Gálvez-Martín P (2020) Role of mesenchymal stromal cells as therapeutic agents: potential mechanisms of action and implications in their clinical use. J Clin Med 9(2):445. https://doi.org/10.3390/jcm9020445

Johnson V, Webb T, Norman A, Coy J, Kurihara J, Regan D, Dow S (2017) Activated mesenchymal stem cells interact with antibiotics and host innate immune responses to control chronic bacterial infections. Sci Rep 7(1):1–18. https://doi.org/10.1038/s41598-017-08311-4

Kadam S, Shai S, Shahane A, Kaushik KS (2019) Recent advances in non-conventional antimicrobial approaches for chronic wound biofilms: have we found the "chink in the armor"? Biomedicine 7(2). https://doi.org/10.3390/biomedicines7020035

Kanji S, Das H (2017) Advances of stem cell therapeutics in cutaneous wound healing and regeneration. Mediat Inflamm 2017. https://doi.org/10.1155/2017/5217967

Kaufman DS, Hanson ET, Lewis RL, Auerbach R, Thomson JA (2001) Hematopoietic colony-forming cells derived from human embryonic stem cells. Proc Natl Acad Sci U S A 98(19):10716–10721. https://doi.org/10.1073/pnas.191362598

Kia NA, Bahrami AR, Ebrahimi M, Matin MM, Neshati Z, Almohaddesin MR, Aghdami N, Bidkhori HR (2011) Comparative analysis of chemokine receptor's expression in mesenchymal stem cells derived from human bone marrow and adipose tissue. J Mol Neurosci 44(3):178–185. https://doi.org/10.1007/s12031-010-9446-6

Kitaori T, Ito H, Schwarz EM, Tsutsumi R, Yoshitomi H, Oishi S, Nakano M, Fujii N, Nagasawa T, Nakamura T (2009) Stromal cell-derived factor 1/CXCR4 signaling is critical for the recruitment of mesenchymal stem cells to the fracture site during skeletal repair in a mouse model. Arthritis Rheum 60(3):813–823. https://doi.org/10.1002/art.24330

Klymiuk N, Aigner B, Brem G, Wolf E (2010) Genetic modification of pigs as organ donors for xenotransplantation. Mol Reprod Dev 77(3):209–221. https://doi.org/10.1002/mrd.21127

Krampera M, Glennie S, Dyson J, Scott D, Laylor R, Simpson E, Dazzi F (2003) Bone marrow mesenchymal stem cells inhibit the response of naive and memory antigen-specific T cells to their cognate peptide. Blood 101(9):3722–3729. https://doi.org/10.1182/blood-2002-07-2104

Krasnodembskaya A, Song Y, Fang X, Gupta N, Serikov V, Lee JW, Matthay MA (2010) Antibacterial effect of human mesenchymal stem cells is mediated in part from secretion of the antimicrobial peptide LL-37. Stem Cells 28(12):2229–2238. https://doi.org/10.1002/stem.544

Kriston-Pál É, Czibula Á, Gyuris Z, Balka G, Seregi A, Sükösd F, Süth M, Kiss-Tóth E, Haracska L, Uher F, Monostori É (2017) Characterization and therapeutic application of canine adipose mesenchymal stem cells to treat elbow osteoarthritis. Can J Vet Res 81(1):73–78

Lara E, Rivera N, Cabezas J, Navarrete F, Saravia F, Rodríguez-Alvarez L, Castro FO (2018) Endometrial stem cells in farm animals: potential role in uterine physiology and pathology. Bioengineering 5(3). https://doi.org/10.3390/bioengineering5030075

Le Blanc K, Tammik C, Rosendahl K, Zetterberg E, Ringdén O (2003) HLA expression and immunologic properties of differentiated and undifferentiated mesenchymal stem cells. Exp Hematol 31(10):890–896. https://doi.org/10.1016/S0301-472X(03)00110-3

Lennartsson J, Rönnstrand L (2012) Stem cell factor receptor/c-Kit: from basic science to clinical implications. Physiol Rev 92(4):1619–1649. https://doi.org/10.1152/physrev.00046.2011

Liang G, Zhang Y (2013) Embryonic stem cell and induced pluripotent stem cell: an epigenetic perspective. Cell Res 23(1):49–69. https://doi.org/10.1038/cr.2012.175

Lin G, Liu G, Banie L, Wang G, Ning H, Lue TF, Lin CS (2011) Tissue distribution of mesenchymal stem cell marker stro-1. Stem Cells Dev 20(10):1747–1752. https://doi.org/10.1089/scd.2010.0564

Lu T, Song Z, Li Q, Li Z, Wang M, Liu L, Tian K, Li N (2017) Overexpression of histone deacetylase 6 enhances resistance to porcine reproductive and respiratory syndrome virus in pigs. PLoS One 12(1):1–21. https://doi.org/10.1371/journal.pone.0169317

MacDonald ES, Barrett JG (2020) The potential of mesenchymal stem cells to treat systemic inflammation in horses. Front Vet Sci 6(January):1–14. https://doi.org/10.3389/fvets.2019.00507

Maga EA, Cullor JS, Smith W, Anderson GB, Murray JD (2006) Human lysozyme expressed in the mammary gland of transgenic dairy goats can inhibit the growth of bacteria that cause mastitis and the cold-spoilage of milk. Foodborne Pathog Dis 3(4):384–392. https://doi.org/10.1089/fpd.2006.3.384

Maleki M, Ghanbarvand F, Behvarz MR, Ejtemaei M, Ghadirkhomi E (2014) Comparison of mesenchymal stem cell markers in multiple human adult stem cells. Int J Stem Cells 7(2):118–126. https://doi.org/10.15283/ijsc.2014.7.2.118

Malik HN, Singhal DK, Saini S, Malakar D (2020) Derivation of oocyte-like cells from putative embryonic stem cells and parthenogenetically activated into blastocysts in goat. Sci Rep 10(1):1–14. https://doi.org/10.1038/s41598-020-66609-2

Mambelli LI, Winter GHZ, Kerkis A, Malschitzky E, Mattos RC, Kerkis I (2013) A novel strategy of mesenchymal stem cells delivery in the uterus of mares with endometrosis. Theriogenology 79(5):744–750. https://doi.org/10.1016/j.theriogenology.2012.11.030

Marrazzo P, Crupi AN, Alviano F, Teodori L, Bonsi L (2019) Exploring the roles of MSCs in infections: focus on bacterial diseases. J Mol Med 97(4):437–450. https://doi.org/10.1007/s00109-019-01752-6

Menchaca A, dos Santos-Neto PC, Mulet AP, Crispo M (2020) CRISPR in livestock: from editing to printing. Theriogenology 150:247–254. https://doi.org/10.1016/j.theriogenology.2020.01.063

Mira H, Andreu Z, Suh H, Chichung Lie D, Jessberger S, Consiglio A, Emeterio JS, Hortigüela R, Marqués-Torrejón MÁ, Nakashima K, Colak D, Götz M, Fariñas I, Gage FH (2010) Signaling through BMPR-IA regulates quiescence and long-term activity of neural stem cells in the adult hippocampus. Cell Stem Cell 7(1):78–89. https://doi.org/10.1016/j.stem.2010.04.016

Moritz CT, Ambrosio F (2017) Regenerative rehabilitation: combining stem cell therapies and activity-dependent stimulation. Pediatr Phys Ther 29:S10–S15. https://doi.org/10.1097/PEP.0000000000000378

Najar M, Raicevic G, Boufker HI, Fayyad-Kazan H, De Bruyn C, Meuleman N, Bron D, Toungouz M, Lagneaux L (2010) Adipose-tissue-derived and Wharton's jelly-derived mesenchymal stromal cells suppress lymphocyte responses by secreting leukemia inhibitory factor. Tissue Eng Part A 16(11):3537–3546. https://doi.org/10.1089/ten.tea.2010.0159

Nakada D, Levi BP, Morrison SJ (2011) Integrating physiological regulation with stem cell and tissue homeostasis. Neuron 70(4):703–718. https://doi.org/10.1016/j.neuron.2011.05.011

Nauta AJ, Fibbe WE (2007) Immunomodulatory properties of mesenchymal stromal cells. Blood 110(10):3499–3506. https://doi.org/10.1182/blood-2007-02-069716

Nemeth K, Keane-Myers A, Brown JM, Metcalfe DD, Gorham JD, Bundoc VG, Hodges MG, Jelinek I, Madala S, Karpati S, Mezey E (2010) Erratum: Bone marrow stromal cells use TGF-ß to suppress allergic responses in a mouse model of ragweed-induced asthma (Proc Natl Acad Sci U S A (2010) 107, (5652–5657) https://doi.org/10.1073/pnas.0910720107). Proc Natl Acad Sci U S A 107(17):8041. https://doi.org/10.1073/pnas.1003664107

Ota KI (2008) Fuel cells: past, present and future. IEEJ Trans Fundam Mater 128(5):329–332. https://doi.org/10.1541/ieejfms.128.329

Pan GJ, Chang ZYI, Schöler HR, Pei D (2002) Stem cell pluripotency and transcription factor Oct4. Cell Res 12(5–6):321–329. https://doi.org/10.1038/sj.cr.7290134

Peralta OA, Carrasco C, Vieytes C, Tamayo MJ, Muñoz I, Sepulveda S, Tadich T, Duchens M, Melendez P, Mella A, Torres CG (2020) Safety and efficacy of a mesenchymal stem cell intramammary therapy in dairy cows with experimentally induced Staphylococcus aureus clinical mastitis. Sci Rep 10(1):1–12. https://doi.org/10.1038/s41598-020-59724-7

Ren G, Zhang L, Zhao X, Xu G, Zhang Y, Roberts AI, Zhao RC, Shi Y (2008) Mesenchymal stem cell-mediated immunosuppression occurs via concerted action of chemokines and nitric oxide. Cell Stem Cell 2(2):141–150. https://doi.org/10.1016/j.stem.2007.11.014

Ren G, Zhao X, Zhang L, Zhang J, L'Huillier A, Ling W, Roberts AI, Le AD, Shi S, Shao C, Shi Y (2010) Inflammatory cytokine-induced intercellular adhesion molecule-1 and vascular cell adhesion molecule-1 in mesenchymal stem cells are critical for immunosuppression. J Immunol 184(5):2321–2328. https://doi.org/10.4049/jimmunol.0902023

Renzi S, Riccò S, Dotti S, Sesso L, Grolli S, Cornali M, Carlin S, Patruno M, Cinotti S, Ferrari M (2013) Autologous bone marrow mesenchymal stromal cells for regeneration of injured equine ligaments and tendons: A clinical report. Res Vet Sci 95(1):272–277. https://doi.org/10.1016/j.rvsc.2013.01.017

Rosado MM, Bernardo ME, Scarsella M, Conforti A, Giorda E, Biagini S, Cascioli S, Rossi F, Guzzo I, Vivarelli M, Dello Strologo L, Emma F, Locatelli F, Carsetti R (2014) Inhibition of B-cell proliferation and antibody production by mesenchymal stromal cells is mediated by T cells. Stem Cells Dev 24(1):93–103. https://doi.org/10.1089/scd.2014.0155

Rubin LL, Haston KM (2011) Stem cell biology and drug discovery. BMC Biol 9. https://doi.org/10.1186/1741-7007-9-42

Saalbach A, Anderegg U (2019) Thy-1: more than a marker for mesenchymal stromal cells. FASEB J 33(6):6689–6696. https://doi.org/10.1096/fj.201802224R

Sampaio RV, Chiaratti MR, Santos DCN, Bressan FF, Sangalli JR, Sá ALA, Silva TVG, Costa NN, Cordeiro MS, Santos SSD, Ambrosio CE, Adona PR, Meirelles FV, Miranda MS, Ohashi OM (2015) Generation of bovine (Bos indicus) and buffalo (Bubalus bubalis) adipose tissue derived stem cells: isolation, characterization, and multipotentiality. Genet Mol Res 14(1):53–62. https://doi.org/10.4238/2015.January.15.7

Sasaki A, Mizuno M, Mochizuki M, Sekiya I (2019) Mesenchymal stem cells for cartilage regeneration in dogs. World J Stem Cells 11(5):254–269. https://doi.org/10.4252/wjsc.v11.i5.254

Sato K, Ozaki K, Oh I, Meguro A, Hatanaka K, Nagai T, Muroi K, Ozawa K (2007) Nitric oxide plays a critical role in suppression of T-cell proliferation by mesenchymal stem cells. Blood 109(1):228–234. https://doi.org/10.1182/blood-2006-02-002246

Seita J, Weissman I (2010) Published in final edited form as: hematopoietic stem cell: self-renewal versus differentiation. Wiley Interdiscip Rev Syst BiolMed 2(5):1–20. https://doi.org/10.1002/wsbm.86.Hematopoietic

Shah K, Drury T, Roic I, Hansen P, Malin M, Boyd R, Sumer H, Ferguson R (2018) Outcome of allogeneic adult stem cell therapy in dogs suffering from osteoarthritis and other joint defects. Stem Cells Int 2018. https://doi.org/10.1155/2018/7309201

Shi X, Chen Q, Wang F (2019) Mesenchymal stem cells for the treatment of ulcerative colitis: a systematic review and meta-analysis of experimental and clinical studies. Stem Cell Res Ther 10(1):1–12. https://doi.org/10.1186/s13287-019-1336-4

Singh AM (2019) An efficient protocol for single-cell cloning human pluripotent stem cells. Front Cell Dev Biol 7(January):1–7. https://doi.org/10.3389/fcell.2019.00011

Spaggiari GM, Capobianco A, Becchetti S, Mingari MC, Moretta L (2006) Mesenchymal stem cell-natural killer cell interactions: evidence that activated NK cells are capable of killing MSCs, whereas MSCs can inhibit IL-2-induced NK-cell proliferation. Blood 107(4):1484–1490. https://doi.org/10.1182/blood-2005-07-2775

Spaggiari GM, Abdelrazik H, Becchetti F, Moretta L (2009) MSCs inhibit monocyte-derived DC maturation and function by selectively interfering with the generation of immature DCs: central role of MSC-derived prostaglandin E2. Blood 113(26):6576–6583. https://doi.org/10.1182/blood-2009-02-203943

St. Jean G, Anderson DE (2014) Decision analysis for fracture management in cattle. Vet Clin North Am Food Anim Pract 30(1):1–10. https://doi.org/10.1016/j.cvfa.2013.11.011

Stanton MM, Tzatzalos E, Donne M, Kolundzic N, Helgason I, Ilic D (2019) Prospects for the use of induced pluripotent stem cells in animal conservation and environmental protection. Stem Cells Transl Med 8(1):7–13. https://doi.org/10.1002/sctm.18-0047

Štefková K, Procházková J, Pachernfk J (2015) Alkaline phosphatase in stem cells. Stem Cells Int 2015(2006). https://doi.org/10.1155/2015/628368

Takahashi K, Yamanaka S (2006) Induction of pluripotent stem cells from mouse embryonic and adult fibroblast cultures by defined factors. Cell 126(4):663–676. https://doi.org/10.1016/j.cell.2006.07.024

Tam BYFY, Wei K, Rudge JS, Hoffman J, Holash J, Park S, Yuan J, Hefner C, Chartier C, Lee J, Jiang S, Nayak NR, Kuypers FA, Ma L, Sundram U, Wu G, Garcia JA, Schrier SL, Maher JJ, et al (2009). Corrigendum: VEGF modulates erythropoiesis through regulation of adult hepatic erythropoietin synthesis. Corrigendum: Bone marrow stromal cells attenuate sepsis via prostaglandin E 2 – dependent reprogramming of host macrophages to increase their interle. 15(4), 2009

Tan W, Proudfoot C, Lillico SG, Whitelaw CBA (2016) Gene targeting, genome editing: from Dolly to editors. Transgenic Res 25(3):273–287. https://doi.org/10.1007/s11248-016-9932-x

Tang XP, Zhang M, Yang X, Chen LM, Zeng Y (2006) Differentiation of human umbilical cord blood stem cells into hepatocytes in vivo and in vitro. World J Gastroenterol 12(25):4014–4019. https://doi.org/10.3748/wjg.v12.i25.4014

Ullah M, Liu DD, Thakor AS (2019) Mesenchymal stromal cell homing: mechanisms and strategies for improvement. IScience 15:421–438. https://doi.org/10.1016/j.isci.2019.05.004

Ungerer C, Quade-Lyssy P, Radeke HH, Henschler R, Königs C, Köhl U, Seifried E, Schüttrumpf J (2014) Galectin-9 is a suppressor of T and B cells and predicts the immune modulatory potential of mesenchymal stromal cell preparations. Stem Cells Dev 23(7):755–766. https://doi.org/10.1089/scd.2013.0335

van Thuan N, Kishigami S, Wakayama T (2010) How to improve the success rate of mouse cloning technology. J Reprod Dev 56(1):20–30. https://doi.org/10.1262/jrd.09-221A

Vassiliev I, Vassilieva S, Beebe LFS, Mcilfatrick SM, Harrison SJ, Nottle MB (2010) Development of culture conditions for the isolation of pluripotent porcine embryonal outgrowths from in vitro

produced and in vivo derived embryos. J Reprod Dev 56(5):546–551. https://doi.org/10.1262/jrd.09-197A
Voga M, Adamic N, Vengust M, Majdic G (2020) Stem cells in veterinary medicine—current state and treatment options. Front Vet Sci 7(May):1–20. https://doi.org/10.3389/fvets.2020.00278
Wall RJ, Powell AM, Paape MJ, Kerr DE, Bannerman DD, Pursel VG, Wells KD, Talbot N, Hawk HW (2005) Genetically enhanced cows resist intramammary Staphylococcus aureus infection. Nat Biotechnol 23(4):445–451. https://doi.org/10.1038/nbt1078
Wang H, Wang X, Xu X, Kyba M, Cooney AJ (2016) Germ cell nuclear factor (GCNF) represses Oct4 expression and globally modulates gene expression in human embryonic stem (hes) cells. J Biol Chem 291(16):8644–8652. https://doi.org/10.1074/jbc.M115.694208
Wang J, Liu C, Fujino M, Tong G, Zhang Q, Li X-K, Yan H (2019) Stem cells as a resource for treatment of infertility-related diseases. Curr Mol Med 19(8):539–546. https://doi.org/10.2174/1566524019666190709172636
White J, Dalton S (2005) Cell cycle control of embryonic stem cells. Stem Cell Rev 1(2):131–138. https://doi.org/10.1385/SCR:1:2:131
Wolfe JH (2009) Gene therapy in large animal models of human genetic diseases. ILAR J 50(2):107–111. https://doi.org/10.1093/ilar.50.2.107
Wolfe DF (2018) Review: abnormalities of the bull – occurrence, diagnosis and treatment of abnormalities of the bull, including structural soundness. Animal 12(s1):s148–s157. https://doi.org/10.1017/S1751731118000939
Wu G, Schöler HR (2014) Role of Oct4 in the early embryo development. Cell Regen 3(1):3:7. https://doi.org/10.1186/2045-9769-3-7
Xu J, Yu L, Guo J, Xiang J, Zheng Z, Gao D, Shi B, Hao H, Jiao D, Zhong L, Wang Y, Wu J, Wei H, Han J (2019) Generation of pig induced pluripotent stem cells using an extended pluripotent stem cell culture system. Stem Cell Res Ther 10(1):1–16. https://doi.org/10.1186/s13287-019-1303-0
Yang SH, Park MJ, Yoon IH, Kim SY, Hong SH, Shin JY, Nam HY, Kim YH, Kim B, Park CG (2009) Soluble mediators from mesenchymal stem cells suppress T cell proliferation by inducing IL-10. Exp Mol Med 41(5):315–324. https://doi.org/10.3858/emm.2009.41.5.035
Yu J, Thomson JA (2014) Embryonic stem cells: derivation and properties In: Essentials of stem cell biology, 3rd edn. Elsevier Inc. https://doi.org/10.1016/B978-0-12-409503-8.00027-5
Zhao W, Ji X, Zhang F, Li L, Ma L (2012) Embryonic stem cell markers. Molecules 17(6):6196–6236. https://doi.org/10.3390/molecules17066196
Zhu Z, Huangfu D (2013) Human pluripotent stem cells: an emerging model in developmental biology. Development (Cambridge) 140(4):705–717. https://doi.org/10.1242/dev.086165

Chapter 5
Metabolomics and Proteomics Signatures in Feed-Efficient Beef and Dairy Cattle

Ahmed A. Elolimy, Mohamed Zeineldin, Mohamed Abdelmegeid, Alzahraa M. Abdelatty, Abdulrahman S. Alharthi, Mohammed H. Bakr, Mona M. M. Y. Elghandour, Abdelfattah Z. M. Salem, and Juan J. Loor

Abstract Feed accounts for 40–60% of total expenses of beef and dairy cattle production costs. Therefore, feed-efficient cattle have a great potential to reduce production costs without compromising meat or milk production levels, resulting in a greater profit margin for producers. Many approaches for measuring feed efficiency are available with residual feed intake being one of the most common. The residual feed intake is defined as the difference between actual dry matter intake and expected dry matter intake based on animal size and production level. Therefore, compared with a least-efficient animal, the most-efficient animal would have a negative resid-

A. A. Elolimy (✉)
Department of Animal Production, National Research Centre, Giza, Egypt

M. Zeineldin
Department of Animal Medicine, College of Veterinary Medicine, Benha University, Banha, Egypt

M. Abdelmegeid
Department of Animal Medicine, College of Veterinary Medicine Kafr-Elsheikh University, Kafrelsheikh, Egypt

A. M. Abdelatty
Department of Nutrition and Clinical Nutrition, Faculty of Veterinary Medicine, Cairo University, Giza, Egypt

A. S. Alharthi
Department of Animal Production, College of Food and Agriculture Sciences, King Saud University, Riyadh, Saudi Arabia

M. H. Bakr
Animal Production Department, Faculty of Agriculture, Cairo University, Giza, Egypt

Mona M. M. Y. Elghandour · A. Z. M. Salem (✉)
Facultad de Medicina Veterinaria y Zootecnia, Universidad Autonoma del Estado de Mexico, Toluca, Mexico

J. J. Loor
Department of Animal Sciences, University of Illinois, Urbana, IL, USA

V. K. Yata et al. (eds.), *Sustainable Agriculture Reviews 54*, Sustainable Agriculture Reviews 54, https://doi.org/10.1007/978-3-030-76529-3_5

ual feed intake coefficient value, indicating that it consumed less dry matter intake while maintaining the same level of production. Recent studies have focused on investigating changes in key metabolites and proteins that would shift metabolic pathways to support better feed efficiency. Recent reports highlighted that in most-efficient cattle metabolic pathways associated with energy, vitamins, and amino acid metabolism in rumen and skeletal muscle are upregulated to provide extra energy, thus, allowing for a similar level of production despite lower dry matter intake. Other studies demonstrated that most-efficient cattle reduce protein turnover in skeletal muscle including upregulation of key protein synthesis pathways, such as mechanistic target of rapamycin signaling, and the downregulation of key proteins in protein degradation such as ubiquitin-proteasome pathway, resulting in greater protein deposition in muscle. In this chapter, we discuss applications of novel comprehensive techniques for protein and metabolite profiling in rumen, intestine, blood, liver, and skeletal muscle to elucidate adaptive biological functions that support better feed efficiency in beef and dairy cattle.

Keywords RFI · Cow · Calves · Metabolomics · Proteomics · Rumen · Blood · Liver · Muscles · Hindgut

5.1 Introduction

Feed costs in the beef and dairy cattle industry are the most expensive inputs, and represent on average 40–60% of total expenses (Montaño-Bermudez et al. 1990). Therefore, enhancing feed efficiency would dramatically decrease overall costs, raise producer profitability, and increase animal protein availability for consumers (Clemmons et al. 2020). Hence, discovering robust biomarkers for selecting the most-efficient beef and dairy cattle is crucial.

Residual feed intake (RFI) is a commonly-used measurement of feed efficiency in beef and dairy cattle (Li et al. 2020; Zhang et al. 2020). The RFI is defined as the divergence of predicted dry matter intake (DMI) for maintenance and meat or milk production from the actual DMI after adjusting DMI for the level of production through a linear regression model (Xi et al. 2016). The predicted DMI is calculated as a function of changes in body weight (BW) and production level (Potts et al. 2017). The regression model defines which animal is below (negative) or above (positive) the predicted DMI (Durunna et al. 2011). The fact that RFI is a trait independent of body size and production level renders it a reliable measurement of feed efficiency (Gomes et al. 2012). Most-efficient (M-eff) cattle, i.e. with the favorable negative RFI coefficient, utilize less DMI than predicted to cover both maintenance and production requirements (Potts et al. 2017). Therefore, M-eff cattle are biologically- and economically-efficient compared with their least-efficient (L-eff) counterparts having an undesirable positive RFI coefficient (Gomes et al. 2012; Lawrence et al. 2013). Despite the proven value of the RFI, the underlying biology associated with this trait is still not well known, hence, supporting the use of modern technologies in an effort to uncover putative biomarkers.

Metabolomics profiling provides a novel approach for rapidly-identifying M-eff animals. These techniques focus on detecting, identifying, and quantifying available free metabolites in a given biological sample (Fontanesi 2016). The origin of these metabolites could be either endogenous (i.e. derived from the animal) or xenobiotic (i.e. metabolites from plants or microbes) (Fontanesi 2016). Metabolites are intermediates or products of metabolic pathways (Clemmons et al. 2020) involved in energy, protein, and vitamin metabolism, all of which are of particular interest for feed efficiency divergence in cattle (Nafikov and Beitz 2007; Ferrell and Jenkins 1984). Liquid chromatography-mass spectrometry (LC-MS) techniques have been used to detect differences in ruminal, intestinal and circulating metabolite abundance between beef and dairy cattle divergent in RFI (Clemmons et al. 2017; Elolimy et al. 2020), with several metabolites identified as having a key contribution to better feed efficiency (summarized in Table 5.1). These alterations in metabolome profiles between M-eff and L-eff cattle could be used as predictive biomarkers for feed efficiency (Clemmons et al. 2019b; Artegoitia et al. 2017; Novais et al. 2019).

The study of protein profiles in a biosample collected at a certain physiological state is called "proteome" (Reinhardt et al. 2012).Proteomics data provide a wealth of information that gene expression analysis could not. Therefore, proteomics analyses provide a better understanding of the biological functions in cattle. Few studies in recent years have applied proteomic approaches to unmask alterations in protein profiles in blood and skeletal muscle associated with RFI divergence that would help our understanding of the biology behind this trait (summarized in Table 5.2).

5.2 Metabolomic Signature in Feed-Efficient Cattle

In a recent study, Clemmons et al. (2020) used DionexUltiMate 3000 ultra-high performance liquid chromatography (UHPLC) system with an Exactive Plus Orbitrap MS to identify the metabolic signature in ruminal fluid between M-eff and L-eff beef steers. Authors used 50 purebred Angus steers at the age of 7 months and 264 ± 2.7 kg of BWfed individually using the GrowSafe System for 70 days to evaluate RFI divergent groups. At day 70 of the trial, the authors selected the extreme M-eff (n = 14) and the top L-eff (n = 15) steers. Rumimal fluid samples were collected on day 70 of the study using stomach tubing. Results revealed 8 DEM between M-eff and L-eff steers, with M-eff steers having greater succinate. In the rumen, bacteria such as *Selenomasruminantium* metabolize succinate in the presence of several bacterial enzymes and coenzyme A (CoA) to propionate (Wirth et al. 2018), a vital volatile fatty acid (VFA) used in the liver for gluconeogenesis (Wirth et al. 2018). Therefore, the greater ruminal succinate in M-eff group indicates a better capacity for hepatic gluconeogenesis to maintain similar meat

Table 5.1 Metabolomics studies in residual feed intake (RFI) divergent beef and dairy cattle

Breed type	Breed name	Sex	Age (months)	Average body weight (kg)	Sample type	Approach and platform	Changes in M-eff	References
Beef	Angus	Steers	7	264	Rumen fluid	Untargeted meabolomics (UHPLC-MS)	↑ succinate ↑ Uracil ↑ Thymine ↑ Hypoxanthine ↑ Pyridoxic acid ↑ Citraconate ↑ Xylose	Clemmons et al. (2020)
Beef	Nellore	Bulls	16-20	376	Blood (serum)	Untargeted meabolomics (LC-MS)	↓ Retinal ↓ Progesterone ↓ Stearic acid ↓ Vomifoliol, 2,3 ↓ Dihydroflavone ↓ Limonoate ↓ Phytanic acid	Novais et al. (2019)
Beef	Angus	Steers	7	264	Blood (serum)	Untargeted meabolomics (LC-MS)	↑ Pantothenate	Clemmons et al. (2019a, b)
Beef	Nellore	Cows	36	484	Blood	Targeted metabolites (enzymatic kits)	↑ Cholesterol	Broleze et al. (2020)
Beef	Nellore	Bulls	7	207	Blood (serum)	Targeted metabolites (enzymatic kits)	↓ Cortisol	Bonilha et al. (2017)
Dairy	Holstein	Heifers	At birth	43	Feces	Untargeted meabolomics (LC-MS)	↑ Cholesterol ester ↑ Biotin ↑ L-Tryptophan	Elolimy et al. (2020)
Dairy	Holstein	Cows	584	Lactating	Blood (serum)	Targeted metabolites (enzymatic kits)	↑ NPY ↓ Leptin ↓ NEFA	Xi et al. (2016)
Dairy	Murrah buffalo	Calves	4-6	70	Blood (plasma)	Targeted metabolites (enzymatic kits)	↑ IGF-1 ↑ Triiodothyronine T3 ↓ Thyroxin	Sharma et al. (2016)

Table 5.2 Proteomics studies in residual feed intake (RFI) divergent beef and dairy cattle

Breed type	Breed name	Sex	Age (months)	Average body weight (kg)	Sample type	Approach and platform	Changes in M-eff	References
Beef	Red Angus	Steers and heifers	Finishing stage	836	Ruminal epithelium	Targeted proteins (western blot)	↑ p-EEF2K ↑ p-EEF2K:EEF2K ↑ p-EIF2A:EIF2A ↓ UBA1 ↓ NEDD4 ↓ STUB1 ↓ MDM2	Elolimy et al. (2019)
Beef	Nellore	Bulls	7	239	Skeletal muscles	2D-PAGE	↑ HSPB1 ↓ 14-3-3 epsilon	Carvalho et al. (2019)
Beef	Nellore	Bulls	24-26	557	Liver	2D-PAGE	↓ HBB ↓ Aldehyde dehydrogenase ↓ Aspartate aminotransferase ↓ Glycine amidinotransferase	Baldassini et al. (2018)

production levels relative to L-eff animals. In support of this notion, Myer et al. (2015) reported that M-eff cattle had more succinate- and propionate-producing bacteria in the rumen such as *Succiniclasticum spp*. Similarly, M-eff steers had a greater concentrations of pantothenate in the serum, a precursor of CoA (Clemmons et al. 2017). These results indicate greater hepatic energy production in in M-eff steers, which likely contribute to maintain growth in M-eff steers despite lower DMI (Fan et al. 2015).

Clemmons et al. (2020) also reported that M-eff steers had more abundant nucleic acids and nucleic acid derivatives in the rumen including uracil, thymine, and hypoxanthine, indicating an increased production of microbial protein in M-eff steers (Leng and Nolan 1984). Additionally, Clemmons et al. (2020) highlighted that M-eff steers had greater pyridoxic acid, a byproduct of vitamin B6 catabolism (Linkswiler and Reynolds 1950), suggesting better protein and muscle accretion in the M-eff group (Clemmons et al. 2020). Other key metabolites associated with carbohydrate metabolism such as citraconate was more abundant in ruminal fluid in M-eff steers (Clemmons et al. 2020). Citraconate is a metabolite generated through TCA cycle activity suggesting higher energy production took place in the rumen of M-eff steers (Clemmons et al. 2020). Collectively, data indicate that M-eff cattle have greater capacity for energy production in the rumen and liver to maintain similar growth performance despite lower DMI.

A recent study useduntargeted metabolomics via liquid chromatography-mass spectrometry (LC-MS) to uncover differencs in serum metabolomic profiles between young M-eff and L-eff Nellore bulls (Novais et al. 2019). In this study, serum samples from 98 Nellore bulls at 16–20 months of age and 376 ± 29 kg BW were collected 21 days before the start of a 70 day RFI evaluation period (Novais et al. 2019). Authors detected 7808 DEM between M-eff and L-eff groups (Novais et al. 2019). Seven metabolites had lower concentrations in the M-eff group including retinal, progesterone, stearic acid, vomifoliol, 2,3 dihydroflavone, limonoate and phytanic acid (Novais et al. 2019). Retinal is involved in the retinol pathway, previously reported to be downregulated in feed-efficient beef cattle (de Almeida Santana et al. 2016). Similarly, progesterone (a key metabolite in steroid hormone biosynthesis) was suppressed in the liver of M-eff Jersey cows (Salleh et al. 2017). Interestingly, this study revealed metabolites exclusively produced by bacteria or plants inclduing vomifoliol, 2,3 dihydroflavone, limonoate and phytanic acid that were lower in the M-eff group (Novais et al. 2019), likely due to lower DMI.

Clemmons et al. (2019a) conducted a study to discover differences in serum metabolome between M-eff and L-eff beef cattle. In this study, they used LC-MS analysis for untargeted metabolomics of serum samples collected from M-eff (n = 14) and L-eff (n = 15) weaned Angus steers at 7 months old and 264 ± 2.7 kg of BW. The GrowSafe system was used to monitor individual DMI for each steer during the 70-day RFI trial (Clemmons et al. 2019a). Weekly, 9 mL of blood samples were collected via venipuncture from the coccygeal vein to separate serum. Results indicated that pantothenate was greater in the M-eff group (Clemmons et al. 2019a). Pantothenate, a substrate for CoA synthesis, is produced by ruminal bacteria such as Flavobacteriiathen absorbed via ruminal epithelium to reach the circulation (Clemmons et al. 2019a). Interestingly, Clemmons et al. (2019a)reported that Flavobacteriia were more abundant in M-eff steers, a result that is in line with previous studies revealing a better capacity for energy production from lower DMI in M-eff cattle.

Using commercial enzymatic kits, Broleze et al. (2020) evaluated differences in targeted metabolites in blood between M-Eff and L-eff beef cows. In this study, DMI of 53 primiparous Nellore beef cows at 36.8 ± 1.23 months of age and 484 ± 40.9 kg of BW was monitored individually using the GrowSafe System for 168 days between 22 and 190 days in milk (DIM; early-tomid-lactation stage) to calculate RFI coefficients for each cow (Broleze et al. 2020). Blood samples were collected from all animals for analysis of glucose, cholesterol, triglycerides, andß-hydroxybutyrate (Broleze et al. 2020). Cholesterol was the only metabolite that differed in concentrationse, being greater in M-eff beef cows (204 mg/dL vs. 192 mg/dL) (Broleze et al. 2020). Because another study reported increased plasma cholesterol in feed-restricted dairy cows (Gross et al. 2015), authors suggested that lower DMI in M-eff cows (consumed 11.5% DMI) partly explaind the response observed (Broleze et al. 2020). In another study, Bonilha et al. (2017) used commercial enzymatic kits to evaluate differences in specific serum metabolites including insulin, non-esterified fatty acids (NEFA), insulin-like growth factor I (IGF-I), and cortisol between M-eff (n = 13) and L-eff (n = 12) Nellore bulls at 210 days of age and 207 kg BW monitored for individual DMI for 70 days. Cortisol, a biomarker of stress, was lower in M-eff bulls indicating a lower degree of systemic stresss (Bonilha et al. 2017) and providing support to previous studies reporting lower circulating cortisol in M-eff cattle (Gomes et al. 2013; Richardson et al. 2004).

In a recent study, Elolimy et al. (2020) detected shifts in hindgut metabolomics profiles between M-eff and L-eff dairy preweaned calves. In this study, DMI in 26 neonatal Holstein heifer calves was individually monitored from birth to weaning at 42 days of age. Calves were retrospectively classified into two groups: M-eff (n = 13) and L-eff (n = 13) heifers based on individual RFI coefficient (Elolimy et al. 2020). Fecal samples were collected every two weeks throughout the study to perform untargeted metabolomics using an LC-MS approach (Elolimy et al. 2020). At birth, M-eff calves had an enrichment of metabolites belonging to energy-producing pathways including pyruvate metabolism, gluconeogenesis, TCA cycle, and biotin suggesting a greater availability of energy for growth and development during the preweaning period (Akram 2014; Vailati-Riboni et al. 2016; Elolimy et al. 2020). Furthermore, the M-eff group upregulated vitamin (biotin metabolism), fatty acid (arachidonic acid metabolism), and amino acid (alanine metabolism) related pathways that would likely enhance gut function and development (Elolimy et al. 2020; León-Del-Río 2019). During the preweaning period, Elolimy et al. (2020) demonstrated that M-eff calves had greater supply of B vitamins in the hindgut including vitamins B6, B7 (biotin) and B9 (folate). Vitamin B6 is essential for metabolism of fatty acids, amino acids, and glucose (Rodriguez-Melendez and Zempleni 2003). Vitamin B7 is important for mucosal immune responses (Jenkins et al. 2017). M-eff calves also had greater capacity for metabolism of amino acids such as tyrosine, tryptophan, and phenylalanine (Elolimy et al. 2020) likely contributing to the similar growth achieved (Elolimy et al. 2020).

Another study measured the concentrations of leptin, prolactin, neuropeptide Y (NPY), insulin-like growth factor 1 (IGF-1), ghrelin, insulin, β-hydroxybutyrate, glucose, NEFA and growth hormone (GH) in serum between M-eff and L-eff dairy cows (Xi et al. 2016). The authors selected 29 lactating Holstein cows from a total of 84 based on their RFI coefficients to end up with two groups including M-eff (n = 15) and L-eff (n = 14) dairy cows (Xi et al. 2016). Blood samples were collected from all 29 cows through jugular venipuncture on day 1, 25 and 50 during the feeding period (Xi et al. 2016). The authors reported no differences in serum prolactin, IGF-1, ghrelin, insulin, β-hydroxybutyrate, glucose and GH (Xi et al. 2016). However, M-eff cows had greater concentration of NPY and lower leptin and NEFA (Xi et al. 2016) indicating lower propensity for fat mobilization, suggesting that M-eff cows likely had sufficient energy supply and experienced a lesser degree of negative energy balance.

Sharma et al. (2016) calculated RFI coefficients for 18 growing male Murrah buffalo calves at 4–6 months old and 70 ± 1.0 kg of BW after 99 days of a feeding trail, resulting in M-eff (n = 7) and L-eff calves (n = 11). Blood samples were collected at the start and end of the feeding trial by venipuncture of the anterior *vena cava* to evaluate plasma content of IGF-1, GH, creatinine, insulin, albumin, hydroxyproline, triio-dothyronine (T3), thyroxin (T4), and total protein using commercial kits (Sharma et al. 2016). No differences in plasma concentrations of creatinine insulin, albumin, hydroxysproline and total protein were detected (Sharma et al. 2016). However, M-eff calves had greater plasma IGF-1, and T3, but lower T4 (Sharma et al. 2016).

Overall, the above findings of alterations in several metabolites associated with better feed efficiency in beef and dairy cattle provide a list of robust biomarkers that could be studied in the future for their potential as physiological indicators predictive of feed-efficient cattle.

5.3 Proteomics Signature in Feed-Efficient Cattle

Elolimy et al. (2019) investigated changes in ruminal epithelium protein abundance between M-eff (n = 6) and L-eff (n = 6) Red Angus heifers and steers using the western blot approach. In this study, Elolimy et al. (2019) evaluated 29 proteins involved in protein synthesis (MTOR signaling) and degradation (ubiquitin-proteasome pathways) in ruminal epithelium collected after slaughter at the end of 70 days of a feeding trail. The M-eff group had greater abundance of proteins crucial for cellular protein synthesis such as phosphorylated eukaryotic elongation factor 2 kinase (p-EEF2K), phosphorylated eukaryotic elongation factor 2 kinase:total eukaryotic elongation factor 2 kinase (p-EEF2K:EEF2K, and phosphorylated eukaryotic translation initiation factor 2A:total eukaryotic translation initiation factor 2A (p-EIF2A:EIF2A)(Elolimy et al. 2019). On the other hand, M-eff cattle had lower abundance of proteins involved in protein degradation pathways such as total ubiquitin like modifier activating enzyme 1 (UBA1), total neural precursor cell expressed, developmentally downregulated 4, E3 ubiquitin protein ligase (NEDD4), total STIP1 homology and U-box containing protein 1 (STUB1), and total MDM2 proto-oncogene (MDM2) (Elolimy et al. 2019). No differences were detected in plasma insulin and ruminal epithelium insulin signaling proteins (Elolimy et al. 2019). These data indicated that M-eff beef cattle have a greater rate of protein synthesis relative to protein degradation in ruminal epithelium. These changes likely result in better growth of ruminal epithelium to absorb more VFA produced from the anaerobic microbial fermentation of plant fiber in M-eff beef cattle.

Carvalho et al. (2019) employed a proteomics approach to unmask differences in key proteins associated with energy metabolism in skeletal muscle of RFI divergent beef cattle. Daily DMI was recorded for 129 young Nellore bulls at 7 months old and 239 ± 30.1 kg of BW during 98 days of RFI evaluation period (Carvalho et al. 2019). At the end of the study, Carvalho et al. (2019) selected 9 bulls for M-eff group and another 9 bulls for L-eff group. After slaughter, longissimus muscle was sampled for protein profiling using a two-dimensional electrophoresis (2D-PAGE) with mass spectrometry (ESI-MS) (Carvalho et al. 2019). Heat shock protein beta 1 (HSPB1), a key protein for cellular development and differentiation (Zhang et al. 2014; Carvalho et al. 2014) and inhibitor of protein degradation in muscle fibers, was greater in the M-eff group (Carvalho et al. 2019) suggesting greater protein synthesis. Therefore, this adaptation likely contributed to better feed efficiency in M-eff cattle through decrease protein turnover in skeletal muscle.

Another proteomics study conducted by Baldassini et al. (2018) used 2D-PAGE and ESI-MS techniques to profile hepatic proteins in RIF divergent Nellore bulls. In this study, Baldassini et al. (2018) used 18 Nellore bulls at 24–26 months of age during the finishing period (M-eff = 9 and L-eff = 9). After slaughter, liver samples were collected from the 18 animals for protein extraction and proteomic profiling (Baldassini et al. 2018). Results indicated that hemoglobin subunit beta protein (HBB) was downregulated in the M-eff group (Baldassini et al. 2018) likely from lower numbers of red blood cells since hemoglobin binds to oxygen to form oxyhemoglobin inside the red blood cells (Hsia 1998). The data indicated a need to determine blood hemoglobin concentrations between M-eff and L-eff groups in future studies. Baldassini et al. (2018) reported that the M-eff group had lower abundance of oxidative stress-associated proteins such as aldehyde dehydrogenase, aspartate aminotransferase, and glycine amidino transferase proteins, highlighting a lower degree of hepatic oxidative stress and less reactive oxygen species (ROS) content in liver of feed-efficient cattle.

Davis et al. (2016) investigated differences in oxygen uptake by mitochondria in muscle and respiratory chain proteins between M-eff and L-eff beef cattle. They calculated RFI coefficients in 92 Hereford-crossbreed steers in63-day feeding period using the individual feed intake system. The top 10 M-eff and top 8 L-eff steers based on RFI coefficient ranking were used for subsequent analysis. Mitochondrial complex I (CI), II (CII), and III (CIII) protein concentration in M-eff and L-eff groups was assessed using bicinchoninic acid colorimetric procedures.

Lymphocytes were isolated from blood samples collected via jugular venipuncture from both groups (Davis et al. 2016). Results indicated that M-eff steers did not differ in CI, CII, and CIII protein concentration. Therefore, mitochondrial proteins do not seem to play a key role in the RFI divergence between M-eff and L-eff steers.

5.4 Conclusions

Comprehensive metabolome and proteome studies revealed associations between RFI divergent cattle and shifts in metabolome/proteome profiles that might explain the biology behind superior feed efficiency in cattle. Overall, M-eff cattle are characterized by metabolite and protein profiles that would provide extra energy and nutrients to help maintain similar levels of production despite a lower DMI. Further studies are warranted to expand our understanding of the biological contribution of key metabolites and proteins to RFI divergence in cattle at different production stages. Additionally, the relationship between metabolome/proteome profiling among different tissue in the same individual animals such as rumen, blood, skeletal muscles and milk should be examined in order to provide a more holistic overview into feed-efficient cattle.

References

Akram M (2014) Citric acid cycle and role of its intermediates in metabolism. Cell Biochem Biophys 68(3):475–478. https://doi.org/10.1007/s12013-013-9750-1

Artegoitia VM, Foote AP, Lewis RM, Freetly HC (2017) Rumen fluid metabolomics analysis associated with feed efficiency on crossbred steers. Sci Rep 7(1):2864. https://doi.org/10.1038/s41598-017-02856-0

Baldassini WA, Bonilha SFM, Branco RH, Vieira JCS, Padilha PM, Lanna DPD (2018) Proteomic investigation of liver from beef cattle (Bos indicus) divergently ranked on residual feed intake. Mol Biol Rep 45(6):2765–2773. https://doi.org/10.1007/s11033-018-4341-2

Bonilha SFM, Branco RH, Mercadante MEZ, Dos Santos Goncalves Cyrillo JN, Monteiro FM, Ribeiro EG (2017) Digestion and metabolism of low and high residual feed intake Nellore bulls. Tropl Anim Health Prod 49(3):529–535. https://doi.org/10.1007/s11250-017-1224-9

Broleze DF, Souza LL, Zorzetto MF, Savegnago RP, Negrao JA, Bonilha SFM, Mercadante MEZ (2020) Feed efficiency and maternal productivity of Bos indicus beef cows. PLoS One 15(6):e0233926. https://doi.org/10.1371/journal.pone.0233926

Carvalho ME, Gasparin G, Poleti MD, Rosa AF, Balieiro JCC, Labate CA, Nassu RT, Tullio RR, LCdA R, Mourão GB, Coutinho LL (2014) Heat shock and structural proteins associated with meat tenderness in Nellore beef cattle, a Bos indicus breed. Meat Sci 96(3):1318–1324. https://doi.org/10.1016/j.meatsci.2013.11.014

Carvalho EB, Gionbelli MP, Rodrigues RTS, Bonilha SFM, Newbold CJ, Guimaraes SEF, Silva W, Verardo LL, Silva FF, Detmann E, Duarte MS (2019) Differentially expressed mRNAs, proteins and miRNAs associated to energy metabolism in skeletal muscle of beef cattle identified for low and high residual feed intake. BMC Genomics 20(1):501. https://doi.org/10.1186/s12864-019-5890-z

Clemmons BA, Mihelic RI, Beckford RC, Powers JB, Melchior EA, McFarlane ZD, Cope ER, Embree MM, Mulliniks JT, Campagna SR, Voy BH, Myer PR (2017) Serum metabolites associated with feed efficiency in black angus steers. Metabolomics 13(12):147. https://doi.org/10.1007/s11306-017-1282-z

Clemmons BA, Martino C, Powers JB, Campagna SR, Voy BH, Donohoe DR, Gaffney J, Embree MM, Myer PR (2019a) Rumen Bacteria and serum metabolites predictive of feed efficiency phenotypes in beef cattle. Sci Rep 9(1):19265. https://doi.org/10.1038/s41598-019-55978-y

Clemmons BA, Martino C, Schneider LG, Lefler J, Embree MM, Myer PR (2019b) Temporal stability of the ruminal bacterial communities in beef steers. Sci Rep 9(1):9522. https://doi.org/10.1038/s41598-019-45995-2

Clemmons BA, Powers JB, Campagna SR, Seay TB, Embree MM, Myer PR (2020) Rumen fluid metabolomics of beef steers differing in feed efficiency. Metabolomics 16(2):23. https://doi.org/10.1007/s11306-020-1643-x

Davis MP, Brooks MA, Kerley MS (2016) Relationship between residual feed intake and lymphocyte mitochondrial complex protein concentration and ratio in crossbred steers. J Anim Sci 94(4):1587–1591. https://doi.org/10.2527/jas.2015-9843

de Almeida Santana MH, Junior GA, Cesar AS, Freua MC, da Costa GR, da Luz ESS, Leme PR, Fukumasu H, Carvalho ME, Ventura RV, Coutinho LL, Kadarmideen HN, Ferraz JB (2016) Copy number variations and genome-wide associations reveal putative genes and metabolic pathways involved with the feed conversion ratio in beef cattle. J Appl Genet 57(4):495–504. https://doi.org/10.1007/s13353-016-0344-7

Durunna ON, Mujibi FD, Goonewardene L, Okine EK, Basarab JA, Wang Z, Moore SS (2011) Feed efficiency differences and reranking in beef steers fed grower and finisher diets. J Anim Sci 89(1):158–167. https://doi.org/10.2527/jas.2009-2514

Elolimy AA, Abdel-Hamied E, Hu L, McCann JC, Shike DW, Loor JJ (2019) Rapid communication: residual feed intake in beef cattle is associated with differences in protein turnover and nutrient transporters in ruminal epithelium. J Anim Sci 97(5):2181–2187. https://doi.org/10.1093/jas/skz080

Elolimy A, Alharthi A, Zeineldin M, Parys C, Loor JJ (2020) Residual feed intake divergence during the preweaning period is associated with unique hindgut microbiome and metabolome profiles in neonatal Holstein heifer calves. J Animal Sci & Biotechnol 11:13. https://doi.org/10.1186/s40104-019-0406-x

Fan H, Wu Y, Zhou X, Xia J, Zhang W, Song Y, Liu F, Chen Y, Zhang L, Gao X, Gao H, Li J (2015) Pathway-based genome-wide association studies for two meat production traits in Simmental cattle. Sci Rep 5:18389. https://doi.org/10.1038/srep18389

Ferrell CL, Jenkins TG (1984) Energy utilization by mature, nonpregnant, nonlactating cows of different types. J Anim Sci 58(1):234–243. https://doi.org/10.2527/jas1984.581234x

Fontanesi L (2016) Metabolomics and livestock genomics: insights into a phenotyping frontier and its applications in animal breeding. Anim Front 6(1):73–79. https://doi.org/10.2527/af.2016-0011

Gomes RC, Sainz RD, Silva SL, César MC, Bonin MN, Leme PR (2012) Feedlot performance, feed efficiency reranking, carcass traits, body composition, energy requirements, meat quality and calpain system activity in Nellore steers with low and high residual feed intake. Livest Sci 150(1):265–273. https://doi.org/10.1016/j.livsci.2012.09.012

Gomes RC, Sainz RD, Leme PR (2013) Protein metabolism, feed energy partitioning, behavior patterns and plasma cortisol in Nellore steers with high and low residual feed intake. Rev Bras Zootec 42:44–50

Gross JJ, Kessler EC, Albrecht C, Bruckmaier RM (2015) Response of the cholesterol metabolism to a negative energy balance in dairy cows depends on the lactational stage. PLoS One 10(6):e0121956. https://doi.org/10.1371/journal.pone.0121956

Hsia CC (1998) Respiratory function of hemoglobin. N Engl J Med 338(4):239–247. https://doi.org/10.1056/nejm199801223380407

Jenkins TP, Rathnayaka Y, Perera PK, Peachey LE, Nolan MJ, Krause L, Rajakaruna RS, Cantacessi C (2017) Infections by human gastrointestinal helminths are associated with changes in faecal microbiota diversity and composition. PLoS One 12(9):e0184719. https://doi.org/10.1371/journal.pone.0184719
Lawrence P, Kenny DA, Earley B, McGee M (2013) Intake of conserved and grazed grass and performance traits in beef suckler cows differing in phenotypic residual feed intake. Livest Sci 152(2):154–166. https://doi.org/10.1016/j.livsci.2012.12.024
Leng RA, Nolan JV (1984) Nitrogen metabolism in the rumen. J Dairy Sci 67(5):1072–1089. https://doi.org/10.3168/jds.S0022-0302(84)81409-5
León-Del-Río A (2019) Biotin in metabolism, gene expression, and human disease. J Inherit Metab Dis 42(4):647–654. https://doi.org/10.1002/jimd.12073
Li B, VanRaden PM, Guduk E, O'Connell JR, Null DJ, Connor EE, VandeHaar MJ, Tempelman RJ, Weigel KA, Cole JB (2020) Genomic prediction of residual feed intake in US Holstein dairy cattle. J Dairy Sci 103(3):2477–2486. https://doi.org/10.3168/jds.2019-17332
Linkswiler H, Reynolds MS (1950) Urinary and fecal elimination of B6 and 4-pyridoxic acid on three levels of intake. J Nutr 41(4):523–532. https://doi.org/10.1093/jn/41.4.523
Montaño-Bermudez M, Nielsen MK, Deutscher GH (1990) Energy requirements for maintenance of crossbred beef cattle with different genetic potential for milk. J Anim Sci 68(8):2279–2288. https://doi.org/10.2527/1990.6882279x
Myer PR, Smith TP, Wells JE, Kuehn LA, Freetly HC (2015) Rumen microbiome from steers differing in feed efficiency. PLoS One 10(6):e0129174. https://doi.org/10.1371/journal.pone.0129174
Nafikov RA, Beitz DC (2007) Carbohydrate and lipid metabolism in farm animals. J Nutr 137(3):702–705. https://doi.org/10.1093/jn/137.3.702
Novais FJ, Pires PRL, Alexandre PA, Dromms RA, Iglesias AH, Ferraz JBS, Styczynski MP, Fukumasu H (2019) Identification of a metabolomic signature associated with feed efficiency in beef cattle. BMC Genomics 20(1):8. https://doi.org/10.1186/s12864-018-5406-2
Potts SB, Boerman JP, Lock AL, Allen MS, VandeHaar MJ (2017) Relationship between residual feed intake and digestibility for lactating Holstein cows fed high and low starch diets. J Dairy Sci 100(1):265–278. https://doi.org/10.3168/jds.2016-11079
Reinhardt TA, Lippolis JD, Nonnecke BJ, Sacco RE (2012) Bovine milk exosome proteome. J Proteomics 75(5):1486–1492. https://doi.org/10.1016/j.jprot.2011.11.017
Richardson EC, Herd RM, Archer JA, Arthur PF (2004) Metabolic differences in Angus steers divergently selected for residual feed intake. Aust J Exp Agric 44(5):441–452
Rodriguez-Melendez R, Zempleni J (2003) Regulation of gene expression by biotin (review). J Nutr Biochem 14(12):680–690. https://doi.org/10.1016/j.jnutbio.2003.07.001
Salleh MS, Mazzoni G, Höglund JK, Olijhoek DW, Lund P, Løvendahl P, Kadarmideen HN (2017) RNA-Seq transcriptomics and pathway analyses reveal potential regulatory genes and molecular mechanisms in high- and low-residual feed intake in Nordic dairy cattle. BMC Genomics 18(1):258. https://doi.org/10.1186/s12864-017-3622-9
Sharma VK, Kundu SS, Prusty S, Datt C, Kumar M (2016) Nutrient utilisation, growth performance and blood metabolites in Murrah buffalo calves (Bubalus bubalis) divergently selected for residual feed intake. Arch Anim Nutr 70(6):455–469. https://doi.org/10.1080/1745039X.2016.1233678
Vailati-Riboni M, Meier S, Burke CR, Kay JK, Mitchell MD, Walker CG, Crookenden MA, Heiser A, Rodriguez-Zas SL, Roche JR, Loor JJ (2016) Prepartum body condition score and plane of nutrition affect the hepatic transcriptome during the transition period in grazing dairy cows. BMC Genomics 17(1):854. https://doi.org/10.1186/s12864-016-3191-3
Wirth R, Kádár G, Kakuk B, Maróti G, Bagi Z, Szilágyi Á, Rákhely G, Horváth J, Kovács KL (2018) The planktonic core microbiome and core functions in the cattle rumen by next generation sequencing. Front Microbiol 9:2285. https://doi.org/10.3389/fmicb.2018.02285

Xi YM, Wu F, Zhao DQ, Yang Z, Li L, Han ZY, Wang GL (2016) Biological mechanisms related to differences in residual feed intake in dairy cows. Animal: Int J Animal Biosci 10(8):1311–1318. https://doi.org/10.1017/S1751731116000343

Zhang Q, Lee HG, Kang SK, Baik M, Choi YJ (2014) Heat-shock protein beta 1 regulates androgen-mediated bovine myogenesis. Biotechnol Lett 36(6):1225–1231. https://doi.org/10.1007/s10529-014-1489-2

Zhang Y, Li F, Chen Y, Wu H, Meng Q, Guan LL (2020) Metatranscriptomic profiling reveals the effect of breed on active rumen eukaryotic composition in beef cattle with varied feed efficiency. Front Microbiol 11:367. https://doi.org/10.3389/fmicb.2020.00367

Chapter 6
Biotechnological Applications in Dairy Products and Safety

Ayushi Kapoor, Monica Yadav, Aparna Verma, and Kiran Ambatipudi

Abstract From harvest to distribution, the food industry's application is as old as the industry itself. Conventional technologies such as pasteurization, homogenization, and direct microscopic count are well established and practiced mainly in the dairy industries. However, there is a need to evolve these technologies due to food security, changing consumer needs, environmental concerns, economic viability, and policy reforms. Apart from this, today's food industry is also facing several challenges related to food safety, preservation, nutrition, and allergies. Therefore, the presently used technologies in the production, processing, packaging, and safety of various dairy products need to evolve further to overcome these challenges. Here we review the potential applications of food safety tools like hazard analysis and critical control point, sustainable packaging materials like polylactic acid, and non-conventional technologies such as ultrasound to solve current problems. This chapter discusses the role of various biotechnologies in food packaging, processing, and safety, focusing on different dairy products, from indigenous to by-products. The food industry also needs to identify and overcome socioeconomic challenges to proceed with these non-conventional, relatively new technologies from laboratory to industry.

Keywords Food industry · Pasteurization · Dairy products · Food safety · Food Packaging · Nutrition · Non-conventional technologies · Non-thermal processing · Policy reform · Biotechnological applications

A. Kapoor · M. Yadav · A. Verma · K. Ambatipudi (✉)
Department of Biosciences and Bioengineering, Indian Institute of Technology Roorkee, Roorkee, Uttarakhand, India
e-mail: akapoor@bt.iitr.ac.in; myadav@bt.iitr.ac.in; averma@bt.iitr.ac.in; kiran.ambatipudi@bt.iitr.ac.in

V. K. Yata et al. (eds.), *Sustainable Agriculture Reviews 54*, Sustainable Agriculture Reviews 54, https://doi.org/10.1007/978-3-030-76529-3_6

6.1 Introduction

Food technology deals with various applications used in the production, packaging, safety, storage, and distribution of food products. Such technologies enhance the shelf life of perishable food, covert the comparatively bulky food to a convenient form, offer more food choices and make food available during off-seasons, thus fulfilling consumer demand. Once harvested from agricultural fields, post-harvest technologies such as threshing, drying, cleaning, and milling are applied to the agricultural produce to protect it until further processing begins. From the age-old techniques such as smoking, salting, drying, and pickling to recent developments like irradiation, canning, freeze-drying, today's food industry uses all sorts of technologies for processing. With technological advancements, it's not only the main product processed, but several by-products are also manufactured with further processing. The large-scale production and availability of food are also advantageous in ensuring food security, especially for developing nations. Biotechnological applications like enzyme and fermentation technology are responsible for increasing the shelf life, enhancing the flavor, texture, and nutritive value of food products (FAO International Technical Conference, 2010). Fermentation technology is highly employed in the dairy industry, especially in yogurt, cheese, cultured buttermilk, kefir, and sour cream. More recently, the kefir grains have been used to ferment and develop soybean beverages supplemented with inulin (dos Santos et al. 2019). Similarly, enzyme technology is well established in fruit juice for liquefaction and clarification. For example, the addition of protease in cherry juice and immobilizing polygalacturonase in calcium alginate microspheres in apple juice reduces juice's turbidity levels (Pinelo et al. 2010; Deng et al. 2019). The advances taking place in these technologies from time to time further improve the product, thus meeting evolving market demand requirements.

Food packaging is one of the most crucial manufacturing steps since it is responsible for protecting its content, conveying information through labels, appealing to consumers, and maintaining product quality. The food packaging industry has come a long way from coating cheese in grease and lard to biobased packaging material and edible films. For instance, active food packaging with polymer material is functionalized with synthetic peptides to form antimicrobial packaging (Agrillo et al. 2019). The packaging is closely associated with the safety aspect of food. The food safety tools are implemented to control physical, chemical, and biological hazards and assure that food upon consumption shall not cause harm. Therefore, it becomes equally essential to ensure that processing methods must comply with food safety norms. Not just nutritious, food must be safe and fit for consumption to prevent foodborne illness and outbreak. Food scares are the repercussions of pandemics and public health emergencies as they cause safety concerns among consumers, and the COVID-19 pandemic is a recent example. Even though the researchers reported no relation between the spread of the virus through food, the pandemic has created awareness and raised significant concern for safe food consumption.

The dairy industry is an integral part of the whole food system, with annual global milk output of 843 million tons in 2018, and India as the largest producer and consumer. In recent times, various dairy by-products have started being explored for their functional properties, for example, phospholipids isolated from buttermilk, butter serum, and cheese whey confer health benefits in humans (Verardo et al. 2017). Utilizing dairy industry by-products is economical and adds to the existing products, and reduces the environmental load. The dairy industry, in particular, uses most of the applications, which are used in food industries. This includes, but is not limited to, fermentation technology, membrane filtration, microbial testing, etc. The chapter consists of various dairy industry biotechnological applications, especially in dairy products (e.g., cheese, yogurt, paneer, and infant formula). The chapter starts with the dairy industry description and further explains various production, processing, packaging, and safety technologies applied in the dairy industry. The current limitations of food industries and their potential solutions are also discussed, including socioeconomic constraints and possible solutions.

6.2 Historical Perspective

Due to limited knowledge, technological advancements, and demand, most dairy products prepared until the mid-nineteenth century were farm-level with the challenges such as non-uniformity in product quality, low quality of raw material, lack of proper market, improper packaging. But with increasing population and subsequently increasing demand, rapid industrialization, improving transportation facilities, and development of technology, dairy production transited from farm to factory level production (De 1980). Pasteurization was invented in 1862 for processing wine, but in the early 1900s, it found application in the dairy industry for pasteurizing milk (Smith-Howard 2017). From vat pasteurization, high-temperature short-time pasteurization around the 1930s to ultra-high-temperature pasteurization in 1948, the technology has evolved because of the advancements of the core concept of killing unwanted microorganisms. Cooling milk is a crucial step in milk processing since it is highly responsible for keeping milk quality. The cooling cans used earlier were replaced due to the advent of mechanical refrigeration in the mid-twentieth century. Homogenization, the pressure-driven process of reducing the size of milk fat globule, was first used in commercialized milk in 1919 but gained consumer acceptance around 1940–1950 (Cano-Ruiz and Richter 1997; Weimar 1994).

It was not just the technological development in terms of processing, but the food safety norms also progressed with time. Processing methods are highly interlinked and have a significant impact on product quality and microbial safety. This relation could be better understood, for example, pasteurization kills pathogenic microorganisms, and refrigeration of milk helps in keeping the microbial load low, therefore, processing methods like these directly contribute to ensuring food safety. Similarly, penicillin, whose discovery was unrelated to the dairy industry, impacted milk processing (Gaynes 2017). Even the packaging of dairy products has come a long way. For example, earlier cheese was packaged in lard, cheese grease, or paraffin, which then shifted to plastic materials like polyethylene terephthalate and now progressing towards edible films and coatings. The commercialized milk was earlier available in returnable quart bottles, then in plastic-coated milk cartons, and later in plastic containers. With rising environmental concerns, plastic is being replaced with environment-friendly and sustainable solutions such as biobased packaging material. The industrial concern is transitioning from mere food preservation to safety, food security, and sustainability. This is apparent with stricter regulations where nutrition labeling was introduced and emphasized in the 1970s, acts such as the food safety modernization act are the need of the hour. Prevention of Food Adulteration Act, 1954 was a prominent act in the Indian food system until 2011, when food safety and standards rule came into force. Prevention of Food Adulteration Act and various such acts were repealed. This itself exemplifies the shift from food adulteration to food safety. The rising concern of the environmental impact of food processing, waste treatment, and the quest for greener and cleaner solutions is shifting scientists' focus on long-term food processing methods. Novel food processing methods such as high hydrostatic pressure and pulsed electric fields offer great potential in this regard (Fig. 6.1).

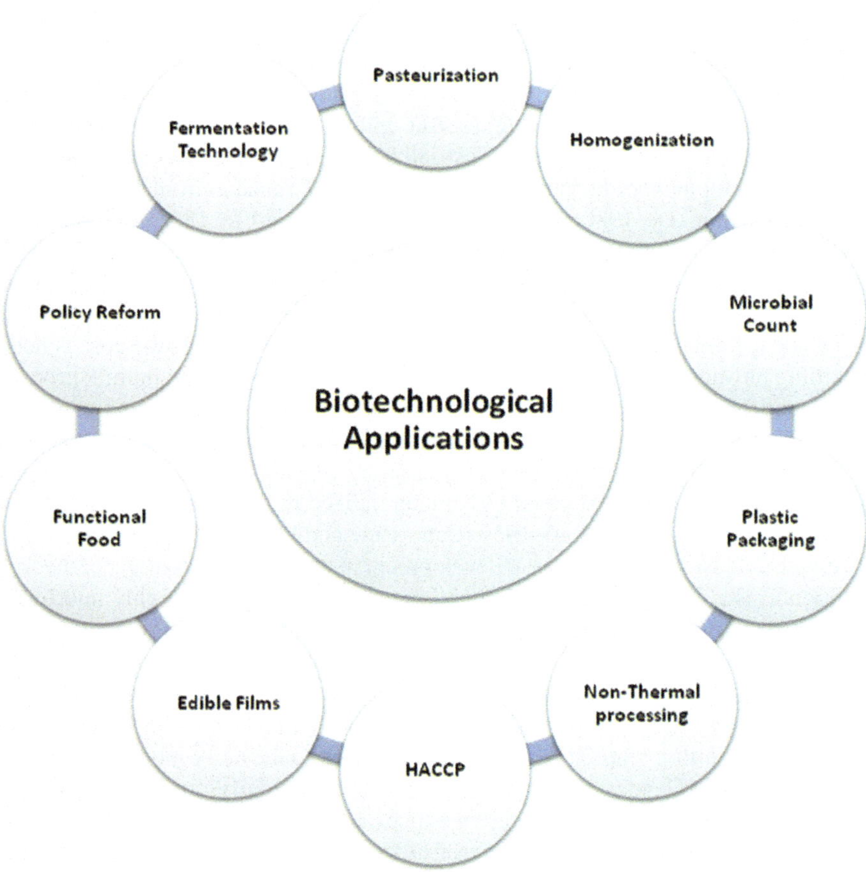

Fig. 6.1 Schematic overview of biotechnological applications in the food industry

6.3 Dairy Products

6.3.1 Cheese

Cheese is made from milk using rennet enzymes and/or lactic acid bacteria for casein coagulation followed by separation of developed curd. Initially, it was made at the farmhouse level in small quantities with low milk quality as the major challenge in large-scale production (Johnson 2017).

6.3.1.1 Production and Processing

The advancement of microbiology and technology in refrigeration, commercial starters, and pasteurized milk has improved cheese quality and initiated factory level production in the latter half of the nineteenth century (Johnson 2017; De 1980). The early challenges in cheese production were overcome gradually by applying various processing methods such as milk in cans, methods to cool milk, direct microscopic counting of bacteria, and, most importantly, by pasteurization. These processing steps contributed to bringing uniformity to the end product, promoting commercial production. Milk naturally contains many bacteria, such as coliforms hindering the fermentation process, resulting in undesirable acidification and gassy cheese. In contrast, pasteurization kills most bacteria that ferment lactose and/or cause uncontrolled fermentation thus giving more control over the cheese production. Even though pasteurization interferes with the flavor development of cheese and a costly technique, it had an advantage in quality improvement with higher cheese yield (Price 1927; Wilson et al. 1945). Presently, a high-temperature short-time pasteurization method is used where milk is heated using a heat exchanger to 72 °C for 15 s, followed by cooling (Singh and Heldman 2001). This heating-holding-cooling method enables the production of cheese with uniform quality.

Milk composition and cheese yield are the key factors that determine profit, milk price, and cheese making efficiency, with yield primarily depending on moisture, fat and casein content, and retention of each component during the processing (Wedholm et al. 2006; Lopez-Fandino et al. 1996). Variation in milk composition is a relatively common phenomenon, attributed to alterations across seasons, breed, feed, health conditions of the animal, etc. Therefore, to maintain uniformity in commercial cheese production, it is imperative to ensure consistency in the raw milk used to produce cheese. Membrane technology (e.g., microfiltration, nanofiltration, and osmosis) is a separation process where pressure is applied to facilitate the separation of individual components in the form of retentate while allowing the permeate to flow through. This pressure-driven process is used in cheese manufacturing because of its role in improving the nutritive value, cheese yield and consistency, whey utilization, and relative reduction in the use of rennet and starter culture (Kumar et al. 2013). Membrane filtration has been strongly advocated to remove lactose and water for concentrating the milk fat and casein. With the membrane pore size ranging from 1 to 100 nm, the widely used ultrafiltration technology is primarily used to eliminate lactose and increase fat and casein content to improve cheese yield. The ultrafiltration technology is further categorized based on the concentration factor: low, medium, and high-concentration ultrafiltration. The low-concentration ultrafiltration is mainly used to standardize milk protein, while medium-concentration factor ultrafiltration is primarily used to concentrate milk (Soodam and Guinee 2018). The microfiltration process efficiently removes bacteria and spores from milk due to its membrane pore size of 0.2 to 2μm. Thus, the microfiltration of pre-treated milk improves curd firmness and reduces the number of additives such as calcium dichloride to promote ripening (Caron et al. 1997).

Membrane technology is also applied in the processing of whey, the nutritionally rich by-product obtained during cheese production. The large-scale manufacturers retrieve the different nutrients such as lactose, protein, and minerals from whey, while small-scale processors consider it as a waste as further processing for valorization is regarded as uneconomical, especially in developing nations (Chen et al. 2019). Membrane filtration of whey leads to the production of high-quality products such as whey protein isolates (>90% protein) and concentrates (30–80% protein) for foaming, gelling and emulsifying agents (Kumar et al. 2013). Recently, Samtlebe et al. (2017) designed a tubular membrane filtration system for controlling the phage before processing cheese whey to develop different whey products. A 0.1μm membrane filtration model was found better suited for phage reduction as well as protein transmission. The application of membrane technology has become a well-established technology in the cheese industry, including the processing of its by-product whey with fouling as one of the significant limitations of the technology. Fouling occurs mainly due to pore blockage, particle adsorption on pores, the formation of cake and protein, and mineral deposition (James et al. 2003), resulting in a significant drop in membrane efficiency and economic loss. Therefore, fouling is reduced by regularly cleaning membranes, using low-fouling membranes, and applying high pressure.

Fermentation is an integral part of cheese making, and commercial starters proved to be a milestone in factory level cheese production. Pasteurization of raw milk kills most of the bacteria, resulting in unwanted fermentation, enabling to better control cheese's fermentation characteristics, and developed acidity is due to added starter culture to maintain cheese quality. The discovery of plasmid DNA encoding proteolytic activity and lactose fermentation was a significant advancement in starter cultures (McKay et al. 1976; Romero and Klaenhammer 1993). For example, commonly used starters in cheese consist of mesophilic lactic acid bacteria, predominantly *Lactococcus lactis* species which is responsible for acidification, proteolytic activity, flavor production, and phage insensitivity. Starter cultures are also associated with flavor development in cheese. Ayad et al. (2000) reported wild strains of *Lactococci* spp. in developing new flavors in cheese. Additionally, permeabilization, autolysis, and intracellular enzyme release of starter culture are also impacted by the alterations in cheese components such as fat and salt content (Yanachkina et al. 2016). Rennet is a set of enzymes used to coagulate milk during cheese making, and chymosin is the active enzyme in rennet. Before 1980, most of the rennet was extracted from calf stomach, but the advent of calf genes' insertion led to chymosin production from molds. And today, the majority of the rennet used is fermentation-produced chymosin (Johnson 2017). Various advances have been made in recent studies researching the parameters associated with the rennet in cheese making. For instance, the milk-clotting enzyme BY-1 developed from *Bacillus subtilis* PNG27 exhibited better flavor and acceptability in cheese, indicating its potential use in cheese production (Meng et al. 2018). More recently, Dussault-Chouinard et al. (2019) reported the enhanced cheese-making characteristics due to lowered renneting pH. Additionally, processing of the milk is performed using ultrasound, different inoculation methods, starter cultures, and ripened molds to influence and enhance the rennet coagulation properties and sensory qualities during cheese making (Ragab et al. 2020; Galli et al. 2016). Cheese ripening is the development of flavor, texture, and other characteristics, such as aroma and specific taste. It is the enzymes and microbiota in specific which contributes to the flavor development of cheese. Flavor in cheese plays a crucial role and often distinguish one variety of cheese from the other, for example, without the addition of lipase enzyme, the flavor of Provolone and Romano would be similar to mozzarella (Johnson 2017).

6.3.1.2 Packaging

Initially, cheese was coated with cheese grease, paraffin, or lard to prevent drying, but with the large-scale production and increasing demand, newer ways were developed to protect the product and prevent contamination. Significant challenges in cheese commercialization are the losses due to contamination by yeast, moulds, and bacteria during storage, often leading to off-flavor and reducing the quality (Costa et al. 2018). The improper packaging causes moisture loss in a few cheese varieties, resulting in increased hardness and low organoleptic properties. The majority of the currently used food packaging materials are composed of plastic materials, including polyethylene terephthalate, polystyrene, polypropylene, polyvinylchloride, and polyethylene, which is primarily associated with environmental concerns because they cannot be recycled entirely or biodegraded (Piscopo et al. 2019). Thus, researchers focus on developing newer technologies such as edible films and coatings, nanoparticle embedded packaging materials, engineered nanofibers for packaging, and treating food with pulsed light to prevent post-process contamination. Biobased packaging materials, such as polylactic acid, are primarily derived from renewable resources produced through chemical synthesis of biomass monomers, have relatively low water vapor barrier properties limiting their use in cheese packaging. For instance, polylactic acid in packing semi-hard cheese resulted in moisture loss and surface drying (Holm et al. 2006). The edible films and coatings comprising polysaccharides, proteins, and lipids are another recent advancement in food packaging where films can be consumed with the food, thereby avoiding waste generation. To curb the bacterial contamination in cheese, researchers have recently developed nanofibers containing nisin and loaded with poly-g-glutamic acid/

chitosan and moringa-oil chitosan nanoparticles, which has shown promise as a potential active packaging material in cheese preservation (Cui et al. 2017; Lin et al. 2019). Pulsed light treatment, a non-thermal processing technology, also finds a potential application in reducing cheese contamination. More recently, it has been reported that pulsed light's post-packaging application could be used successfully to minimize post-processing contamination in cheese.

6.3.1.3 Food Safety

The quality of the milk used for cheese production has drastically improved over time, mainly due to the application of technologies such as pasteurization. Today, the bacterial count in pasteurized milk is significantly fewer (e.g., 100 bacteria/mL) compared to the previous cheesing making procedure (5 billion CFU/mL) (Kelly 1939). As cheese is highly susceptible to contamination at various processing and post-processing steps, it demands strict cleaning and sanitation facilities. For instance, water activity, pH of cheese, and competitiveness with other bacteria determine the growth and survival of pathogens present due to contamination during manufacturing, storage, and cheese ripening. Similarly, few surface-ripened kinds of cheese like Limburger and mold-ripened cheese such as Camembert and Brie are more prone to pathogenic growth because of their high-water activity and tendency to lose acidity during ripening (Johnson 2017). Although pasteurization has proved to produce better quality cheese, negligence in proper cleaning and sanitization and prolonged use of the equipment has often led to developing a complex microbial consortium called biofilms. Recently, Lee et al. (2017) reported the biofilm-producing ability of *Listeria monocytogenes* on stainless steel and polystyrene microplates in cheese plants. Antibiotics residue is another primary safety concern in cheese manufacturing as it could inhibit acid production by starter cultures that leads to technological and economic loss (Chiesa et al. 2020). In contrast, Silva et al. (2020) reported no effect of fermentation and physicochemical characteristics of Minas Frescal cheese by an antibiotic (e.g. monensin) residue in milk. Nonetheless, antibiotic residues remain a matter of concern due to their effect on product and consumer health.

6.3.2 Paneer

Paneer is an indigenous dairy food product rich in animal protein and possesses a high biological value of 80–86. It is obtained by heat treatment of milk, followed by coagulation, where the end product has a mildly acidic flavor, spongy texture, and whitish color. Both cow and buffalo milk are suitable to prepare paneer; however, the latter is preferred because of its high fat, minerals, and calcium content and whitish appearance, as desirable. In contrast, paneer made from cow milk has a soft texture with a yellowish appearance and does not appeal to consumers. Paneer has been classified as one of the soft cheese varieties highly used in culinary dishes (Kumar et al. 2014).

6.3.2.1 Production and Processing

To maintain uniformity in the final product, it is necessary to ensure consistency in the raw materials obtained, processing conditions, quality analysis of the product, proper storage, and optimum conditions throughout the supply chain. For instance, membrane technology, the widely used filtration technology, has been extensively used to concentrate the standardized milk for paneer preparation. The ultrafiltration concentrates standardized milk to 27% total solids, resulting in more whey proteins for paneer manufacturing, leading to the total solids recovery up to 95% and higher paneer yield. Kanawjia and Rizvi (2000) reported that adding 0.15% calcium chloride to the membrane filtration retentate improves the paneer's textural and organoleptic properties. Reverse osmosis has also been used for concentration, but its main advantage lies in low energy consumption. The recent advancement of utilizing reverse osmosis to obtain milk solids from dairy rinse water is an efficient environmental strategy for the dairy industry (Brião et al. 2019). In paneer processing, milk is coagulated after filtration, heating, and cooling. The most commonly used coagulants in paneer preparation are lactic acid, citric acid, alum, tartaric acid, or sour whey, including several coagulants that impart different properties to the quality of the final product. The use of citric and tartaric acid at a 2% concentration has been reported to make paneer with the most suitable properties from reconstituted milk (Khan et al. 2014).

In contrast, malic acid was found as the most suitable coagulant for better flavor and soft body, while calcium lactate for higher yield and whiter color, and citric acid for overall acceptability of paneer prepared from cow milk (Kumar et al. 2019; Shanaziya et al. 2018). Recently, Amini et al. (2019) reported fermented rice milk as a novel coagulant for paneer preparation. Sharma et al. (2002) reported that solids-not-fat, total solids, and paneer yield were highest in winter (8.983%, 13.639%, 15.5.1% respectively), summer, and rainy season, emphasizing not only coagulants but seasons also influencing paneer composition and yield. Interestingly, sour fruit juices such as lemon and amla (Indian gooseberry) juice have also been used as paneer coagulants. The sensory evaluation to analyze the appearance, flavor, body, and texture found sour juice coagulated paneer as a suitable product (Ahmed and Bajwa 2019). Like further processing, paneer whey, similar to cheese whey, is utilized as a potential product for human consumption. Lactose, also called milk sugar, accounts for approximately 70% of total solids present in paneer whey. Recovering lactose post further processing is economical and environment friendly since lactose reduces the biochemical oxygen demand of whey by 80%. Bund and Pandit (2007) studied sonocrystallization to fasten lactose recovery from paneer whey and optimized the recovery rate at 90.3% in 20 min. Paneer whey has also been explored to prepare functional pineapple beverages and has been reported to exhibit higher mineral content and better shelf life than cheese whey (Baba et al.

2016). Additionally, paneer whey has also been reported for inulinase production as a cost-effective substrate (Singh et al. 2019).

6.3.2.2 Packaging

Food packaging is mainly done to enhance the product's shelf life, maintain quality, provide mechanical, chemical, and microbial protection, appeal costumer, and provide details on labels. Paneer is a perishable product, highly susceptible to physico-chemical and microbial contamination, and therefore, packaging correctly is critical for consumption. The most commonly used packaging material for paneer includes coextruded films, polythene sachets, laminates, parchment paper, etc. (Kumar et al. 2014). Similarly, modified atmospheric packaging is a food packaging technique with a modified atmosphere inside the package to enhance the shelf life and maintain food quality. This is achieved either by flushing desirable gases inside the package, or it occurs due to food's respiration process. The application of modified atmospheric packaging in paneer packaging is relatively new, but the technology has gained significance since the primary concern associated with paneer packaging is its shelf life. The modified atmospheric packaging has been studied in paneer packaging to enhance the product shelf life (along with hurdle technology for paneer processing), to evaluate paneer quality based on its color, and to study its influence on chemical properties of paneer such as moisture, pH, titratable acidity and free fatty acid (Thippeswamy et al. 2011; Shrivastava et al. 2013; Rai et al. 2008). The application of sequential treatment such as supercritical-carbon dioxide in association with food-grade acetic acid works as a sustainable, non-thermal method to enhance paneer shelf life up to 30 days (Kapoor et al. 2013). Edible films and coatings are also a potential packaging material for paneer as it not only increases paneer's shelf life but also fortifies it, thereby enhancing its nutritional value. This includes the usage of whey protein concentrate-based edible coatings consisting of ferrous sulfate heptahydrate, starch, and carboxymethyl cellulose-based edible film treated with clove oregano essential oil (Jotarkar et al. 2018; Karunamay et al. 2020). Vacuum packaging is another advancement in food packaging where the air is removed from the package and sealed hermetically. The primary advantages include shelf life extension, flexible film packages like shrink film, and oxygen reduction to limit aerobic bacteria's growth. Paneer is usually vacuum packed in co-extruded and laminated films. The linear low-density polyethylene and the binding agent have been proved as an effective packaging combination for vacuum packing and paneer tikka storage for 40 days (Ahuja and Goyal 2013).

6.3.2.3 Food Safety

Sensory evaluation is a discipline that evaluates consumer food products by using human senses such as smell, sight, touch, taste, and hear. The food should taste not only acceptable but also look appealing. For example, no matter how nutritious orange juice is, consumers will not accept the product unless it appears to be 'orange.' Therefore, the sensory evaluation of products is conducted to analyze the overall acceptability. Since sensory evaluation is all about perceiving human senses, mostly human panels are involved as trained and consumer panels. Schlossareck and Ross (2020) evaluated the intensity of aftertaste and liking of spicy paneer by employing a consumer panel and observed that the aftertaste stayed was directly proportional to the spice level of paneer. Nonetheless, error and inaccuracy are still high with human panelists in sensory evaluation because of each panelist's varying experience. Therefore, more recently, the electronic tongue, an analytical tool, has drawn attention for better accuracy as a potential tool in qualitative analysis of food products like paneer.

Paneer is a highly perishable food product with short shelf life, and the high moisture content of approximately 70% increases the chance of microbial contamination. Although refrigeration is suggested to store it, there are microorganisms like *Escherichia coli* that jeopardize the safety of food. Sharma et al. (2019) demonstrated the use of phytochemicals, including gallic acid, piperine, menthol, quercetin, and eugenol, with antimicrobial properties against *Escherichia coli* in paneer. Another critical parameter in paneer preservation is maintaining a low temperature. The cold-chain facilities in India are still in the developing phase where most areas lack proper electricity supply, thus hampering the product quality and limiting the shelf life of various dairy products. The paneer's shelf life is a significant restriction in its proper utilization, and therefore, efforts are made to enhance the paneer's shelf life. One such attempt was made by Pandey et al. (2019), where paneer whey was used for the production of paneer making powder with bio-preservative potential. The paneer obtained exhibited a lower count of the total plate, *Staphylococcus*, lactic acid bacteria, and coliforms, thereby increasing the shelf life to 75% compared to control.

6.3.3 Yogurt

Milk is the most desired food consumed to fulfill the essential nutritional requirement of humans. The dairy products such as yogurt and other fermented dairy products (e.g., cheese, kefir) impart high nutritional value, sensory properties, and overall health (Cuevas-Gonzalez et al. 2020). The incorporation of lactic acid bacteria in fermented products due to their beneficial health benefits has led to probiotic's evolution, with the ingestion of live microorganisms to restore or improve the host's gut microflora. Yogurt (popular fermented dairy product) is consumed worldwide because of its crucial role in improving intestinal health, immune system (Weerathilake et al. 2014; Pei et al. 2017), maternal health, and pregnancy outcomes (Kok C. R and Hutkins R 2018; He A et al., 2020). It acts as a suitable carrier for delivering functional ingredients like probiotics and prebiotics (Castro et al. 2015; Hill et al. 2017). Yogurt is relatively more nutritious than milk as an excellent source of calcium, magnesium, potassium, milk fat, and vitamins (e.g., B_2, B_6, B_{12}) (Ayar et al. 2014), including milk proteins and amino acid with a higher biological value for sustaining good health (Lourens-Hattingh 2001). Ndife et al (2014) reported a crucial role of yogurt in improving lactose tolerance, immune enhancement, metabolic, and gastrointestinal disorders (e.g., inflammatory bowel disease). Furthermore, several non-nutrient components present in yogurts, such as sphingolipids, conjugated linoleic acid, and butyric acid, have been reported to have anti-cancerous properties (Sanders et al. 2007; Gahruie et al. 2015). Interestingly, fortified yogurt consumption with iron, fibers, and seed oil has been reported to effectively ameliorates nutritional diseases such as osteoporosis, rickets, malnutrition, and atherosclerosis (Gahruie et al. 2015). Yogurt has also found an important place in the older population's diet due to its accessibility and beneficial role in improving their nutrition and health profiles (El-Abbadi et al. 2014). Thus, due to increased awareness of the numerous beneficial effects of yogurt, there has been a steady rise in the demand and popularity of yogurt globally, and in turn, has driven scientists to develop innovative strategies to improve further the quality and quantity of yogurt production (Chandan et al. 2017). In the market, multiple yogurt varieties such as plain, stirred/blended, vanilla flavored, smoothies/whips/mousses, frozen, organic, and natural are available for consumption. These yogurts are incorporated with different additives such as stabilizers, colors, exopolysaccharides, flavors, fruits, preservative, and sweetening agents to improve their sensory and texture characteristics (Fazilah et al. 2018).

6.3.3.1 Production and Processing

Bacterial exopolysaccharides are natural stabilizers and have gained momentum as safe during yogurt manufacturing, thereby eliminating the commercial stabilizers (Gawai et al. 2017). To generate novel functional food, yogurt-based products have been enriched with plant-derived sterols/olive or carrot juice, including many other products (e.g., cereal yogurt-like beverages, synbiotic yogurt) to impact human health positively (Buyuktuncer et al. 2017; Zoidou et al. 2017). To this end, emerging fields of biotechnology, microbiology, and food bioengineering have contributed significantly to improve the overall quality and safety of different dairy products. For example, biotechnological techniques have been used to incorporate microorganisms in food to enhance the aroma, taste, flavor, shelf life, and nutritional value of foods. Additionally, biotechnological tools have been used to produce a range of products (e.g., enzymes, vitamins, and food ingredients), including selection and manipulation of microorganisms to upgrade the product quality, safety, consistency, and yield (Ghoshal et al. 2018). Yogurt has been defined as a mixture of milk (whole, reduced-fat, low fat) produced by culturing lactic-acid producing bacteria such as *Lactobacillus delbrueckii* spp. *Bulgaricus* and *Streptococcus thermophiles*. Additionally, *Acidophilus* spp. culture has also been reported to produce yogurt and considered as "Generally recognized as safe" (Weerathilake et al. 2014). The microbial cultures decrease the pH and coagulate the milk proteins *via* lactic acid obtained through lactose fermentation, providing a viscous gel-like structure. The milk components converted to volatile and non-volatile acids (e.g., acetone, acetaldehyde) within the yogurt give a characteristic flavor (Beshkova et al. 1998; Dan T et al. 2019).

Starter culture strains such as *Streptococcus thermophiles* zlw TM11 secrete a high amount of exopolysaccharide to improve the texture and viscosity (Folkenberg et al. 2006). Of note, flavor, texture, and taste vary depending on the lactic acid bacteria starters and metabolites secreted by the bacteria. Subsequently, the different manufacturing processes (e.g., homogenization, heat treatment, pasteurization) and types of ingredients (e.g., fruit pieces, stabilizers, bacterial cultures, flavors, and colors) have been reported to alter yogurt's quality (Nagaoka et al. 2019). The most commonly available yogurts are the set and strained, but frozen and drinking yogurt has gained significant popularity more recently. To date, yogurt manufacturing has been mainly from bovine milk (e.g., sheep, goat, water buffalo), but other mammals' (e.g., camels) milk has recently aroused interest. These dairy products are carriers of probiotic bacteria, which enter the human intestine by maintaining their survivability through their stomach and modulating intestinal microflora. It requires a thorough understanding to increase their usage and application in dairy science and medicine (Elizaquivel et al. 2011). The manufacturing of yogurt begins with the standardization based on varying fat amount in milk, and proteins (5–15%), followed by physical treatment such as heat (90–95°C, 3–7 min), homogenization (20–25 Mpa at 70°C), deareation (70°C), cooling (42°C), and inoculation of microbial culture (10^6–10^7 CFU/ml). Subsequently, after fermentation, yogurt production proceeds with harvesting, post-treatment, and packaging. The formation of a few yogurts such as set-type, stirred, drinking, and concentrated depends on the steps (e.g. number) during the manufacturing influences the yogurt's final quality (Lopes R P et al. 2019). After the final packaging, yogurt is stored at low temperature (4° C or 5°C) to limit the post-acidification and improve safety. The fermentation of yogurt is a batch process, but few operations (e.g., treatment, cooling, packaging) are designed and carried out in a continuous and semi-continuous process. Acceptable and good manufacturing practices are followed to maintain control of industrial manufacturers' microbial risk (Corrieu and Beal 2015).

6.3.3.2 Genetic improvement of starter cultures

The Fermentation bioprocess makes use of microbial inoculants to produces a wide range of microbial products that functions as preservatives, stabilizers, flavoring, and coloring agents. Several traditional and molecular-based approaches have been applied to improve microorganisms' metabolic properties and quality found in fermented food.

Traditional approaches

Traditional methods such as mutagenesis and natural gene transfer methods have formed the basis for developing bacterial starter cultures. For instance, mutants are produced by exposure of the microbial strain to UV rays or mutagenic chemicals (e.g., alkylating and deaminating agents) by inducing changes in their genomic sequence in classical mutagenesis. Subsequently, the strains with improved quality are selected based on flavor or ability to resist bacterial viruses. In contrast, gene transfer methods such as transduction, conjugation, and transformation have led to a better understanding of microbial starter cultures' genetics and used for strain improvement (Harlander et al. 1992). Additionally, the introduction of DNA into bacteria through electroporation has revolutionized bacteria's transformation (Chassy et al. 1988; Hisatsune et al. 2016). Recently, Welker et al. (2019) reported that multi-pulse electroporation increases the transformation efficiency in bacteria (e.g., *Lactococcus lactis)* compared to that obtained withsingle-pulse electroporation (Welker et al. 2019).

Molecular Approaches

The different molecular approaches used to improve starter cultures are genetic modification, characterization, and genomics-based platforms. For example, recombinant DNA technology has been extensively used to modify bacteria, yeast, and mould to either promote, alter, or inactive the expression of a specific gene (Shinde et al. 2018). To make this approach more successful, specifically in the application of food processing, requires the construction of food-grade cloning vectors with plasmids devoid of antibiotic-resistant marker carrying DNA sequence from generally recognized as a safe organism (Landete et al. 2017). Additionally, the research work on the genetic characterization of microbial organisms has accelerated due to the progress made in information collection of the genome sequence, high-throughput RNA and protein analysis, and high-throughput comparative genome bioinformatics tools. For instance, molecular diagnostic techniques such as polymerase chain reaction (PCR) and fluorescent in situ hybridization (FISH) tools have been extensively used to identify, detect and characterize different microbial strains to improve the fermentation process. Similarly, the application of high-throughput approaches such as the Next-Generation Sequencing (NSG) tool, Loop-Mediated Isothermal Amplification (LAMP), and microbial marker development could be applied to understand food biotechnological processes better. For instance, the techniques above have been used to study the ecological communication between bacterial strains thoroughly, appearance, competition, prevalence, food-microbe, microbe-microbe interaction in the gut, and development of the final product (De Vos et al. 2001). The era of microbial genomics also holds an important place in food biotechnology for the conversion of raw materials to edible form. As microorganisms are an integral part of food-grade ingredients and fermented food due to their interaction within the intestine, functional and comparative genomics of food-grade organisms are crucial to developing functional and safe food. More recently, functional genomics has been used to regulate the gene expression pattern and interactions based on the availability of the complete genome sequence of a particular organism, such as understanding the response of microorganisms to environmental influences (Parkhill et al. 2000). As we move to proteins from a gene level, proteomics' application has led to the identification and characterization of protein expression and protein-protein interaction in a microorganism to improve the fermentation technology (Parkhill et al. 2000). Similarly, genetic and protein engineering techniques have also been used to increase enzyme production (e.g., chymosin) in the dairy industry to manufacture fermented products such as cheese (Teuber et al. 1993).

6.3.4 Buttermilk

Buttermilk is the liquid phase by-product that is obtained during the processing of butter. It has high nutritional value and is explored as a functional food. When the cream is churned in butter making, it yields 2 streams – butter and buttermilk. Upon churning, the milk fat globule membrane breaks down, and while most of the milk fat goes to butter, phospholipids present on the milk fat globule membrane (MFGM) fragment migrates to the buttermilk. This liquid phase primarily consists of water-soluble components, including lactose, milk proteins, and minerals, making it a nutritionally rich by-product (Morin et al. 2007).

6.3.4.1 Production and Processing

When the cream is churned to produce butter, a stream of buttermilk is obtained. Although ideally, the cream should yield 50% butter and 50% buttermilk, various conditions such as fat content of the cream, churning temperature, churning speed, aging of cream, etc., are responsible for the yield of both products. Therefore, just like any other dairy product, processing parameters are crucial in buttermilk yield as well. Not just the churning but the initial steps in buttermilk processing also affects the composition of the product. Morin et al. (2007) studied the effects of processing such as cream pasteurization and spray drying on the buttermilk composition and microstructure. Cream pasteurization was a pivotal step in determining higher recovery and composition of milk fat globule membranes, whereas spray drying had a pronounced effect on phospholipids content. Buttermilk has applications in various other dairy products, such as yogurt, cheese, fermented beverage, etc. The development of fermented drink based on buttermilk, where barley and fructooligosaccharide were used as a functional ingredient, is an example of buttermilk utilization in manufacturing other products (Sheth and Hirdyani 2015). Incorporating buttermilk in the preparation of fermented milk beverages reduced excessive post-acidification during storage, thus highlighting the acceptability of buttermilk in a fermented beverage (De Bassi et al. 2012). Using buttermilk in the making of a variety of cheese has been reported by several studies. Poduval and Mistry (1999) prepared Mozzarella cheese using 0.5% fat milk with various combinations of cream and ultrafiltered sweet buttermilk and found alterations in functional properties such as spongy and open protein matrix due to the addition of sweet buttermilk. A similar pizza cheese was also studied for its compositional changes due to the inclusion of unconcentrated sweet cream buttermilk and was found responsible for improved cheese yield (Govindasamy-Lucey et al. 2006). On the other hand, utilizing buttermilk powder in the processing of cheese similar to that of Cheddar was a successful attempt since the process resulted in softer cheese with better quality and potential health benefits (Hickey et al. 2018).

As far as yogurt is concerned, adding buttermilk is responsible for enhancing reduction in undesirable sourness and an increase in odor and flavor, thereby improving the product's overall acceptability (Zhao et al. 2018). Similarly, employing buttermilk powder for the milk fat globule membrane's isolation and then making yogurt indicated functional product development (Le et al. 2011). Apart from being nutritionally rich, buttermilk is also known to have good emulsifying properties. Wong and Kitts (2003) explained how buttermilk solids exhibit antioxidant properties. Therefore, it is a potential ingredient for value addition in stabilizing food matrix against lipid peroxidation reactions (Fig. 6.2).

6.3.4.2 Food Safety

Being an integral part of the food industry, food safety plays a crucial role, especially in dairy products like buttermilk, which are highly susceptible to spoilage. Buttermilk is a perishable product, and therefore, its proper packaging is of utmost importance whether it is meant to be on the shelves of supermarkets or transported to another industry to be used in processing. The food product itself, but the manufacturing facility is also responsible for ensuring food safety. Manufacturing units where food safety tools such as hazard analysis and critical point are not implemented were found to produce cultured buttermilk contaminated with microorganisms such as *Escherichia coli* and coliforms (Paraffin et al. 2019).

6.3.5 Milk

Milk is a significant source of proteins, carbohydrates, vitamins, and minerals (potassium, magnesium, phosphorous, calcium, iodine, sodium). Milk comprises total solids (12–13%)and water (87%). Total solids consist of fat (4%) and solids-not-fat (9%), e.g., proteins, lactose, and minerals and vitamins. The fat amount is the primary factor that categorizes milk drinks into different types. On average, fresh milk has a milk fat of 3.8%, while low-fat milk has a fat of less than 1.5%. Milk is further categorized based on processing such as pasteurization and

Fig. 6.2 Timeline diagram of major technological advancements of the dairy industry. The pink color indicates production and processing; blue color indicates food packaging; and green color indicates food safety

Foot note:

Production and Processing	Food Packaging	Food Safety

homogenization. Pasteurized kills any potentially harmful bacteria, while homogenized milk reduces the size of MFGM and uniformly blends to prevent them from rising with cream. Similarly, fortified milk with extra nutrients such as calcium, vitamin D, or omega-3 fatty acids and modified milk like lactose-free milk are also popular products. Furthermore, according to usage, milk has been categorized into Grades A and B. For instance, grade A milk is produced under appropriate sanitary conditions for usage as fluid milk, while grade B milk is acceptable for manufactured products such as certain cheeses and requiring further processing. In contrast, certified milk is produced under exceedingly high sanitary standards and sold at a higher price than Grade A milk.

Besides, specialty milk includes flavored milk, such as chocolate milk with a flavoring syrup. Furthermore, concentrated milk products have varying degrees of water removed from evaporated milk, condensed milk, and dry milk. Biotechnological intervention can add more value to milk by extending the shelf life, applying better packaging, storage facilities, sterilization, and processing procedures. Milk plants process raw milk collected from farms into marketable milk, requiring standard and clean processing procedures. Maintaining aseptic and clean processes directly affect the quality of milk. Furthermore, the application of rapid detection methods would prevent spoilage of milk, thus making sure the safety and quality of milk.

6.3.5.1 Extending Shelf Life (ESL) Milk

Processing technology and packaging material can extend milk's shelf life. Limited shelf life has been a matter of concern for dairy retailers and producers, and investing in technology that can extend the shelf life can enhance the longevity of dairy products allowing milk to be transported longer distances. The three steps to extend the shelf life of milk include:

Reducing bacterial count

Microbial contamination in raw milk occurs from milk-secreting alveoli cells, the udder's exterior surface, and the surface of milking equipment (Guerra et al. 2013), causing undesirable changes affecting the physicochemical properties of milk (Pereda et al. 2007). In contrast, healthy udder, hygienic procedures for milking, aseptic storage, significant, and transportation practices significantly affect milk's microbial quality. Thus, bacterial count in milk should be reduced to minimal to make heat treatment methods more efficient. There are two ways to reduce bacteria: bactofugation and microfiltration.

Bactofugation

This technique separates the bacteria from the milk by centrifugation. The technique is done in an airtight mode, thereby eliminating any bacteria entering into milk throughout the bactofugation process and preventing oxidation to preserve milk quality. Bactofugation reduces the microbial milk load to 89.66% by eliminating pasteurization-resistant microorganisms and psychrotrophic bacteria like *Lysinibacillus fusiformis*, *Bacillus invictae*, *Enterococcus faecalis*, and *Kurthia gibsonii*. Milk psychrotrophics, e.g., Gram-negative (e.g., Pseudomonas, Aeromonas) and Gram-positive (e.g., Bacillus, Clostridium), are majorly responsible for milk deterioration (Sørhaug and Stepaniak 1997) and significantly reduce the quality and shelf life of fluid such as milk by producing lipases and proteases. (Dogan and Boor 2003). Bactofugation has proven to be an efficient way to remove spores and vegetative forms of bacteria from milk, thereby prolonging the shelf life by 3–5 days.

Microfiltration

This technique is very efficient in removing bacteria (99.999 %) and preserving the shelf life of milk. In this technique, pressure is used to push the milk through a membrane for the particles to go through it, while majority of bacteria are left behind. More recently, ceramic membranes have been used for microfiltration to reduce the mesophilic microflora to 4 logs and 2 logs, thereby improving the quality of ovine and bovine milk (Panopoulos et al. 2020).

Preventing Re-introduction of Bacteria

After removing bacteria, additional steps should be taken as precautionary measures to avoid recontamination of milk, which can happen during handling and pre-processing activities. Residual milk deposited on the surfaces of milking tools and equipment supports the growth of various microorganisms such as coliforms, pseudomonas, and streptococci (Poghossian et al. 2019, Naing et al. 2019). Additionally, improper refrigeration conditions allow psychotrophs to multiply rapidly before milk is processed (Paludetti et al. 2018). Thus, to ensure recontamination, a well-designed milk plant with no exposure to air and proper cleaning methods will prevent bacterial growth making it more suitable for preservation.

Preservation

Milk components make it a suitable medium for microorganisms to grow. Pathogenic and spoilage bacteria (e.g., *Escherichia coli*, staphylococcus, streptococci) cause the spoilage of milk by releasing the extracellular and intracellular enzymes, which makes it unfit for consumption. Enzymes (e.g., oxidases, proteases, lipases) cause degradation of milk components, facilitating bacteria's growth (Poghossian et al.

2019). Thus, preserving milk and screening it to be microbe-free in all becomes an essential step. Preserving the milk can be done either by heat treatment (Pasteurization) or chemical treatment.

Pasteurization

Pasteurization's main objective is to eliminate pathogenic bacteria and enzymes that cause spoilage and extend the milk's shelf life. Pasteurization has been classified in four ways:

High-Temperature Short Time (HTST) Method

This is the most popular way of preserving milk. In this process, milk is heated to 72° C (or 161.5°F) for 15 s to kill any pathogenic bacteria. It is also known as Flash pasteurization or continuous flow pasteurization. Milk pasteurized by this method is labeled as HTST milk and has a refrigerated shelf life of two or three weeks.

Ultra-High Temperature (UHT)

In this method, milk is heated at 138 °C (280 °F) for at least 2 s. The UHT pasteurized milk has a refrigerated shelf life of a month and is the fastest and most economical way of preserving milk.

Vat or Batch Pasteurization

It is the longest and expensive method of pasteurization. Temperature controlled closed vats are used in batch pasteurization, with the milk heated to71° C (160°F) for 30 min, followed by cooling to 4°C (40°F).

Solar Milk Pasteurization

This method is based on the utility of solar intensity as a source of energy for pasteurization. The solar pasteurizing unit consists of a parabolic solar dish, energy conversion system, heat exchanger, and cooling unit. The system can pasteurize milk at 73°C for 30 min and subsequently transferred it to the cooling unit. However, this method is directly dependent on the availability of solar intensity. Development of novel solar pasteurization systems with advancements in higher temperature using photovoltaic integrated concentrator (Meraj et al. 2021) and portable pasteurization plants (Sur et al. 2020) for remote areas can be an efficient alternative for pasteurization.

Chemical Treatment

Chemical preservatives have been added to milk to improve shelf-life; however, they are a challenge to milk safety. Chemicals like hydrogen peroxide, salicylic acid, benzoic acid, mercuric chloride, and formalin. These preservatives are bactericidal and have a health/environmental hazard, including unsuitable for storage. Studies conducted to preserve raw milk using hydrogen peroxide (H_2O_2) showed an absence of coliform (e.g., *Escherichia coli*) growth in the treated milk and was found safe for human consumption. Furthermore, the application of transglutaminase enhanced the micellar stability in skim milk on the addition of ethylene diamine tetraacetic acid (EDTA) and did not affect the dynamics between salts and serum phase. Additionally, the use of thyme-free nano emulsified essential oil is also known for inhibiting microbes' growth in milk (Jemma et al. 2017).

Proper Packaging

The longer shelf life of dairy products is achieved through packaging in a sealed and sterilized environment. Milk can be packed into different types of packages such as cartons, glass, pouches, Polyethylene terephthalate (PET) bottles, C-enamelled tinplate cans, etc. The introduction of different plastic materials, combined with paper, was developed for packaging dairy products. Petrus et al. 2010 studied storage temperature evaluation on pasteurized milk with different packaging materials like polyethylene (PE) monolayers specked with titanium dioxide (TiO2). The study showed better storage of pasteurized milk when high-density polyethylene (PE-HD) bottles were used compared to low-density polyethylene (PE-LD) pouches. Similarly, Polyethylene terephthalate (PET), another packaging material, has shown better results by protecting milk from light-induced lipid oxidation (Kontominas 2010; Van Aardt et al. 2001).

The growing demand for safe and fresh products has led to the development of technologies that assure food quality and improve shelf life through cold chain distribution (Rysstad and Kolstad 2006). The enhanced shelf life of dairy products depends on the biochemical interactions between dairy products, packaging, and the existing environment (Haghighi-Manesh, 2017). Therefore, modern food preservation methods must ensure packaging to enhance the shelf life, fresh characteristics, and quality of food. Newer technologies used in packaging include nanotechnology, modified atmosphere, active and intelligent/smart packaging. The active form of packaging has been introduced to the market to enhance products' safety and quality. It is an innovative food packaging system that includes additives or agents that absorb oxygen, moisture, carbon dioxide, ethylene, flavors, and odours or release carbon dioxide, antimicrobial agents, and antioxidants flavours (Wong and Goddard 2014). Some popular active packaging systems include moisture-scavenging (desiccation), oxygen-scavenging, carbon dioxide-absorbing, and antimicrobial systems (Jalilzadeh et al. 2015). Furthermore, non-thermal food preservation technologies such as pulsed electric fields (PEF) and high-pressure technology (HPP) are also useful in extending the shelf life of the pasteurized product (Abida et al. 2014).

Novel Preservation Technologies

The use of increasing temperatures in pasteurization impacts the taste of milk and is not sufficient enough as an extended shelf life product. Different technologies have been adopted by milk plants to extend the shelf life without impacting the taste of pasteurized milk. New technologies include non-thermal technologies like ultrasound, irradiation, and ultraviolet, while thermal technologies include microwave, ohmic heating, and radiofrequency pasteurization methods. Similarly, microwave and radiation processing have also evolved as novel pasteurization techniques. These new technologies exhibit great potential in maintaining the pasteurized milk product's shelf life and nutritional properties.

Ultrasound Technology

In this technique, low-frequency ultrasound (power ultrasound) waves are applied to milk, causing vibration of molecules and the generation of bubbles, which disrupts the cells, leading to inactivation of microorganisms. Application of low-frequency ultrasound waves (20–100 kHz) generates micro-currents of implosion and explosion, causing physical damage to cell membrane and breakage of cell walls (Carcel et al. 1998). These waves have a lethal effect on cells, thereby inactivating the pathogenic and spoilage bacteria (BermuÂdez-Aguirre et al. 2009). A study reported by Noci et al. (2009) thermo-sonication of milk and the pulsed electric field application enhanced the inactivation of *Listeria innocua* in milk significantly. This technology renders shorter processing time and imparts significant enhancements to final milk quality after processing. For instance, the milk color turns whiter due to fat globules' homogenization (BermuÂdez-Aguirre et al. 2009).

Irradiation Technology

Irradiation (transfer of energy) in milk is used to inactivate pathogenic microorganisms (Osaili et al. 2008). Microorganisms like Salmonella, Listeria, and *Escherichia coli* have been inactivated with irradiation technology (Crawford and Ruff 1996). The ionizing radiation includes gamma rays, X-rays, and electron beams (Satin 1996) and generates secondary electrons causing ionization, dissociation, and excitation. The transfer of energy kills the pathogenic bacteria; however, changes the physical and chemical properties of milk components. For example, in a study, lower vitamin A and carotene contents were observed on the application of gamma-irradiation with thermal treatment to pasteurized milk ((Naghmoush et al. 1983).

Ohmic Heating

Ohmic heating (electro-pasteurization) involves the application of electrical current to pasteurize milk. The technique of Ohmic heating is based on electroporation with pore-forming mechanisms leading to the inactivation of cells. The current generation generates heat of low frequency (50–60 Hz), mainly responsible for bacterial death. Nevertheless, some studies showed permanent membrane damage on the application of high voltage discharges, while the killing with lower voltages needs further investigation. As a novel technology, there are few studies reported on electro-pasteurization of milk. In a study reported by Sun et al. 2008, ohmic heating pasteurization inactivated *Streptococcus thermophilus* faster than thermal treatment processing.

6.3.6 Infant Formulas

Human milk is a complete food for the development, nourishment, and protection of infants. The mother's milk's unavailability has driven a new platform for manufacturing artificial formulas to nurture developing infants. Although inferior to natural milk, these formulas are designed to provide early nutrition to infants and long-lasting metabolic influence during their life (Thompson et al. 2012). The most commonly used breast milk substitute is cow or soya milk, which is marketed as ready-to-feed, liquid, and powder forms. The specialized formulas depending on the infant's health conditions (e.g., gastrointestinal disorders, allergy), are also designed and mentioned as "safe to use" (Green corkins et al. 2016). The government closely monitors infant formulas' composition, and the manufacturers should adhere to strict guidelines to maintain nutrients' self-life (Martin et al. 2016). The recent advances in biotechnology have opened up numerous unique food industry opportunities to produce formulas with bioactives, targeting the infant's gut, nutritional, and functional composition to simulate human milk. For example, the production of recombinant milk proteins (e.g., lactoferrin) through genetic engineering techniques to enhances the nutritional value of synthetic infant formulas and human milk fortifiers have improved the diet and feeding of very-low-birth-weight and premature infants (Gaul 2000; Hartmann et al. 2008; Aggarwal and Aggarwal 2020). Notably, a functional ingredient such as human milk oligosaccharides, long-chain polyunsaturated fatty acids (PUFA), nucleotides, probiotic bacteria, synbiotics, and fat globule membrane is being incorporated to humanize infant formula to provide nutrition, including enhancing immunity, microbiome, and cognitive development (Pande and Akoh 2016; Ahern et al. 2019;). Formulas with prebiotics and probiotics could generate epigenetic modification by modulating gut microbiota's composition and activity (Lemaire et al. 2018).

Human milk is dynamic in composition, especially during the initial six months of lactation compared to the static composition of infant formulas, making it noteworthy to prepare age-based formulas to meet the infant's desirable nutritional requirement of the infant (Lönnerdal et al. 2016; Spalinger et al. 2017). Additionally, during formula preparation, it is essential to maintain a hygienic condition and acceptable manufacturing practices to avoid microbial contamination and adulteration due to the neonates' fragile gut and immature immune systems. In the future, efforts should be directed to develop fast and effective technologies to increase the prepared formulas' microbiological safety, such as bacteriophage therapy, as natural weapons to control pathogens in the food industry (Gutierrez et al. 2016). Similarly, natural antimicrobial (e.g., lactic acid) has effectively prevented microbial spoilage (Al-Holy et al. 2010, 2015; Yemiş G. P and Delaquis P 2020). Furthermore, strategies to increase milk's shelf life and reduce microbial spoilage without hampering the bioactive milk components have been employed through different techniques such as microfiltration, ultrafiltration, microfiltration, and reverse osmosis (Ahern et al. 2019). The use of partially fermented formulas has recently gained momentum in restoring infant health with components derived from a microorganism such as microbial cells, metabolites (postbiotics), and non-viable cells (parabiotics) (Wegh et al. 2019; Salminen et al. 2020). These postbiotics create an appropriate environment for the proper colonization and development of intestinal microbiota, vital for the infant's future well-being. Recent studies have also supported the application of postbiotics as suitable and safest alternative agents than other 'biotics' to improve infant's health with fewer challenges to product storage and shelf life (Deshpande et al. 2018; Rad et al. 2020). Most importantly, these unique features make postbiotics a promising tool for food, biotechnology, and the pharmaceutical industry to promote health benefits as a functional food (Cuevas-González et al. 2020).

The new era of 'biotic' research (e.g., postbiotics, parabiotics) has been increasingly used for the development of biotechnology-related products (e.g., Bactistatin ®, Cytoflora ®, Lacteol ® Fort) with functional elements (e.g., metabolites, cell lysate, non-viable cells) for the nutraceutical industry. The evolving concept of postbiotics brings food, microbiology, and personalized therapies closer (Olle et al. 2013). A particular focus on the impact of host-gut microbe interaction on potential benefits and risk factors is: essential. Therefore, future studies based on bioinformatics, epigenetic and omics-based approaches (e.g., metatranscriptomics, metabolomics, lipidomics) will provide an insight to understand better the complex interplay between postbiotics, host biology, and the gut microbiome, including measurement of the composition and functions of gut microbiota (Wegh et al. 2019). Despite these biotics' tremendous positive impact, researchers are looking forward to developing novel bioengineered postbiotics with desirable features (Cuevas-González et al. 2020). Interestingly, the postbiotics could be tailored towards the prevention and treatment of pandemic disease such as SARS-CoV-2 as the changes in structure and metabolic activity of microbiota residing in the intestine from a healthy state may lead to the occurrence of a microbial biomarker for the prediction of coronavirus (Gou et al. 2020). Taken together, such progress may pave the route to reproduce future infant formula mimicking human milk. Of note, postbiotics' mechanism and signaling pathways need further studies in promoting health, including their usage in infant-based formulas.

6.4 Biotechnology in Food Diagnosis and Safety

Milk production has tremendously increased in India, with sufficient quantities available to prepare different dairy products with acceptable microbial load and quality. For human consumption, the safety of milk and processed products (e.g., yogurt) are ensured through the use of different molecular techniques such as gene probing and immunological assays (e.g., monoclonal and polyclonal antibodies) to detect the foodborne pathogens (Sundarraj et al. 2018). The characterization of individual pathogens and their products is crucial to identify the pathogen source involved in diseases. The application of molecular techniques provides valuable information to understand how pathogens pass through the food chain and cause infections (Macori and Cotter 2018). For example, molecular techniques such as Polymerase Chain Reaction (PCR), an extremely versatile, sensitive, and reliable technique, have been extensively used to rapidly detect, identify, and determine pathogens. Additionally, they have been extensively used in DNA cloning, genetic fingerprinting, and detection and diagnosis of infectious diseases. Through this technique, the 16S rRNA gene of bacteria is amplified using universal or specific primers, followed by sequencing to identify not known or novel bacteria species (Magistrado et al. 2001). Furthermore, modified PCR technique such as Multiplex PCR uses multiple primers within the one PCR mixture in order to rapidly detect, identify and differentiate various bacterial species and strains (Shi et al. 2010), leading to amplification, more than one target sequence in a reaction and for different DNA sequences.

Similarly, modified PCR technique such as real-time PCR (qPCR) amplifies and quantifies the target DNA simultaneously within a reaction. It uses a specified primer set, one or two probes, or fluorescent dye for improving the detection signals (Dhanasekaran et al. 2010). As the reaction progresses, the amplified DNA is detected rather than towards the end and shortens detection time compared to standard PCR, including resolving the relative or an absolute number of bacteria across various samples (Shi et al. 2010; Agrimonti C et al. 2019). This technique has also been used to detect *Enterobacter* in the infant formula and *Clostridium* spore quantification in dairy products (Liu et al. 2006; Lopez-Brea 2018). It even encompasses the capability for quantifying target organisms in complex matrices and hence is regarded as a promising tool in improving the safety and quality of food (Martínez et al. 2011). Various studies have recently studied the utilization of droplet digital polymerase chain reaction (ddPCR) that involves eight individual TaqMan™ reactions for the simultaneous detection of, without selective enrichment of *Listeria* spp, *L. monocytogenes*, *Salmonella* spp., and *Campylobacter* spp. in cheese. The ddPCR is a third-generation PCR that quantifies target DNA without standard curve and higher sensitivity than qPCR (Cremonesi et al. 2016; Pan et al. 2020). Additionally, High Resolution Melting quantitative-PCR (HRM-qPCR), a development of PCR technique, has been implemented to identify foodborne pathogens (Forghani et al. 2015; Agrimonti C et al., 2019). In contrast, in the reverse-transcription PCR, the target RNA is reverse transcribed to complementary DNA (cDNA) to detect only viable pathogenic cells (Choi S.H. and Lee S. B, 2011).

The molecular typing methodologies such as restriction fragment length polymorphism (RFLP), random amplified polymorphic DNA (RAPD), and pulsed-field gel electrophoresis (PFGE) has been used to monitor, characterize, detect, and diagnose spoilage flora (Tabit F T 2016; Maryam et al. 2017). The PFGE is classified as a third-generation technique and is considered the 'gold standard' typing method by many researchers to investigate foodborne pathogens and other epidemiological studies (Alonso R et al., 2005; Parizad EG et al., 2016; Neoh HM et al., 2019). It enables research on the entire food supply chain and also traces, including estimating the strain responsible for food contamination, from animal raising to foodborne illness in a human. The PFGF is applied to correlate the strain isolated from different sites to resolve the relationship between strains (Frasao, B. D. S et al. 2017). The RAPD-based method is used for rapid detection of target DNA using arbitrary primers (typically 10-mer primers) under low-stringency PCR conditions (Wassenaar T.M. and Newell D. G, 2000). Furthermore, RFLP uses variation in homologous DNA sequences for bacteria's characterization (Foley SL et al., 2009; Adzitey F et al., 2013). Recently, biosensor-based nanomaterial has been gaining interest to detect allergenic and toxicant compounds and food authentication (Gomez-Arribas et al. 2018; Nikoleli et al. 2018). The application of such approaches and methods has significantly improved human health and hygiene (Meshram et al. 2018; Patel et al. 2016). Newer techniques such as high-resolution melting HRM-qPCR, ddPCR, Loop-mediated isothermal amplification (LAMP), and Next-generation sequencing (NGS) have been used for food labeling and authentication due to their speed, specificity, sensitivity, and multiplexing (Bohme et al., 2019). Notably, the applications of culture-independent nucleic acid-based food microbiology techniques have solved newer emerging problems associated with culture-dependent methodologies. Furthermore, the application of foodomics (e.g., metagenomics, metatranscriptomics) has been extensively used to analyze genes linked with characteristics properties and functionalities such as flavor formation, taste, and probiotic activity, including understanding the behavior of microbes in the food ecosystem (Ceuppens et al. 2014). The membrane-based technologies that are non-thermal environment-friendly, greener, and energy-saving technologies have gained broader applications in the food and dairy sector (Dhinesh Kumar et al., 2017). Different types of membranes, such as microfiltration, ultrafiltration, nanofiltration, and reverse osmosis, have been used to enhance shelf life, yield, texture, compositional properties, and quality of milk products (Kumar et al., 2013). Therefore, it is essential to continuously develop superior membranes that could be extensively used to improve other dairy products for human consumption (Kumar et al., 2013). From the burgeoning data, it is evident to acknowledge biotechnology's immense contribution to reshaping the dairy industry. However, we envision and confident that newer technologies will be developed to improve the quality, nutrition, and safety of different dairy products to impact human health and introduce newer products in the market.

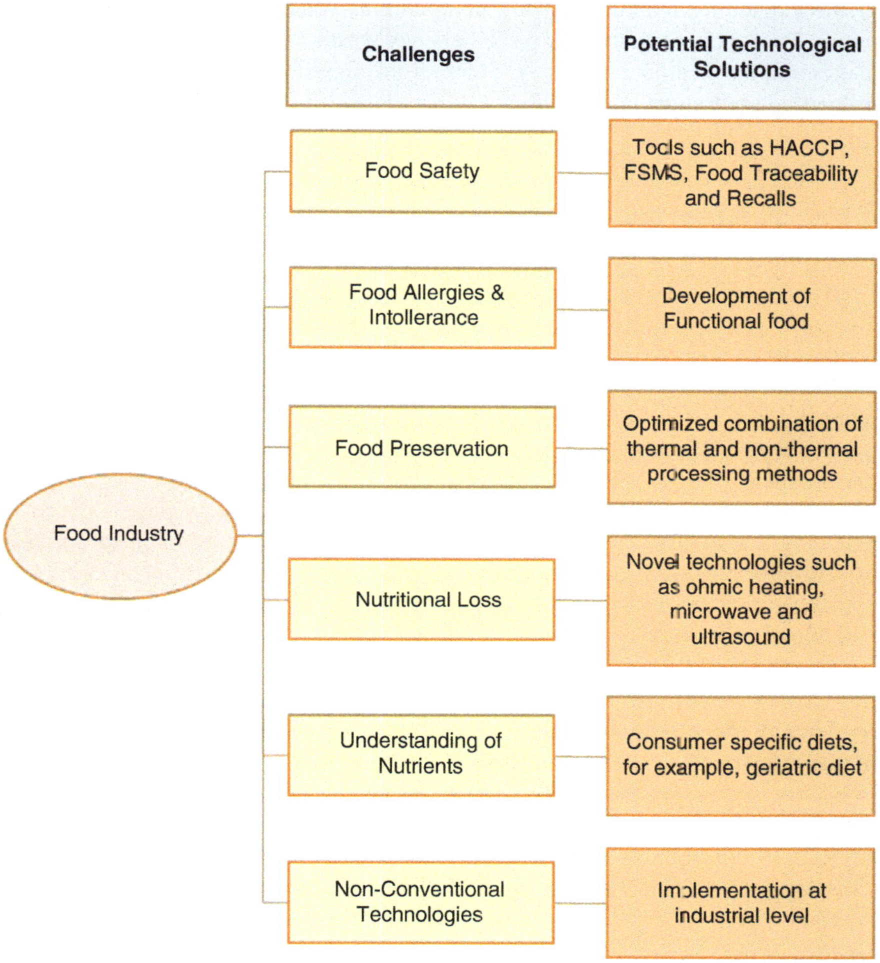

Fig. 6.3 Overview of challenges and potential technological solutions. The present challenges of the food industry could be resolved by respective technological solutions, as presented. HACCP: Hazard analysis, critical control point. FSMS: Food safety management system

6.5 Challenges in Application of Technologies in the Food Industry

Food is an indispensable requirement for life, and therefore, the food industry is always in demand. The food industry has evolved from using fire for cooking to thermal processing, sprinkling salt for preservation to brine, sun drying to solar drying, the focus has shifted from merely satiating the hunger to safe food consumption. Nevertheless, the food industry has been experiencing multiple challenges due to the expanding market, growing competition, volatile demand, and changing trends. Food technologists and researchers worldwide are trying to solve these challenges and have come a long way. Some of the significant challenges are listed as follows and presented in Fig. 6.3.

6.5.1 Food safety

Food safety is an assurance that the food has been prepared, handled, and stored in a way to avoid any food-borne illness and suitable for consumption. The concerns about safe food consumption have been there since the advent of factory-level food production. Although processing technologies such as pasteurization were implemented at the commercial level in the late nineteenth and early twentieth century, food safety has remained a significant concern even today primarily because of the development of new products, and knowledge about microorganisms keeps on emerging. Various tools and techniques in assuring food safety have focused on detecting, monitoring, and preventing physical, chemical, and biological hazards. For instance, hazard analysis critical control point (HACCP) is a systematic approach to identify, evaluate, and control food safety hazards (FDA HACCP Guidelines, 2017). From heating, chilling, drying, and fermentation, and HACCP can be successfully implemented in all cases. Food traceability and recall, the capability to follow the development of food at any stage, and the action of removing food at any stage from the market, respectively, are other fundamental approaches in managing risk as a response to food safety events (FAO, Traceability and Recalls). Food scares are the repercussions of pandemics and public health emergencies as they cause safety concerns among consumers. Although there is no evidence of COVID-19 transmission by food, the pandemic has raised concern about safe food consumption among consumers and increased attention on zoonotic disease, antimicrobial resistance, and food fraud among scientists (FAO, Food Safety and Quality, 2020).

6.5.2 Food Allergies and Intolerances

Allergens and intolerances are a significant challenge for the dairy industry and the whole food industry. The main concern associated with food allergies is the industry's limited knowledge of dealing with them, mainly to avoid cross-contamination (Augusto 2020).Cow milk allergy arising from a particular immune response is undoubtedly the most common type of milk allergy that affects non-exclusively breastfed infants (Vandenplas 2017). Another dairy-associated allergy is casein sensitivity, where the body mistakenly produces antibodies to fight against the milk protein. The third most common dairy intolerance is lactose intolerance, which is often wrongly referred to as an allergen. Lactose intolerance is associated with the inability to digest milk sugar and lactose. Avoiding milk intake due to allergies and intolerances is undoubtedly not the solution since it may lead to adverse health effects such as nutritional rickets, low bone mineral density, and increased fracture risk (Heine et al. 2017). Therefore, functional food development is a potential solution; for example, fermented dairy products consisting of probiotics reduce lactose intolerance symptoms (Hasan et al. 2014).

6.5.3 Food preservation

It is the technique of preserving food from physical, chemical, biological, and biochemical spoilage. The main objectives of food preservation include, but not limited to, are making food safe for consumption, enhancing its shelf life, improving texture, flavor, and consistency, and increasing its acceptability. The age-old practice of drying, curing, pickling, fermentation, and newer technologies of canning, irradiation, and freezing are all food preservation forms. Despite food preservation being practiced for long, various gaps still exist, such as a lack of thorough understanding of microbial and enzymatic inactivation. Thermal inactivation of microbial spores, which forms the basis of calculating thermal preservation processes' safety, rarely follows first-order kinetics (Peleg and Cole 1998). Another case is enzymatic inactivation in commercially available coconut water. The drink's thermal processing damages its flavor, while non-thermal treatment such as high-pressure processing cannot inactivate the peroxidase and polyphenol oxidase, resulting in pink coloration. Therefore, an optimized combination of thermal and non-thermal processing methods could be a potential solution (Augusto 2020).

6.5.4 Nutritional Loss

Over-processing food items lead to the nutritional loss. Conventional thermal processing such as pasteurization, sterilization, and canning are widely used techniques to control microbial spoilage that heavily contributes to the loss of nutrients, including loss of color, texture, and appearance of the food products. There are two approaches to avoid nutritional loss during processing and/or enhancing food's nutrient content. Firstly, through a process known as food enrichment, lost nutrients during processing are added to the food that comes with an added cost and is not feasible in every case. However, a more reliable long-term solution is in novel thermal techniques like ohmic heating, microwave, non-thermal technologies, for example, irradiation and high-pressure processing, and pre-treatments like an ultrasound for food drying. Novel thermal processing, for example, ohmic heating, microwave, radio-frequency heating, is efficient in minimizing nutrient and quality loss due to heating (Ling et al. 2015). Employing an electron beam for pasteurization is another advancement in non-thermal milk processing affecting raw milk's nutrient and aroma profile. Rojas et al. (2019) studied ultrasound technology for incorporating microencapsulated nutrients such as iron and carotenoids during pre-treatments of food drying. Surprisingly, it has been reported that thermal treatments responsible for the loss of natural antioxidants in tomato and coffee derivatives also aids in the formation of antioxidants due to Maillard reaction products (Nicoli et al. 1997).

6.5.5 Better Understanding of Nutrients

Although food safety and quality are often the prime focus while developing a new product or improving existing ones, its nutritional requirement throughout the processing is still overlooked. Also, it is equally crucial to understand the presence of nutrients in food and bioavailability. Verkempinck et al. (2020) reported how various food processing methods could enhance nutrient accessibility. For example, non-thermal techniques like high-pressure homogenization and ultrasound can be used to improve *in vitro* carotenoids. Similarly, the physicochemical properties of food matrices such as hardness, compactness, elasticity responsible for nutrient absorption of dairy products like yogurt (semi-solid), milk (liquid), and cheese (solid) are significantly affected (Fardet et al. 2019). Another essential aspect to understand is a customer requirement, which has to evolve due to consumer awareness, purchasing power, health issues, personal choices, etc. For instance, the nutrient requirement of the elderly is entirely different from a young person as it affects their mental, physical and social health (Kaur et al. 2019). Whey protein consists of β-lactoglobulin, bovine serum albumin, lactoferrin, α-lactalbumin are nutritionally rich and confer benefit in geriatric nutrition (Sreeja et al. 2013). Therefore, food technologists must work closely with nutrition scientists for wholesome development.

6.5.6 Potential of Non-conventional Technologies

Non-conventional technologies such as ultrasound, ohmic heating, pulsed electric fields, etc. offer great potential, as they are relatively environmentally friendly, economical, and consume less energy for operations. Nonetheless, these food-processing methods are used in industrial production. Although the concrete reasons behind this are unclear, ohmic heating has been studied for its application in improving the fermentation process to develop the probiotic dairy products like yogurts, fermented milk, and dairy-based drinks (Pereira et al. 2018). Similarly, non-conventional cell disruption technologies (e.g., high-pressure homogenization, microwave-assisted extraction, pulsed electric fields) have been reported to offers sustainable chemistry solutions in the extraction of food additives (Barba et al. 2015).

6.6 Socioeconomic Constraints and Potential Ways to Overcome Them

The local government's role in developing core sectors like food and agriculture cannot be neglected. The policies and regulations must be drafted to promote biotechnological developments in the food and dairy industry. The government's support in the form of favorable, flexible policies plays a crucial role in encouraging technological advancements. At the same time, it is necessary to ensure that the developed policies follow the global system. The flexibility of trade flow plays a crucial role in determining economic progress. Fostering startups and micro, small and medium enterprises could prove beneficial in the long run, especially when these businesses could quickly be promoted by providing tax incentives, elasticity in loans, etc. Since the developing nations have limited resources, they need to identify the key areas and prioritize their technological development work plan. Food safety is one such domain that should be the prime concern for every country, however, the extent and approach to deal with this issue could vary.

Food security is still an obstacle yet to overcome for many low-income countries. In most of these cases, agricultural production is sufficient to meet the population's demand. However, due to the poor food supply chain, the food wastage while reaching from farm to fork is countless. Another major drawback in the food supply chain is the proper maintenance of the cold chain due to disrupted power supply. Therefore, a consistent power supply needs to be achieved. There is a strong need to boost intellectual property rights since most indigenous products' traditional knowledge is passed through cultural channels. Disseminating information regarding patents and other intellectual property rights would promote and inculcate a sense of motivation among researchers and scientists. Another crucial aspect that is often overlooked while drafting policies is the consumer's needs, which could be achieved through effective communication between various stakeholders. The information and opinions must be shared between consumers and food industry stakeholders. Public participation in decision-making is crucial for implementing new biotechnological applications and working towards safe food consumption.

6.7 Conclusion

Biotechnological applications offer the food industry potential solutions to food processing, production, packaging, and safety challenges. The age-old pasteurization process is still relevant with certain modifications and finds its application in almost all dairy products. Notably, technologies such as biobased packaging material are prospective solutions towards a greener and sustainable environment. The utilization of dairy industry by-products such as whey and buttermilk are economical and confer health benefits.

Presently, the food industry faces multiple challenges covering various aspects of the industry, including food safety, allergy and intolerances, preservation, nutritional loss, and unexplored potential of non-conventional. Technology advances with time, and it is worthwhile to envision existing technologies to evolve based on requirements and demand. For instance, recent technologies such as non-thermal food processing need to find their way from the laboratory scale to the industry. This will further help meet various challenges that the food industry is facing today. Technology alone cannot be implemented without the support of policies and regulations. With evolving technology, the regulatory structures have to evolve to support the widespread adoption and rollout of the technology in the food industry. Technology, policy, and regulatory infrastructure should sync and complement each other to develop the food market. The ability of biotechnological applications in various food industry domains has enormous potential and further needs proper implementation.

Acknowledgments This work is supported by the Department of Biotechnology (DBT), Government of India, grant no. BT/PR28190/BBE/117/276/2018. We also sincerely appreciate the help of Mr. Gaurav Gupta with the figures.

References

Abida J, Rayees B, Masoodi FA (2014) Pulsed light technology: a novel method for food preservation. Int Food Res J 21(3):839–848

Adzitey F, Huda N, Ali GRR (2013) Molecular techniques for detecting and typing of bacteria, advantages and application to foodborne pathogens isolated from ducks. 3. Australas Biotechnol 3(2):97–107. https://doi.org/10.1007/s13205-012-0074-4

Aggarwal R, Aggarwal S (2020) Dry powdered human milk fortifier. U.S. Patent Application 16/781,149

Agrillo B, Balestrieri M, Gogliettino M, Palmieri G, Moretta R, Proroga YT et al (2019) Functionalized polymeric materials with bio-derived antimicrobial peptides for "active" packaging. Int J Mol Sci 20(3):601. https://doi.org/10.3390/ijms20030601

Agrimonti C, Bottari B, Sardaro MLS, Marmiroli N (2019) Application of real-time PCR (qPCR) for characterization of microbial populations and type of milk in dairy food products. Crit Rev Food Sci Nutr 59(3):423–442. https://doi.org/10.1080/10408398.2017.1375893

Ahern GJ, Hennessy AA, Ryan CA, Ross RP, Stanton C (2019) Advances in infant formula science. Annu Rev Food Sci Technol 10(1):75–102. https://doi.org/10.1146/annurev-food-081318-104308

Ahmed A, Bajwa U (2019) Composition, texture and microstructure appraisal of paneer coagulated with sour fruit juices. J Food Sci & Technol 56(1):253–261. https://doi.org/10.1007/s13197-018-3484-4

Ahuja KK, Goyal GK (2013) Combined effect of vacuum packaging and refrigerated storage on the chemical quality of paneer tikka. J Food Sci & Technol 50(3):620–623. https://doi.org/10.1007/s13197-012-0688-x

Al-Holy MA, Castro LF, Al-Qadiri HM (2010) Inactivation of Cronobacter spp. (Enterobacter sakazakii) in infant formula using lactic acid, copper sulfate and monolaurin. Lett Appl Microbiol 50(3):246–251. https://doi.org/10.1111/j.1472-765X.2009.02782.x

Al-Holy MA, Sabbah K, Osaili TM, Qatatsheh AA (2015) Inactivation of Cronobacter sakazakii in infant formula and infant cereals using chitosan and lactic acid. J Food Process Preserv 39(6):1229–1234. https://doi.org/10.1111/jfpp.12339

Alonso R, Martin A, Pelaez T, Marin M, Rodriguez-Creixems M, Bouza E (2005) An improved protocol for pulsed-field gel electrophoresis typing of Clostridium difficile. J Med Microbiol 54(2):155–157. https://doi.org/10.1099/jmm.0.45808-0

Amini RK, Islam MZ, Kitamura Y, Kokawa M (2019) Utilization of fermented rice milk as a novel coagulant for development of paneer (soft cheese). Foods 8(8):339. https://doi.org/10.3390/foods8080339

Augusto PE (2020) Challenges, trends and opportunities in food processing. Curr Opin Food Sci. https://doi.org/10.1016/j.cofs.2020.03.005

Ayad EH, Verheul A, Wouters JT, Smit G (2000) Application of wild starter cultures for flavour development in pilot plant cheese making. Int Dairy J 10(3):169–179. https://doi.org/10.1016/S0958-6946(00)00041-8

Ayar A, Gürlin E (2014) Production and sensory, textural, physicochemical properties of flavored spreadable yogurt. Life Sci J 11(4)

Baba WN, Din S, Punoo HA, Wani TA, Ahmad M, Masoodi FA (2016) Comparison of cheese and paneer whey for production of a functional pineapple beverage: Nutraceutical properties and Shelf life. J Food Sci Technol 53(6):2558–2568. https://doi.org/10.1007/s13197-016-2218-8

Barba FJ, Grimi N, Vorobiev E (2015) New approaches for the use of non-conventional cell disruption technologies to extract potential food additives and nutraceuticals from microalgae. Food Eng Rev 7(1):45–62. https://doi.org/10.1007/s12393-014-9095-6

Bermuâ Dez-Aguirre D, Mawson R, Versteeg K, Barbosa-Caâ NOVASGV (2009) Composition parameters, physical±chemical characteristics and shelf-life of whole milk after thermal and thermo-sonication treatments. J Food Quality 32:283–302. https://doi.org/10.1111/j.1745-4557.2009.00250.x

Beshkova D, Simova E, Frengova G, Simov Z (1998) Production of flavour compounds by yogurt starter cultures. J Ind Microbiol Biotechnol 20(3–4):180–186. https://doi.org/10.1038/sj.jim.2900504

Böhme K, Calo-Mata P, Barros-Velázquez J, Ortea I (2019) Review of recent DNA-based methods for main food-authentication topics. J Agric Food Chem 67(14):3854–3864. https://doi.org/10.1021/acs.jafc.8b07016

Brião VB, Salla ACV, Miorando T, Hemkemeier M, Favaretto DPC (2019) Water recovery from dairy rinse water by reverse osmosis: giving value to water and milk solids. Res Conserv & Recyc 140:313–323. https://doi.org/10.1016/j.resconrec.2018.10.007

Bund RK, Pandit AB (2007) Rapid lactose recovery from paneer whey using sonocrystallization: a process optimization. Chem Eng & Proc: Proc Intens 46(9):846–850. https://doi.org/10.1016/j.cep.2007.05.015

Buyuktuncer Z (2017) Chapter 8 – Yogurt with plant sterols and stanols (N. P. B. T.-Y. in H. and D. P. Shah (ed.); pp. 151–169). Academic Press. https://doi.org/10.1016/B978-0-12-805134-4.00008-0

Cano-Ruiz ME, Richter RL (1997) Effect of homogenization pressure on the milk fat globule membrane proteins. J Dairy Sci 80(11):2732–2739. https://doi.org/10.3168/jds.S0022-0302(97)76235-0

Carcel JA, Benedito J, Sanjuan N, Saâ Nchez E (1998) AplicacioÂn de los ultrasonidos en las industrias de productos laÂcteos y derivados. AlimentacioÂn, Equipos y TecnologõÂa 135–141

Caron A, St-Gelais D, Pouliot Y (1997) Coagulation of milk enriched with ultrafiltered or diafiltered microfiltered milk retentate powders. Int Dairy J 7(6-7):445–451. https://doi.org/10.1016/S0958-6946(97)00024-1

Castro J M, Tornadijo M E, Fresno J M, Sandoval H (2015) Biocheese: A food probiotic carrier. In BioMed research international 2015. Hindawi Publishing Corporation. https://doi.org/10.1155/2015/723056.

Ceuppens S, Li D, Uyttendaele M, Renault P, Ross P, Van Ranst M, Cocolin L, Donaghy J (2014) Molecular Methods in Food Safety Microbiology: Interpretation and Implications of Nucleic Acid Detection. Comprehen Rev Food Sci Food Safety 13(4):551–577. https://doi.org/10.1111/1541-4337.12072

Chandan R C, Gandhi A, Shah N P (2017) Chapter 1 – Yogurt: Historical Background, Health Benefits, and Global Trade (N. P. B. T.-Y. in H. and D. P. Shah (ed.); pp. 3–29). Academic Press. https://doi.org/10.1016/B978-0-12-805134-4.00001-8.

Chassy BM, Mercenier A, Flickinger J (1988) Transformation of bacteria by electroporation. Trends Biotechnol 6(12):303–309. https://doi.org/10.1016/0167-7799(88)90025-X

Chen G Q, Leong T S, Kentish S E, Ashok Kumar M, Martin G J (2019) Membrane separations in the dairy industry. In Separation of functional molecules in food by membrane technology (pp. 267–304). Academic Press. https://doi.org/10.1016/B978-0-12-815056-6.00008-5

Chiesa LM, DeCastelli L, Nobile M, Martucci F, Mosconi G, Fontana M, Panseri S (2020) Analysis of antibiotic residues in raw bovine milk and their impact toward food safety and on milk starter cultures in cheese-making process. LWT 131:109783. https://doi.org/10.1016/j.lwt.2020.109783

Choi SH, Lee SB (2011) Improved detection of viable Escherichia coli O157: H7 in milk by using reverse transcriptase-PCR. Korean J Food Sci Anim Resour 31(2):158–165. https://doi.org/10.5851/kosfa.2011.31.2.158

Corrieu G, Béal C (2015) Yogurt: The Product and its Manufacture. In Encyclopedia of Food and Health (pp. 617–624). Elsevier Inc. https://doi.org/10.1016/B978-0-12-384947-2.00766-2

Costa MJ, Maciel LC, Teixeira JA, Vicente AA, Cerqueira MA (2018) Use of edible films and coatings in cheese preservation: Opportunities and challenges. Food Res Int 107:84–92. https://doi.org/10.1016/j.foodres.2018.02.013

Crawford LM, Ruff EH (1996) A review of the safety of cold pasteurization through irradiation. Food Control 7(2):87–97. https://doi.org/10.1016/0956-7135(96)00004-7

Cremonesi P, Cortimiglia C, Picozzi C, Minozzi G, Malvisi M, Luini M, Castiglioni B (2016) Development of a droplet digital polymerase chain reaction for rapid and simultaneous identification of common foodborne pathogens in soft cheese. Front Microbiol 7:1725. https://doi.org/10.3389/fmicb.2016.01725

Cuevas-González P F, Liceaga A M, Aguilar-Toalá J E (2020) Postbiotics and paraprobiotics: From concepts to applications. In Food research international (Vol. 136). Elsevier Ltd. https://doi.org/10.1016/j.foodres.2020.109502

Cui H, Wu J, Li C, Lin L (2017) Improving anti-listeria activity of cheese packaging via nanofiber containing nisin-loaded nanoparticles. LWT-Food Sci & Technol 81:233–242. https://doi.org/10.1016/j.lwt.2017.04.003

Dan T, Chen H, Li T, Tian J, Ren W, Zhang H, Sun T (2019) Influence of Lactobacillus plantarum P-8 on fermented milk flavor and storage stability. Front Microbiol 9:3133. https://doi.org/10.3389/fmicb.2018.03133

Davis J G (1965) Cheese: basic technology (Vol. 1). American Elsevier Publishing Company.

De S (1980) Outlines of dairy technology. Outlines of dairy technology

De Bassi LG, Ferreira GCC, Da Silva AS, Sivieri K, Aragon-Alegro LC, Costa MDR (2012) Evaluation of physicochemical, microbiological and sensorial characteristics of fermented milk beverages with buttermilk addition. Int J Dairy Technol 65(2):282–286. https://doi.org/10.1111/j.1471-0307.2011.00764.x

De Vos W M (2001) Advances in genomics for microbial food fermentations and safety. In Current Opinion in Biotechnology 12(5): 493–498. Elsevier Ltd. https://doi.org/10.1016/S0958-1669(00)00252-4.

Deng Z, Wang F, Zhou B, Li J, Li B, Liang H (2019) Immobilization of pectinases into calcium alginate microspheres for fruit juice application. Food Hydrocoll 89:691–699. https://doi.org/10.1016/j.foodhyd.2018.11.031

Deshpande G, Athalye-Jape G, Patole S (2018) Para-probiotics for preterm neonates—The next frontier. In Nutrients (Vol. 10, Issue 7). MDPI AG. https://doi.org/10.3390/nu10070871.

Dhanasekaran S, Doherty TM, Kenneth J, TB Trials Study Group (2010) Comparison of different standards for real-time PCR-based absolute quantification. J Immunol Methods 354(1-2):34–39. https://doi.org/10.1016/j.jim.2010.01.004

Dhinesh Kumar V, Ramasamy D (2017) Review on membrane technology applications in food and dairy processing. J Appl Biotechnol & Bioeng 3(5):399–407

Dogan B, Boor KJ (2003) Genetic diversity and spoilage potentials among Pseudomonas spp. isolated from fluid milk products and dairy processing plants. Appl Environ Microbiol 69(1):130–138. https://doi.org/10.1128/aem.69.1.130-138.2003

dos Santos DC, de Oliveira Filho JG, Santana ACA, de Freitas BSM, Silva FG, Takeuchi KP, Egea MB (2019) Optimization of soymilk fermentation with kefir and the addition of inulin: Physicochemical, sensory and technological characteristics. LWT 104:30–37. https://doi.org/10.1016/j.lwt.2019.01.030

Dussault-Chouinard I, Britten M, Pouliot Y (2019) Improving rennet coagulation and cheese making properties of reverse osmosis skim milk concentrates by pH adjustment. Int Dairy J 95:6–14. https://doi.org/10.1016/j.idairyj.2019.03.008

El-Abbadi NH, Dao MC, Meydani SN (2014) Yogurt: Role in healthy and active aging. Am J Clin Nutr 99(5). https://doi.org/10.3945/ajcn.113.073957

Elizaquível P, Sánchez G, Salvador A, Fiszman S, Dueñas MT, López P, Fernández de Palencia P, Aznar R (2011) Evaluation of yogurt and various beverages as carriers of lactic acid bacteria producing 2-branched (1,3)-β-d-glucan. J Dairy Sci 94(7):3271–3278. https://doi.org/10.3168/jds.2010-4026

FAO (2010) Current status and options for biotechnologies in food processing and in food safety in developing countries. In Biotechnologies for Agricultural Development: Proceedings of the FAO Technical Conference on 'Agricultural Biotechnologies in Developing Countries: Options and Opportunities in Crops, Forestry, Livestock, Fisheries and Agro-Industry to Face the Challenges of Food Insecurity and Climate Change' (pp. 240–277). Rome: Food and Agriculture Organization of the United Nations. (Accessed 15 November 2020)

Fardet A, Dupont D, Rioux LE, Turgeon SL (2019) Influence of food structure on dairy protein, lipid and calcium bioavailability: A narrative review of evidence. Crit Rev Food Sci Nutr 59(13):1987–2010. https://doi.org/10.1080/10408398.2018.1435503

Fazilah NF, Ariff AB, Khayat ME, Rios-Solis L, Halim M (2018) Influence of probiotics, prebiotics, synbiotics and bioactive phytochemicals on the formulation of functional yogurt. J Funct Foods 48:387–399. https://doi.org/10.1016/j.jff.2018.07.039

Foley SL, Lynne AM, Nayak R (2009) Molecular typing methodologies for microbial source tracking and epidemiological investigations of Gram-negative bacterial foodborne pathogens. Infect Genet Evol 9(4):430–440. https://doi.org/10.1016/j.meegid.2009.03.004

Folkenberg DM, Dejmek P, Skriver A, Skov Guldager H, Ipsen R (2006) Sensory and rheological screening of exopolysaccharide producing strains of bacterial yoghurt cultures. Int Dairy J 16(2):111–118. https://doi.org/10.1016/j.idairyj.2004.10.013

Forghani F, Langaee T, Eskandari M, Seo KH, Chung MJ, Oh DH (2015) Rapid detection of viable Bacillus cereus emetic and enterotoxic strains in food by coupling propidium monoazide and multiplex PCR (PMA-mPCR). Food Control 55:151–157. https://doi.org/10.1016/j.foodcont.2015.02.049

Frasao BDS, Marin VA, Conte-Junior CA (2017) Molecular detection, typing, and quantification of Campylobacter spp. in foods of animal origin. Comprehen Rev Food Sci Food Safety 16(4):721–734. https://doi.org/10.1111/1541-4337.12274

Gahruie HH, Eskandari MH, Mesbahi G, Hanifpour MA (2015) Scientific and technical aspects of yogurt fortification: a review. Food Sci Human Wellness 4(1):1–8

Galli BD, Martin JGP, da Silva PPM, Porto E, Spoto MHF (2016) Sensory quality of Camembert-type cheese: Relationship between starter cultures and ripening molds. Int J Food Microbiol 234:71–75. https://doi.org/10.1016/j.ijfoodmicro.2016.06.025

Gaull GE (2000) Infant formula compositions and nutrition containing genetically engineered human milk proteins. US Patent 6:020,015

Gawai K M, Mudgal S P, Prajapati J B (2017) Chapter 3 – Stabilizers, Colorants, and Exopolysaccharides in Yogurt (N. P. B. T.-Y. in H. and D. P. Shah (ed.); pp. 49–68). Academic Press. https://doi.org/10.1016/B978-0-12-805134-4.00003-1

Gaynes R (2017) The discovery of penicillin – new insights after more than 75 years of clinical use. Emerg Infect Dis 23(5):849. https://doi.org/10.3201/eid2305.161556

Ghoshal G (2018) Chapter 2 – Biotechnology in food processing and preservation: An overview. In A. M. Holban & A. M. B. T.-A. in B. for F. I. Grumezescu (Eds.), Handbook of Food Bioengineering (pp. 27–54). Academic Press. https://doi.org/10.1016/B978-0-12-811443-8.00002-5

Gómez-Arribas L, Benito-Peña E, Hurtado-Sánchez M, Moreno-Bondi M (2018) Biosensing based on nanoparticles for food allergens detection. Sensors 18(4):1087. https://doi.org/10.3390/s18041087

Gou W, Fu Y, Yue L, Chen G, Cai X, Shuai M, Xu F, Yi X, Chen H, Zhu Y, Xiao M, Jiang Z, Miao Z, Xiao C, Shen B, Wu X, Zhao H, Ling W, Wang J, Zheng J-S (2020) Gut Microbiota May Underlie the Predisposition of Healthy Individuals to COVID-19-Sensitive Proteomic Biomarkers. MedRxiv. https://doi.org/10.1101/2020.04.22.20076091

Govindasamy-Lucey S, Lin T, Jaeggi JJ, Johnson ME, Lucey JA (2006) Influence of condensed sweet cream buttermilk on the manufacture, yield, and functionality of pizza cheese. J Dairy Sci 89(2):454–467. https://doi.org/10.3168/jds.S0022-0302(06)72109-9

Green Corkins K, Shurley T (2016) What's in the bottle? A review of infant formulas. Nutr Clin Pract 31(6):723–729. https://doi.org/10.1177/0884533616669362

Guerra MG, do Nascimento Rangel AH, Spyrides MHC, de Lara IAR, de Araújo VM, de Aguiar EM (2013) Factors influencing total bacterial count in tanks: an application of linear mixed-effect models. Italian J Animal Sci 12(4):e75. https://doi.org/10.4081/ijas.2013.e75

Gutiérrez D, Rodríguez-Rubio L, Martínez B, Rodríguez A, García P (2016) Bacteriophages as weapons against bacterial biofilms in the food industry. In Frontiers in Microbiology (Vol. 7, Issue JUN). Frontiers Research Foundation. https://doi.org/10.3389/fmicb.2016.00825

Haghighi-Manesh S, Azizi MH (2017) Active packaging systems with emphasis on its applications in dairy products. J Food Process Eng 40:e12542. https://doi.org/10.1111/jfpe.12542

Harlander SK (1992) Genetic improvement of microbial starter cultures. Applications of Biotechnology to Traditional Fermented Foods. National Academy Press, Washington, DC, pp 20–26

Hartmann PE, Lai CT, Sherriff JL, Simmer KN, Lewis MA, Mitoulas LR, Davis BI, Medela Holding AG (2008) Human milk fortifiers and methods for their production. U.S. Patent Application 11/662,348

Hasan MN, Sultan MZ, Mar-E-Um M (2014) Significance of fermented food in nutrition and food science. J Sci Res 6(2):373–386. https://doi.org/10.3329/jsr.v6i2.16530

He A, Chin J, Lomiguen CM (2020) Benefits of probiotic yogurt consumption on maternal health and pregnancy outcomes: A systematic review. Cureus 12(7). https://doi.org/10.7759/cureus.9408

Heine RG, AlRefaee F, Bachina P, De Leon JC, Geng L, Gong S, Rogacion JM (2017) Lactose intolerance and gastrointestinal cow's milk allergy in infants and children–common misconceptions revisited. World Allergy Org J 10(1):41. https://doi.org/10.1186/s40413-017-0173-0

Hickey CD, O'Sullivan MG, Davis J, Scholz D, Kilcawley KN, Wilkinson MG, Sheehan JJ (2018) The effect of buttermilk or buttermilk powder addition on functionality, textural, sensory and volatile characteristics of Cheddar-style cheese. Food Res Int 103:468–477. https://doi.org/10.1016/j.foodres.2017.09.081

Hill D, Ross R P, Arendt E, Stanton C (2017) Chapter 4 – Microbiology of Yogurt and Bio-Yogurts Containing Probiotics and Prebiotics (N. P. B. T.-Y. in H. and D. P. Shah (ed.); pp. 69–85). Academic Press. https://doi.org/10.1016/B978-0-12-805134-4.00004-3.

Hisatsune J, Sato'o Y, Yu L, Kutsuno S, Hayakawa Y, Sugai M (2016) Efficient transformation of Staphylococcus aureus using multi-pulse electroporation. J Microbiol Methods 130:69–72. https://doi.org/10.1016/j.mimet.2016.08.012

Holm VK, Mortensen G, Risbo J (2006) Quality changes in semi-hard cheese packaged in a poly (lactic acid) material. Food Chem 97(3):401–410. https://doi.org/10.1016/j.foodchem.2005.05.016

http://www.fao.org/food-safety/covid-19-materials/en/ (Accessed 22 November 2020).

http://www.fao.org/food-safety/food-control-systems/supply-chains-and-consumers/traceability-and-recalls/en/ (Accessed 24 November 2020).

https://www.fda.gov/food/hazard-analysis-critical-control-point-haccp/haccp-principles-application-guidelines (Accessed 25 November 2020).

Jalilzadeh A, Tunçtürk Y, Hesari J (2015) Extension shelf life of cheese: a review. Int J Dairy Sci 10(2):44–60

James BJ, Jing Y, Chen XD (2003) Membrane fouling during filtration of milk—a microstructural study. J Food Eng 60(4):431–437. https://doi.org/10.1016/S0260-8774(03)00066-9

Jemaa MB, Falleh H, Neves MA, Isoda H, Nakajima M, Ksouri R (2017) Quality preservation of deliberately contaminated milk using thyme free and nanoemulsified essential oils. Food Chem 217:726–734. https://doi.org/10.1016/j.foodchem.2016.09.030

Johnson ME (2017) A 100-year review: Cheese production and quality. J Dairy Sci 100(12):9952–9965. https://doi.org/10.3168/jds.2017-12979

Jotarkar PS, Panjagari NR, Singh AK, Arora S (2018) Effect of whey protein-iron based edible coating on the quality of Paneer and process optimization. Int J Dairy Technol 71(2):395–407. https://doi.org/10.1111/1471-0307.12478

Kanawjia SK, Rizvi SS (2000) Development of paneer from microfiltered milk. Indian J Dairy & Biosci 11:67–70

Kapoor R, Jash A, Rizvi SS (2013) Shelf-life extension of Paneer by a sequential supercritical-CO2-based process. LWT 135:110060. https://doi.org/10.1016/j.lwt.2020.110060

Karunamay S, Badhe SR, Shulka V, Jaiswal S (2020) Effect of essential oil of clove and oregano treated edible packaging film in extending the shelf life of paneer. Pharm Inno J 9:317–322

Kaur D, Rasane P, Singh J, Kaur S, Kumar V, Mahato DK, Kumar S (2019) Nutritional interventions for elderly and considerations for the development of geriatric foods. Curr Aging Sci 12(1):15–27. https://doi.org/10.2174/1874609812666190521110548

Kelly CD (1939) Studies on the control of gas in Limburger cheese. J Dairy Sci 22(5):299–307. https://doi.org/10.3168/jds.S0022-0302(39)92888-2

Khan SU, Pal MA, Wani SA, Salahuddin M (2014) Effect of different coagulants at varying strengths on the quality of paneer made from reconstituted milk. J Food Sci Technol 51(3):565–570. https://doi.org/10.1007/s13197-011-0525-7

Kok CR, Hutkins R (2018) Yogurt and other fermented foods as sources of health-promoting bacteria. Nutr Rev 76(Supplement_1):4–15. https://doi.org/10.1093/nutrit/nuy056

Kontominas MG (2010) Packaging and the shelf life of milk. In: Food packaging and shelf life: a practical guide (edited by Gordon L. Robertson). Taylor and Francis Group, LLC

Kumar P, Sharma N, Ranjan R, Kumar S, Bhat ZF, Jeong DK (2013) Perspective of membrane technology in dairy industry: a review. Asian Australas J Anim Sci 26(9):1347–1358. https://doi.org/10.5713/ajas.2013.13082

Kumar S, Rai DC, Niranjan K, Bhat ZF (2014) Paneer—an Indian soft cheese variant: a review. J Food Sci & Technol 51(5):821–831. https://doi.org/10.1007/s13197-011-0567-x

Kumar R, Mishra D, Sutariya H, Chaudhary MB, Rao KJ (2019) Effect of different coagulants on the yield, sensory, instrumental colour and textural characteristics of cow's milk Paneer. Int J Dairy Technol 72(4):617–625. https://doi.org/10.1111/1471-0307.12615

Landete J M (2017) A review of food-grade vectors in lactic acid bacteria: from the laboratory to their application. In Critical Reviews in Biotechnology (Vol. 37, Issue 3, pp. 296–308). Taylor and Francis Ltd. https://doi.org/10.3109/07388551.2016.1144044

Le TT, Van Camp J, Pascual PAL, Meesen G, Thienpont N, Messens K, Dewettinck K (2011) Physical properties and microstructure of yoghurt enriched with milk fat globule membrane material. Int Dairy J 21(10):798–805. https://doi.org/10.1016/j.idairyj.2011.04.015

Lee SHI, Barancelli GV, de Camargo TM, Corassin CH, Rosim RE, da Cruz AG et al (2017) Biofilm-producing ability of Listeria monocytogenes isolates from Brazilian cheese processing plants. Food Res Int 91:88–91. https://doi.org/10.1016/j.foodres.2016.11.039

Lemaire M, Le Huërou-Luron I, Blat S (2018) Effects of infant formula composition on long-term metabolic health. In Journal of Developmental Origins of Health and Disease (Vol. 9, Issue 6, pp. 573–589). Cambridge University Press. https://doi.org/10.1017/S2040174417000964

Lin L, Gu Y, Cui H (2019) Moringa oil/chitosan nanoparticles embedded gelatin nanofibers for food packaging against Listeria monocytogenes and Staphylococcus aureus on cheese. Food Packag Shelf Life 19:86–93. https://doi.org/10.1016/j.fpsl.2018.12.005

Ling B, Tang J, Kong F, Mitcham EJ, Wang S (2015) Kinetics of food quality changes during thermal processing: A review. Food Bioproc Tech 8(2):343–358. https://doi.org/10.1007/s11947-014-1398-3

Liu Y, Cai X, Zhang X, Gao Q, Yang X, Zheng Z, Huang X (2006) Real time PCR using TaqMan and SYBR Green for detection of Enterobacter sakazakii in infant formula. J Microbiol Methods 65(1):21–31. https://doi.org/10.1016/j.mimet.2005.06.007

Lönnerdal B, Hernell O (2016) An opinion on "staging" of infant formula: A developmental perspective on infant feeding. In Journal of Pediatric Gastroenterology and Nutrition (Vol. 62, Issue 1, pp. 9–21). Lippincott Williams and Wilkins. https://doi.org/10.1097/MPG.0000000000000806

Lopes RP, Mota MJ, Pinto CA, Sousa S, da Silva JAL, Gomes AM et al (2019) Physicochemical and microbial changes in yogurts produced under different pressure and temperature conditions. LWT 99:423–430. https://doi.org/10.1016/j.lwt.2018.09.074

Lopez-Brea SG, Gómez-Torres N, Arribas MÁ (2018) Spore-forming bacteria in dairy products. Microbiol Dairy Proc 11. https://doi.org/10.1002/9781119115007.ch2

Lopez-Fandino R, Carrascosa AV, Olano A (1996) The effects of high pressure on whey protein denaturation and cheese-making properties of raw milk. J Dairy Sci 79(6):929–936. https://doi.org/10.3168/jds.S0022-0302(96)76443-3

Lourens-Hattingh A, Viljoen B C (2001) Yogurt as probiotic carrier food. In International Dairy Journal (Vol. 11, Issues 1–2, pp. 1–17). Elsevier. https://doi.org/10.1016/S0958-6946(01)00036-X

Macori G, Cotter PD (2018) Novel insights into the microbiology of fermented dairy foods. Curr Opin Biotechnol 49:172–178. https://doi.org/10.1016/j.copbio.2017.09.002

Magistrado PA, Garcia MM, Raymundo AK (2001) Isolation and polymerase chain reaction-based detection of Campylobacter jejuni and Campylobacter coli from poultry in the Philippines. Int J Food Microbiol 70(1-2):197–206. https://doi.org/10.1016/S0168-1605(01)00537-2

Martin C R, Ling P R, Blackburn G L (2016) Review of infant feeding: Key features of breast milk and infant formula. In Nutrients (Vol. 8, Issue 5). MDPI AG. https://doi.org/10.3390/nu8050279

Martínez N, Martín MC, Herrero A, Fernández M, Alvarez MA, Ladero V (2011) qPCR as a powerful tool for microbial food spoilage quantification: Significance for food quality. Trends Food Sci Technol 22(7):367–376. https://doi.org/10.1016/j.tifs.2011.04.004

McKay LL, Baldwin KA, Efstathiou JD (1976) Transductional evidence for plasmid linkage of lactose metabolism in streptococcus lactis C2. Appl Environ Microbiol 32(1):45–52

Meng F, Chen R, Zhu X, Lu Y, Nie T, Lu F, Lu Z (2018) Newly effective milk-clotting enzyme from Bacillus subtilis and its application in cheese making. J Agric Food Chem 66(24):6162–6169. https://doi.org/10.1021/acs.jafc.8b01697

Meraj M, Mahmood SM, Khan ME, Azhar M, Tiwari GN (2021) Effect of N-Photovoltaic thermal integrated parabolic concentrator on milk temperature for pasteurization: A simulation study. Renew Energy 163:2153–2164. https://doi.org/10.1016/j.renene.2020.10.103

Meshram BD, Agrawal AK, Adil S, Ranvir S, Sande KK (2018) Biosensor and its application in food and dairy industry: a review. Int J Curr Microbiol Appl Sci 7(2):3305–3324. https://doi.org/10.20546/ijcmas.2018.702.397

Morin P, Jiménez-Flores R, Pouliot Y (2007) Effect of processing on the composition and microstructure of buttermilk and its milk fat globule membranes. Int Dairy J 17(10):1179–1187. https://doi.org/10.1016/j.idairyj.2007.03.010

Maryam BM, Datsugwai MSS, Shehu I (2017) The role of biotechnology in food production and processing. Industrial Eng 1(1):24–35

Nagaoka S (2019) Yogurt production. In Methods in Molecular Biology (Vol. 1887, pp. 45–54). Humana Press Inc. https://doi.org/10.1007/978-1-4939-8907-2_5

Naghmoush MR, Shehata TE, Girgis ES, Kamal TH, Ibrahim MK (1983) Gamma irradiation effect on some bacteriological and other properties of cow's, buffaloes and goat's milk. Egyptian J Dairy Sci 11:61–69

Naing YW, Wai SS, Lin TN, Thu WP, Htun LL, Bawm S, Myaing TT (2019) Bacterial content and associated risk factors influencing the quality of bulk tank milk collected from dairy cattle farms in Mandalay Region. Food Sci Nutr 7(3):1063–1071. https://doi.org/10.1002/fsn3.945

Ndife J (2014) Production and quality assessment of functional yoghurt enriched with coconut. Int J Nutr & Food Sci 3(6):545. https://doi.org/10.11648/j.ijnfs.20140306.19

Neoh HM, Tan XE, Sapri HF, Tan TL (2019) Pulsed-field gel electrophoresis (PFGE): a review of the "gold standard" for bacteria typing and current alternatives. Infect Genet Evol 74:103935. https://doi.org/10.1016/j.meegid.2019.103935

Nicoli MC, Anese M, Parpinel MT, Franceschi S, Lerici CR (1997) Loss and/or formation of antioxidants during food processing and storage. Cancer Lett 114(1-2):71–74. https://doi.org/10.1016/S0304-3835(97)04628-4

Nikoleli G P, Nikolelis D P, Siontorou C G, Karapetis S, Varzakas T (2018) Novel biosensors for the rapid detection of toxicants in foods. In Advances in Food and Nutrition Research (Vol. 84, pp. 57–102). Academic Press Inc. doi:https://doi.org/10.1016/bs.afnr.2018.01.003

Noci F, Walking-Ribeiro M, Cronin DA, Morgan DJ, Lyng JG (2009) Effect of thermo-sonication, pulsed electric field and their combination on inactivation of Listeria innocua in milk. Int Dairy J 19:30–35. https://doi.org/10.1016/j.idairyj.2008.07.002

Olle B (2013) Medicines from microbiota. Nat Biotechnol 31(4):309–315. https://doi.org/10.1038/nbt.2548

Osaili TM, Al-Nabulsi AA, Shaker RR, Ayyash MM, Olaimat AN, ASA A-H, Kadora KM, Holley RA (2008) Effects of extended dry storage of powdered infant milk formula on susceptibility of Enterobacter sakazakii to hot water and ionizing radiation. J Food Prot 71(5):934–939. https://doi.org/10.4315/0362-028X-71.5.934

Paludetti LF, Kelly AL, O'Brien B, Jordan K, Gleeson D (2018) The effect of different precooling rates and cold storage on milk microbiological quality and composition. J Dairy Sci 101(3):1921–1929. https://doi.org/10.3168/jds.2017-13668

Pan H, Dong K, Rao L, Zhao L, Wu X, Wang Y, Liao X (2020) Quantitative detection of viable but nonculturable state Escherichia coli O157: H7 by ddPCR combined with propidium monoazide. Food Control 112:107140. https://doi.org/10.1016/j.foodcont.2020.107140

Pandey KK, Sood SK, Verma SK, Kumar S, Rani S, Ganguli S (2019) Bioutilization of paneer whey waste for production of paneer making powder containing pediocin PA-1 as a biopreservative to enhance shelf life of paneer. LWT 113:108243. https://doi.org/10.1016/j.lwt.2019.06.010

Panopoulos G, Moatsou G, Psychogyiopoulou C, Moschopoulou E (2020) Microfiltration of Ovine and Bovine Milk: Effect on Microbial counts and biochemical characteristics. Foods 9:284. https://doi.org/10.3390/foods9030284

Paraffin AS, Zindove TJ, Chimonyo M (2019) Effect of structural condition of milk processing facilities and food safety systems on Escherichia coli and Coliforms presence in cultured buttermilk. J Food Quality:2019. https://doi.org/10.1155/2019/7365983

Parizad EG, Parizad EG, Valizadeh A (2016) The application of pulsed field gel electrophoresis in clinical studies. J Clin Diagn Res 10(1):DE01. https://doi.org/10.7860/JCDR/2016/15718.7043

Parkhill J, Wren BW, Mungall K, Ketley JM, Churcher C, Basham D, Chillingworth T, Davies RM, Feltwell T, Holroyd S, Jagels K, Karlyshev AV, Moule S, Pallen MJ, Penn CW, Quail MA, Rajandream MA, Rutherford KM, Van Vliet AHM et al (2000) The genome sequence of the food-borne pathogen Campylobacter jejuni reveals hypervariable sequences. Nature 403(6770):665–668. https://doi.org/10.1038/35001088

Patel S, Nanda R, Sahoo S, Mohapatra E (2016) Biosensors in health care: The milestones achieved in their development towards lab-on-chip-analysis. In Biochemistry Research International (Vol. 2016). Hindawi Publishing Corporation. https://doi.org/10.1155/2016/3130469

Pei R, Martin DA, DiMarco DM, Bolling BW (2017) Evidence for the effects of yogurt on gut health and obesity. Crit Rev Food Sci Nutr 57(8):1569–1583. https://doi.org/10.1080/10408398.2014.883356

Peleg M, Cole MB (1998) Reinterpretation of microbial survival curves. Critic Rev Food Sci 38(5):353–380. https://doi.org/10.1080/10408699891274246

Pereda J, Ferragut V, Quevedo JM, Guamis B, Trujillo AJ (2007) Effects of ultra-high pressure homogenization on microbial and physicochemical shelf life of milk. J Dairy Sci 90(3):1081–1093. https://doi.org/10.3168/jds.S0022-0302(07)71595-3

Pereira RN, Teixeira JA, Vicente AA, Cappato LP, da Silva Ferreira MV, da Silva Rocha R, da Cruz AG (2018) Ohmic heating for the dairy industry: a potential technology to develop probiotic dairy foods in association with modifications of whey protein structure. Curr Opin Food Sci 22:95–101. https://doi.org/10.1016/j.cofs.2018.01.014

Petrus RR, Loiola CG, Oliveira AF (2010) Microbiological shelf life of pasteurized milk in bottle and pouch. J Food Sci 75(1):M36–M40. https://doi.org/10.1111/j.1750-3841.2009.01443.x

Pinelo M, Zeuner B, Meyer AS (2010) Juice clarification by protease and pectinase treatments indicates new roles of pectin and protein in cherry juice turbidity. Food Bioprod Process 88(2-3):259–265. https://doi.org/10.1016/j.fbp.2009.03.005

Piscopo A, Zappia A, De Bruno A, Pozzo S, Limbo S, Piergiovanni L, Poiana M (2019) Use of biodegradable materials as alternative packaging of typical Calabrian Provola cheese. Food Packag Shelf Life 21:100351

Poduval VS, Mistry VV (1999) Manufacture of reduced fat Mozzarella cheese using ultrafiltered sweet buttermilk and homogenized cream. J Dairy Sci 82(1):1–9. https://doi.org/10.3168/jds.S0022-0302(99)75202-1

Poghossian A, Geissler H, Schöning MJ (2019) Rapid methods and sensors for milk quality monitoring and spoilage detection. Biosens Bioelectron 140:111272. https://doi.org/10.1016/j.bios.2019.04.040

Price WV (1927) Cheddar cheese from pasteurized milk. J Dairy Sci 10(2):155–168. https://doi.org/10.3168/jds.S0022-0302(27)93826-0

Rad AH, Abbasi A, Kafil HS, Ganbarov K (2020) Potential pharmaceutical and food applications of postbiotics: a review. Curr Pharm Biotechnol 21. https://doi.org/10.2174/1389201021666200516154833

Ragab ES, Zhang S, Pang X, Lu J, Nassar KS, Yang B et al (2020) Ultrasound improves the rheological properties and microstructure of rennet-induced gel from goat milk. Int Dairy J 104:104642. https://doi.org/10.1016/j.idairyj.2020.104642

Rai S, Goyal GK, Rai GK (2008) Effect of Modified Atmosphere Packaging (MAP) and storage on the chemical quality of paneer. J Dairying Food & Home Sci 27(1):33–37

Report- Dairy and Milk Processing Market in India (2018–2023) ID: 4757746, Report-March 2019, Region: India, Netscribes (India) Pvt Ltd

Rojas ML, Alvim ID, Augusto PED (2019) Incorporation of microencapsulated hydrophilic and lipophilic nutrients into foods by using ultrasound as a pre-treatment for drying: a prospective study. Ultrason Sonochem 54:153–161. https://doi.org/10.1016/j.ultsonch.2019.02.004

Romero DA, Klaenhammer TR (1993) Transposable elements in Lactococci: a review. J Dairy Sci 76(1):1–19. https://doi.org/10.3168/jds.S0022-0302(93)77318-X

Rysstad G, Kolstad J (2006) Extended shelf life milk—advances in technology. Int J Dairy Technol 59(2):85–96. https://doi.org/10.1111/j.1471-0307.2006.00247.x

Salminen S, Stahl B, Vinderola G, Szajewska H (2020) Infant formula supplemented with biotics: current knowledge and future perspectives. Nutrients 12(7):1952. https://doi.org/10.3390/nu12071952

Samtlebe M, Wagner N, Brinks E, Neve H, Heller KJ, Hinrichs J, Atamer Z (2017) Production of phage free cheese whey: design of a tubular laboratory membrane filtration system and assessment of a feasibility study. Int Dairy J 71:17–23. https://doi.org/10.1016/j.idairyj.2017.02.006

Sanders ME, Hamilton J, Reid G, Gibson G (2007) A nonviable preparation of Lactobacillus acidophilus is not a probiotic. Clin Infect Dis 44(6):886

Satin M (1996) Food irradiation. Technomic Publishing Co, Lancaster, PA. https://doi.org/10.1002/food.19960400515

Schlossareck C, Ross CF (2020) Consumer sensory evaluation of aftertaste intensity and liking of spicy paneer cheese. Int J Food Sci Technol. https://doi.org/10.1111/ijfs.14524

Shanaziya ASF, Mangalika ULP, Nayananjalie WAD (2018) Effect of different coagulants on the quality of paneer made from cow milk. Int J Sci Res Publ 8:189–194. https://doi.org/10.29322/IJSRP.8.4.2018.p7627

Sharma RB, Kumar M, Pathak V (2002) Effect of different seasons on cross-bred cow milk composition and paneer yield in sub-himalayan region. Asian Australas J Anim Sci 15(4):528–530. https://doi.org/10.5713/ajas.2002.528

Sharma A, Shivaprasad DP, Chauhan K, Taneja NK (2019) Control of E. coli growth and survival in Indian soft cheese (paneer) using multiple hurdles: Phytochemicals, temperature and vacuum. LWT 114:108350. https://doi.org/10.1016/j.lwt.2019.108350

Sheth M, Hirdyani H (2016) Development and sensory analysis of a buttermilk based fermented drink using barley and fructooligosaccharide as functional ingredients. Int J Home Sci 2:235–239

Shi XM, Long F, Suo B (2010) Molecular methods for the detection and characterization of foodborne pathogens. Pure Appl Chem 82(1):69–79. https://doi.org/10.1351/PAC-CON-09-02-07

Shinde SA, Chavhan SA, Sapkal SB, Shrikhande VN (2018) Recombinant DNA technology and its applications: a review. Int J Medipharm Res 4:79–88

Shrivastava S, Goyal S, Goyal GK (2013) Effect of Modified Atmosphere Packaging on Colour Profile of Paneer Stored at 3±1 °C. J Agric & Sustain 3(2)

Silva FRN, Pereira AD, Baptista DP, Pereira MU, Spisso BF, Gigante ML, Arisseto-Bragotto AP (2020) Monensin residues in the production of Minas Frescal cheese: Stability, effects on fermentation, fate and physicochemical characteristics of the cheese. Food Res Int 137:109440. https://doi.org/10.1016/j.foodres.2020.109440

Singh RP, Heldman DR (2001) Introduction to food engineering. Gulf Professional Publishing

Singh RS, Chauhan K, Pandey A (2019) Influence of aeration, agitation and process duration on fungal inulinase production from paneer whey in a stirred tank reactor. Bioresource Technol Report 8:100343. https://doi.org/10.1016/j.biteb.2019.100343

Smith-Howard K (2017) Pure and modern milk: An environmental history since 1900. Oxford University Press

Soodam K, Guinee TP (2018) The case for milk protein standardisation using membrane filtration for improving cheese consistency and quality. Int J Dairy Technol 71(2):277–291. https://doi.org/10.1111/1471-0307.12502

Sørhaug T, Stepaniak L (1997) Psychrotrophs and their enzymes in milk and dairy products: Quality aspects. Trends Food Sci Technol 8:35–41. https://doi.org/10.1016/S0924-2244(97)01006-6

Spalinger J, Nydegger A, Belli D, Furlano R, Yan J, Tanguy J, Pecquet S, Destaillats F, Egli D, Steenhout P (2017) Growth of infants fed formula with evolving nutrition composition: a single-arm non-inferiority study. Nutrients 9(3):219. https://doi.org/10.3390/nu9030219

Sreeja V, Jana A, Aparnathi K, Prajapati J (2013) Role of whey proteins in combating geriatric disorders. J Sci Food Agric 15(93):3662–3669. https://doi.org/10.1002/jsfa.6345

Sun H, Kawamura S, Himoto J, Itoh K, Wada T, Kimura T (2008) Effects of ohmic heating on microbial couints and denaturation of proteins in milk. Food Sci Tech. https://doi.org/10.3136/fstr.14.117

Sundarraj AA, Rajathi AA, Vishaal SC, Rohit D, Prakash MS (2018) Food biotechnology applications in dairy and dairy products. J Pharm Res 12(4):520–525

Sur A, Sah R P, Pandya S (2020) Milk storage system for remote areas using solar thermal energy and adsorption cooling. Materials Today: Proceedings. https://doi.org/10.1016/j.matpr.2020.05.170

Tabit FT (2016) Advantages and limitations of potential methods for the analysis of bacteria in milk: a review. J Food Sci Technol 53(1):42–49. https://doi.org/10.1007/s13197-015-1993-y

Teuber M (1993) Genetic engineering techniques: in food microbiology and enzymology. Food Rev Intl 9(3):389–409. https://doi.org/10.1080/87559129309540968

Thippeswamy L, Venkateshaiah BV, Patil SB (2011) Effect of modified atmospheric packaging on the shelf stability of paneer prepared by adopting hurdle technology. J Food Sci Technol 48(2):230–235. https://doi.org/10.1007/s13197-010-0155-5

Thompson AL (2012) Developmental origins of obesity: Early feeding environments, infant growth, and the intestinal microbiome. Am J Hum Biol 24(3):350–360. https://doi.org/10.1002/ajhb.22254

Van Aardt M, Duncan SE, Marcy JE, Long TE, Hackney CR (2001) Effectiveness of poly (ethylene terephthalate) and high-density polyethylene in protection of milk flavor. J Dairy Sci 84:1341–1347. https://doi.org/10.3168/jds.S0022-0302(01)70164-6

Vandenplas Y (2017) Prevention and management of cow's milk allergy in non-exclusively breastfed infants. Nutrients 9(7):731. https://doi.org/10.3390/nu9070731

Verardo V, Gómez-Caravaca AM, Arráez-Román D, Hettinga K (2017) Recent advances in phospholipids from colostrum, milk and dairy by-products. Int J Mol Sci 18(1):173. https://doi.org/10.3390/ijms18010173

Verkempinck S, Pallares AP, Hendrickx M, Grauwet T (2020) Processing as a tool to manage digestive barriers in plant-based foods: Recent advances. Curr Opin Food Sci 35:1–9. https://doi.org/10.1016/j.cofs.2019.11.007

Wassenaar TM, Newell DG (2000) Genotyping of Campylobacterspp. Appl Environ Microbiol 66(1):1–9. https://doi.org/10.1128/AEM.66.1.1-9.2000

Wedholm A, Larsen LB, Lindmark-Månsson H, Karlsson AH, Andrén A (2006) Effect of protein composition on the cheese-making properties of milk from individual dairy cows. J Dairy Sci 89(9):3296–3305. https://doi.org/10.3168/jds.S0022-0302(06)72366-9

Weerathilake WADV, Rasika DMD, Ruwanmali JKU, Munasinghe MADD (2014) The evolution, processing, varieties and health benefits of yogurt. 4(4).

Wegh CAM, Geerlings SY, Knol J, Roeselers G, Belzer C (2019) Postbiotics and their potential applications in early life nutrition and beyond. In International Journal of Molecular Sciences (Vol. 20, Issue 19). MDPI AG. https://doi.org/10.3390/ijms20194673

Weimar MR (1994) Landmarks in the US dairy industry (No. 694). US Department of Agriculture, Economic Research Service.

Welker DL, Coburn BM, McClatchy JH, Broadbent JR (2019) Multiple pulse electroporation of lactic acid bacteria Lactococcus lactis and Lactobacillus casei. J Microbiol Methods 166:105741. https://doi.org/10.1016/j.mimet.2019.105741

Wilson HL, Hall SA, Rogers LA (1945) The manufacture of Cheddar cheese from pasteurized milk. J Dairy Sci 28(3):187–200. https://doi.org/10.3168/jds.S0022-0302(45)95162-9

Wong DE, Goddard JM (2014) Short communication: Effect of active food packaging materials on fluid milk quality and shelf life. J Dairy Sci 97(1):166–172. https://doi.org/10.3168/jds.2013-7214. Epub 2013 Nov 13

Wong PYY, Kitts DD (2003) Chemistry of buttermilk solid antioxidant activity. J Dairy Sci 86(5):1541–1547. https://doi.org/10.3168/jds.S0022-0302(03)73739-4

Yanachkina P, McCarthy C, Guinee T, Wilkinson M (2016) Effect of varying the salt and fat content in Cheddar cheese on aspects of the performance of a commercial starter culture preparation during ripening. Int J Food Microbiol 224:7–15. https://doi.org/10.1016/j.ijfoodmicro.2016.02.006

Yemiş GP, Delaquis P (2020) Natural compounds with antibacterial activity against Cronobacter spp. in powdered infant formula: a review. Front Nutr:7. https://doi.org/10.3389/fnut.2020.595964

Zhao L, Feng R, Ren F, Mao X (2018) Addition of buttermilk improves the flavor and volatile compound profiles of low-fat yogurt. Lwt 98:9–17. https://doi.org/10.1016/j.lwt.2018.08.029

Zoidou E, Melliou E, Moatsou G, Magiatis P (2017) Chapter 11 - Preparation of Functional Yogurt Enriched With Olive-Derived Products (N. P. B. T.-Y. in H. and D. P. Shah (ed.); pp. 203–220). Academic Press. https://doi.org/10.1016/B978-0-12-805134-4.00011-0

Zou L, Pande G, Akoh CC (2016) Infant formula fat analogs and human milk fat: new focus on infant developmental needs. Annu Rev Food Sci Technol 7:139–165. https://doi.org/10.1146/annurev-food-041715-033120

Chapter 7
On-Farm Point-of-Care Diagnostic Technologies for Monitoring Health, Welfare, and Performance in Livestock Production Systems

Mohamed Zeineldin, Ahmed A. Elolimy, P. Ravi Kanth Reddy, Mohamed Abdelmegeid, Miguel Mellado, Mona M. M. Y. Elghandour, and Abdelfattah Z. M. Salem

Abstract The occurrence of infectious diseases has a significant adverse effect on livestock health and production efficiency. Current diagnostic approaches in veterinary practice are primarily focused on the observation of changes in physical, clinical, behavioral, or performance of individual or groups of animals. In recent years, these diagnostic approaches have markedly improved livestock profitability during their production cycle. This is mainly achieved by using reliable and conveniently available on-farm and point-of-care diagnostic technologies for the rapid and accurate management of animal health. The availability of on-farm and point-of-care technology is rapidly changing decisions for bovine practitioners on both individual sick cows and herd health level. Early detection of infectious diseases using quick,

M. Zeineldin
Department of Animal Medicine, College of Veterinary Medicine, Benha University, Benha, Egypt

A. A. Elolimy
Department of Animal Production, National Research Centre, Giza, Egypt

P. R. K. Reddy
Veterinary Dispensary, Taticherla, Andhra Pradesh, India

M. Abdelmegeid
Department of Animal Medicine, College of Veterinary Medicine, Kafr-Elsheikh University, Kafrelsheikh, Egypt

M. Mellado
Department of Animal Nutrition, Autonomous Agrarian University Antonio Narro, Saltillo, Mexico

M. M. M. Y. Elghandour · A. Z. M. Salem (✉)
Facultad de Medicina Veterinaria y Zootecnia, Universidad Autonoma del Estado de Mexico, Toluca, Mexico
e-mail: salem@uaemex.mx

V. K. Yata et al. (eds.), *Sustainable Agriculture Reviews 54*, Sustainable Agriculture Reviews 54, https://doi.org/10.1007/978-3-030-76529-3_7

effective, low-cost, automated technologies will allow timely detection of infected animals, thus reducing the economic loss and associated abuse of antimicrobial therapy. Here we review the currently available on-farm and point-of-care diagnostic technologies for health surveillance and disease detection in livestock production systems. We additionally review the advantages and disadvantages of each methodology, concerning their possible effect on the improvement of animal welfare and productivity of farm animals.

Keywords Bovine · Disease · Diagnosis · On-farm · Point-of-care · Technology

7.1 Introduction

Infectious diseases are a major source of livestock production inefficiency and cause significant economic losses to the global livestock industry due to associated morbidity, mortality, and treatment costs (Craft 2015). The presence of infectious agents, and related therapeutic and antibiotic residues, has serious consequences for the public perception of livestock production, food safety, and human health (Daszak et al. 2000). Although disease control and prevention are critical elements for mitigating the effect of disease on animal health and production efficiency, disease-free production systems are unrealistic (Schwabe 1982). Since diseases are an unavoidable component of modern livestock production systems, on-farm and point-of-care diagnostics for the accurate detection and diagnosis of affected animals are vital for minimizing morbidity, optimizing recovery, and maintaining profitability. Early and accurate disease detection plays an important role in reducing mortality, preventing disease transmission, and avoiding the long-term effects of disease on production, welfare, and profitability (Bisson et al. 2015). Early detection and diagnosis of infectious diseases in livestock also help to avoid irreversible pathologies, prevent the production of antibiotic-resistant bacteria, reduce public health concerns and optimize the action of antimicrobial therapy (Hennessy and Wolf 2018).

Compared to human medicine, current diagnostic approaches in veterinary practice have primarily focused on observation of changes in the physical, clinical, behavioral, or performance status of individual or groups of animals. Although these findings have historically been made by animal caretakers and animal health professionals, there has been an increased interest in the use of cow-side diagnostic tests to improve the sensitivity, precision, and timeliness of disease detection (Berckmans 2014a). The investment in rapid, selective, accurate, and cost-effective tests and the recognition among clinicians and animal stakeholders of the value of quick and accurate diagnosis of disease have resulted in rapid progress in point-of-care and on-farm diagnostics for rapid on-site testing of animals (Helwatkar et al. 2014). Similarly, technology adaptation and implementation are being used to

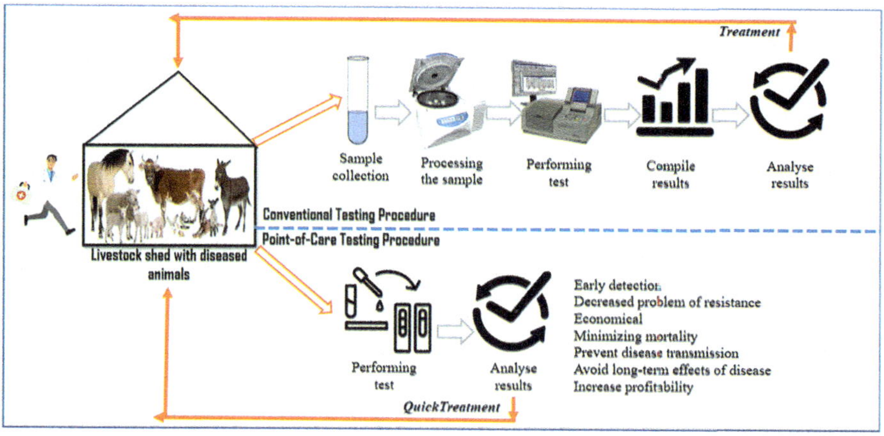

Fig. 7.1 Advantages of point-of-care diagnostic techniques compared to conventional diagnostic methods

create more robust systems for measuring animal growth, production, and the physical environment of animals, all to improve individual or group performance and production efficiency (Lokhorst and Ipema 2010). To date, most of the available diagnostic tests in humans and farm animals focus on the rapid identification of non-infectious conditions (e.g. blood glucose, β-hydroxybutyrate, and hemoglobin concentration, diagnosis of cardiac markers, detection of blood gas and electrolytes, etc.) (Tschandl et al. 2019; Sargeant and O'Connor 2020). On-farm and point-of-care diagnostics also provide an alternative for monitoring livestock health, which has a high potential for increasing the performance of livestock production systems, eliminating subjectivity the management decision-making, and maximizing the efficacy of unskilled or inexperienced labor in animal health management (Helwatkar et al. 2014). Advantages of point-of-care diagnostic techniques compared to conventional diagnostic methods are shown in Fig. 7.1.

In this chapter, we review the currently available on-farm and point-of-care diagnostic technologies for health and performance monitoring in livestock production systems. Additionally, we review the advantages and disadvantages of each approach, in particular regarding its possible effect for improving animal health and productivity.

7.2 Characteristics of an Ideal Point-of-Care and On-Farm Diagnostic Technologies

The list of available point-of-care and on-site diagnostic tests in both human and animal medicine has expanded steadily over the last few years. The global diagnostic market for veterinary diagnostic testing is projected to increase at an average annual growth rate of 8.6% from 2016 to 2021, to hit US$ 6.71 billion by 2021 (Commission 2011). This massive market potential is driven by the availability of quick, portable veterinary diagnostic devices to track animal health disorders (Helwatkar et al. 2014). Ideal technology should be developed as durable, low-cost, reliable, and effective explaining the intrinsic biological process (Frost et al. 2003). Moreover, it should be easily converted into a professionally tested and easy-to-maintain relevant action (Bolboacă 2019). The selection of the appropriate point-of-care and on-farm diagnostic technology is affected by many factors, including the frequency and type of measurement, the labor needed to conduct the test, the conditions of the livestock management system, and the costs of the technology (Berckmans 2014a). The idealized diagnostic technologies are challenged by the sample volumes of the complex structure of biological materials, the processing time of the sampling, and the diagnostic capacity and accuracy of the instrument (Gubala et al. 2012). The design of an ideal point-of-care and on-farm diagnostic should therefore depend not only on the development of technology but also on the design of the biological measurement goals and the surrounding livestock farming system.

7.3 Categories of Point-of-Care and On-Farm Diagnostic Technologies

Successful management of animal diseases depends on the timely identification of clinically infected animals through laboratory and clinical examination, along with an understanding of the factors influencing the epidemiology of the causative agent (Bisson et al. 2015). There is a wide variety of easier and more cost-effective diagnostic techniques, including point-of-care technologies for disease surveillance in both human medicine and veterinary practice (Helwatkar et al. 2014; Yanase and Triantaphyllou 2019; Mohr et al. 2020). Generally, each diagnostic tool measures a specific parameter related to the physiological state of the animal, performance or existence of infectious agents in an individual or group of animals, a change that entails an intervention to alleviate the disorder detected (Lokhorst and Ipema 2010). Besides, there may be relevant algorithms and computer programs necessary to understand an animal's health status, which may need to be coupled with other data (e.g, financial input) before an appropriate and effective decision can be taken (Das et al. 2015). This obstacle can be better tackled by the advancement of point-of-care and on-farm automated technology in a revolutionary way to provide fast and effective identification of animal health threats (Neethirajan et al. 2017). In this context, point-of-care diagnostics are valuable tools for detecting small sample volumes and low concentrations of biological components and infectious agents (Theurer et al. 2013). Practically, all the available technologies for the livestock industry fall into one of four broad categories related to their proximity (attached or non-attached) and association (invasive or non-invasive) with the animals themselves (Helwatkar et al. 2014). Attached technologies include those that can be fixed to the outside the animal's body (non-invasive) (Krieger et al. 2019) or those that are fitted inside the body (invasive) (Rose-Dye et al. 2011). This category includes some of the most reliable technologies for continuously monitoring animal health throughout the day and comprises accelerometers, pedometers, vibration sensors, thermometers, and rumen temperature bolus. Non-attached technologies are those that animals pass by, over, or through for health monitoring (Stone et al. 2017). These devices are often set at fixed locations in the animal's environment (e.g, surveillance cameras (regular or thermographic), and video and audio recording systems). Similarly, point-of-care diagnostic technologies are also achieved with the use of unattached, transportable, and hand-held equipment and test kits to obtain blood samples and provide results in a very short time, so that decisions are taken very rapidly.

7.4 Molecular Diagnostic Technologies for Monitoring Animal Health and Disease

Traditionally, diagnostic approaches in veterinary practice have focused primarily on clinical evaluation of the behavior of individual animals or groups of individuals and the reaction to certain conditions of the disease (Van Veen 1997). These conventional approaches rely on animal history and visual clinical signs and can only be used to diagnose a small percentage of sick animals. Because of the multifactorial disease control and the wide variety and complexity of the livestock farming environment, these conventional methods are not sufficient to improve the sensitivity, precision, and timeliness of disease diagnosis (Saegerman et al. 2011). The usefulness of molecular and high-performance sequencing technologies for tracking the health and disease of livestock has been highlighted for accurate disease control (Reuter et al. 2015; Kumar et al. 2019). The recent application of molecular diagnostic approaches to livestock has shown the importance of these platforms for improving livestock management systems and limiting the spread of diseases (Shirley et al. 2010). Also, the existing use of molecular diagnostic techniques has demonstrated the complexity of disease-causing agents (Zeineldin et al. 2017a, b) and proven associations between early disease detection and livestock health and productivity (Naqvi 2007). Numerous molecular-based techniques are presently used to determine disease-causing agents, including immunohistochemistry, fluorescent in situ hybridization (FISH), marker-assisted selection, cloning, flow cytometry, RNA dot blotting hybridization, ELISA, and quantitative real-time PCR assays (Walker and Subasinghe 2000). Meanwhile, the field of molecular diagnostics has shown dramatic developments over the last few years in the use of low-cost high-density single-nucleotide polymorphism (SNP) technology in the genotyping of individuals at the SNPs level (van Arendonk 2011). Among these technologies, immunohistochemistry and fluorescent in situ hybridization have been used to show the presence and characteristics of species in complex biological samples (Cheon and Chae 2000). Also, qPCR and FISH have been the most effective approaches used to amplify and quantify particular DNA or RNA sequences in complex biological samples from different hosts (Elelu et al. 2016). While these molecular techniques are appropriate for a plethora of different biological samples and provide reliable information on some agents, they do not provide the direct sequence of the infectious agent. Also, the use of these methods includes advanced awareness of the causative agents. These shortcomings eventually led to the development and widespread implementation of high-performance sequencing technologies in veterinary diagnostic reference laboratories, which can sequence all genomic material present in biological samples and produce thousands of sequences of previously unknown biological materials (Zeineldin et al. 2019b). In addition, next-generation technology has the ability to provide an unbiased sequencing platform for several microbial genomes and their antibiotic resistance genes in near-real time, which bring several possible advantages to create diagnostic reference laboratories for rapid diagnostic decision making (Zeineldin et al. 2019a). Recently, the current high-throughput sequencing technologies used in veterinary practice are primarily Illumina sequencing platforms (San Diego, CA, USA), Nanopore technology, and 454 pyrosequencing platforms (Ambardar et al. 2016; Singh et al. 2019). In the coming years, the cost of this sequencing technology is expected to be reduced and sequencing platforms will be available and used on a wide scale in livestock. The entire genome sequencing technology has recently been applied in a variety of studies to improve diagnostic methods accuracy in the veterinary field (Glaser et al. 2016). For example, the use of sequencing technologies for the genetic characterization of viral infectious agents has provided a much deeper and more detailed view of viral capsid and virus genetic material (Chappell et al. 2019). This knowledge is important for a deeper understanding of the genotypes of viruses and will help to identify viruses and the future production of vaccines. More recently, developments in mass spectrometry science, including metaproteomics, metabolomics, and metatranscriptomics, have improved diagnostic technologies to enhance animal health (Elolimy et al. 2020). Although valuable information is given by these techniques, the use of this technology in veterinary practice is minimal. Due to the associated high costs, access to these advanced technologies in developing countries is limited (Zumla et al. 2014). In addition, some enhancements are still required before the next generation and mass spectrometry technologies are used as a point-of-care and on-farm diagnostic testing.

7.5 Electrochemical Point-of-Care Biosensor Technology

The term point-of-care biosensors include instruments that can measure the physiological, immunological, and behavioral responses of animals, as well as the monitoring of the animal environment (Wang 2006). Electrochemical point-of-care is a flexible biosensor with a point-of-care functionality used to generate an electrochemical signal measured using a detector and a data analyzer (Dai and Liu 2019).

These instruments are not only extremely precise and sensitive to the parameters being analyzed, but are also accurate and easy to use and can improve the clinical assessment (Wang 2006). The emerging use of point-of-care biosensors in livestock management offers major benefits and applications in the monitoring of animal productivity, health monitoring, and disease detection, as well as monitoring of animal physiological conditions (Gattani et al. 2019; Robinson et al. 2020). The implementation of this technology is expected to enhance animal health and productivity of the livestock industry in the future, as well as to reduce the impact of the livestock industry on the environment (Gattani et al. 2019). In addition, the implementation of point-of-care biosensors in the livestock sector would contribute to social competitive advantage and organizational benefits for the global economy. The latest existing point-of-care biosensors in the veterinary field focus on the use of the available knowledge of animal physiology, nature, biology, nutrition, and the environment, and the incorporation of this knowledge into an effective and real-time diagnostic method (Neethirajan et al. 2017). Different established sensors and their locations in farm animals are presented in Fig. 7.2.

The key point-of-care biosensing devices rely mainly on identifying the desired biological biomarker (e.g, DNA, RNA, enzymes, hormones, metabolites, tissue, blood, cells, etc.) unique to a particular biological agent using a bioreceptor sensor (Wang 2006). The bioreceptor sensor is a key structure in the point-of-care biosensor device since it is crucial for the distinction between various biomarkers and molecules present in the same biological samples. The association between the bioreceptor sensor and the specific biomarker results in specific signals that allow qualitative and quantitative biomarker measurements (Vidic et al. 2017). A detailed overview of the theory and function of the electrochemical point-of-care biosensors has been reviewed elsewhere (Wang 2006; Dai and Liu 2019; Sun and Hall 2019).

The criteria for an ideal point-of-care biosensor in the veterinary sector is similar to that commonly used in human medicine but may vary concerning the particular needs of livestock farming (Kumar et al. 2020). Although designed for human application, respiratory rate sensors were evaluated successfully in cattle to measure thoracic and abdominal movements (Neethirajan 2020). Laser distance sensor while milking is another system for monitoring the respiratory rate of cows (Pastell and Kujala 2007). Sensor-equipped holters (belt-like device on an animal's chest) have been tested to measure the heart rate in dogs (Lahdenoja et al. 2019). Similarly, den Uijl et al. (2017) analyzed the results of behavioral traits such as headshake, walk, sleep, trot, canter, eat, and drink, using a neck collar-embedded with an accelerometer and validated through clinical setting. Several point-of-care diagnostic platforms are currently available in veterinary practice (Table 7.1), and these tools provide automatic readout results greatly reducing the time of diagnosis. However, the complete adoption and application of such innovations and their effective use in diagnostic procedures in livestock clinical practice imply further progress in the robustness of bioassay production and biomarker detection.

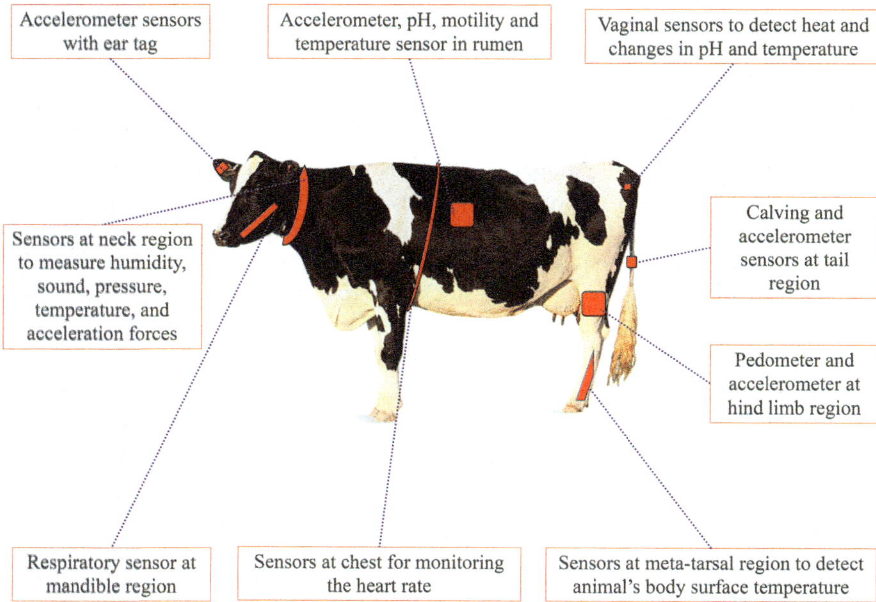

Fig. 7.2 Different established sensors being used in farm animals

7.6 On-Farm Diagnostic Technologies

7.6.1 On-Farm Video Surveillance and Digital Images Analysis Technologies

The frequency of animal behavior changes is often correlated with stress, anxiety, and pathological conditions (Von Holst 1998). These changes in behavior are cost-effective and easy to track via careful observation of individual animals or through the use of video surveillance devices on the farm (Zin et al. 2016). Simultaneous monitoring of these behavioral changes that occur on more than one animal for 24 h a day may be difficult to interpret for the farm observer (Ishiwata et al. 2007). The use of video monitoring systems to track these behavioral activities enables researchers to examine animal activity at their location in real-time (Zin et al. 2016). Further, the availability of digital imaging and the latest innovation of computer programs that can interpret and analyze these data has expanded the use of this technology in the livestock industry to track animal health without distracting the animals (Nilsson et al. 2015). For example, on-farm video monitoring and digital image processing tools have been commonly used for recording particular areas used by individual animals as they get up or lay down to make suggestions on stall size (Ceballos et al. 2004). Moreover, it was also used to track the thermoregulatory clumping of the animal to determine the efficacy of the pen temperature (Shao and Xin 2008). Another example of the use of on-farm video surveillance and digital image

Table 7.1 Some examples of point-of-care diagnostic technologies currently used in livestock husbandry system

Point-of-care test	Health status	Metabolites and agents	Animal species	Sample type	Reference
Protein G-based milk dipstick	Brucellosis	Brucella	Dairy cattle	Blood	Revathi Poonati et al. (2020)
Indirect enzyme-linked immunosorbent (iELISA) assays	Brucellosis	Brucella	Dairy cattle	Blood	Revathi Poonati et al. (2020)
Visual color-based serum phosphorous detection kit	Hypophosphatemia	Phosphorous	Cattle	Blood	D'souza (2020)
Handheld portable lactate analyzer	Bovine respiratory disease	L-lactate concentration	Cattle	Blood	Zeineldin et al. (2017a, b)
Handheld portable lactate analyzer	Healthy	L-lactate concentration	Cattle	Blood	Karapinar et al. (2013)
Portable blood lactate analyzer	Bovine respiratory disease	Plasma lactate	Cattle	Blood	Coghe et al. (2000)
β-hydroxybutyrate hand-held meter	Ketosis	β hydroxybutyrate	Dairy cattle	Blood	Iwersen et al. (2009)
Portable glucometer	Healthy	Blood glucose	Cattle and sheep	Blood	Katsoulos et al. (2011)
Lactate meters	Healthy	L-lactate concentration	Cattle	Blood	Karapinar et al. (2013)
Electrochemical glucose meter	Periparturient healthy cattle	Glucose	Dairy cattle	Blood and plasma	Megahed et al. (2015)
Portable ion-selective electrode meters	Periparturient healthy cattle	Potassium	Dairy cattle	Blood, plasma, milk, and abomasal fluid	Megahed et al. (2016)
Portable somatic cell count-based test	Healthy and sub-clinical mastitis	Somatic cell count	Dairy cows	Milk samples	Iraguha et al. (2017)
FreeStyle Precision Neo™	Healthy	β-hydroxybutyrate and glucose	Dairy cow	Blood samples	Macmillan et al. (2017)

(continued)

Table 7.1 (continued)

Point-of-care test	Health status	Metabolites and agents	Animal species	Sample type	Reference
β-hydroxybutyrate electrochemical meter	Periparturient healthy cattle	β hydroxybutyrate	Dairy cattle	Blood and plasma samples	Megahed et al. (2017)
Point-of-care glucometer	Healthy	Glucose	Goat	Blood samples	Quandt et al. (2018)
Epocal point-of-care analyzer	Healthy	Ionized calcium	Dairy cattle	Blood samples	Mahen et al. (2018)
Erythrocyte osmotic fragility assessment at field level	Stress-related disorders	Erythrocyte membrane	Sheep and goats	Blood samples	Reddy et al. (2019)
Cobalt chloride-impregnated device	Stress-related disorders	Moisture	Cattle	Sweat samples	Pereira et al. (2010)
Leukocyte esterase (LE) test strips	Subclinical endometritis	Leukocyte esterase	Dairy cow	Vaginal discharge	Van Schyndel et al. (2018)
Brix refractometry	Subclinical endometritis	Total solid percentage	Dairy cow	Vaginal discharge	Van Schyndel et al. (2018)
Colorimetric ammonium point-of-care test	Periparturient Holstein-Friesian cows	Ammonium	Dairy cow	Urine sample	Megahed and Constable (2020)

processing technology is the tracking of lameness problems in dairy cows by measuring the shift in animal gait characteristics as they pass to the milking parlor (Berckmans 2014b). The key drawbacks of on-farm video monitoring and digital image processing technologies are the difficulty in detecting the behavior of a large number of animals at the same point over a long period (Von Holst 1998). The challenge of recording in low ambient light and the labor needed to evaluate and display the frequency of such behaviors in individual animals over a long period impedes the application of this technology on a wide scale (Fleishman et al. 1998). Despite current limitations, on-farm video surveillance and digital image processing technologies may be used as a reference framework for other behavior tracking technologies in livestock management system.

7.6.2 *On-Farm Audio Surveillance Systems for Sound Detection*

Animal vocalization has been commonly used as a significant predictor of animal health status (Manteuffel et al. 2004). Sound monitoring technology has been implemented in the livestock industry to track and regulate animal health and welfare (Wathes et al. 2005; Handcock et al. 2009). The animal sound provides details not only about the abnormal state of the animal but also about the personality of the animal (Manteuffel et al. 2004). Automatic detection of irregular animal behaviors using bioacoustics, such as cough sounds, achieved maximum efficiency and minimal costs at a precision level of over 94% in the livestock industry (Anderson et al. 2011). For example, the use of sound recognition devices in the pig and dairy cattle industry provided an earlier diagnosis of respiratory distress (Exadaktylos et al. 2008; Ferrari et al. 2010). Respiratory diseases in the livestock sector have resulted in significant economic losses due to high mortality and morbidity rates (Zeineldin et al. 2016). Coughing is the primary clinical symptom of respiratory diseases in both cattle and pigs (Carpentier et al. 2018). Monitoring coughing as an indicator of respiratory disease can identify early clinical cases before the occurrence of serious complications and thereby reduce the risk of respiratory diseases in the livestock industry (Ferrari et al. 2010). It is well recognized that existing diagnostic methods for respiratory diseases in livestock are expensive and not very precise. Several attempts have therefore been made to classify the coughing characteristics in various animal species as a sign of respiratory diseases. Since coughing sound during respiratory distress has unique characteristics, the clinical use of on-farm audio surveillance systems to investigate cough sounds may be beneficial for real-time monitoring (Chung et al. 2013). Several algorithms have been validated for various cough sounds characterization in the livestock industry (Berckmans 2014b).

The advantage of using on-farm audio monitoring systems for sound detection is that the non-invasive nature of this device does not interfere with the daily activities of animals (Ferrari et al. 2008). In addition, this technology may be used to track multiple individuals at the same time. However, this may result in overlapping and less coherent sounds in a commercial setting (Carpentier et al. 2018). Despite the usefulness of this technology in early disease detection and tracking animal sounds, more consideration needs to be given to the realistic application and avoid drawbacks of these technologies in the veterinary field.

7.6.3 *On-Farm Accelerometers and Pedometer for Walking and Standing Behavior Monitoring*

Several automated technologies have been used in the livestock industry to track animal walking and standing activity as well as to record step frequency, including the use of accelerometers and pedometers placed on animals employing various wearable devices such as collars, ear tags, leg or tail bands (Rushen et al. 2012). Over the last few years, accelerometers are probably the most promising technology for providing accurate data on activity tracking, body orientation, and complex posture behaviors (Sala et al. 2011). These instruments have been used to remotely assess animal walking distance and animal orientation (Rothwell et al. 2011; Ringgenberg et al. 2010; Enstipp et al. 2011). Besides, accelerometers have been used to track lying down activity, rumination, feeding, sleep pattern, approaching parturition, and lameness in dairy cows (Martiskainen et al. 2009). Similarly, pedometers are an inexpensive, simple, portable electronic system used to measure the number of distance traveled and the status of the animal's behavior using a mechanical sensor (Shepley et al. 2017). Recently, pedometer devices have also shown strong predictability when used to classify estrus behavior (Løvendahl and Chagunda 2010). Furthermore, a pedometer can assess several cow activities, including locomotion behavior (Alsaaod and Büscher 2012) and approaching calving time (Felton et al. 2013). The application of accelerometers and pedometers in livestock involves using three main components: the main device, the receiving system, and the software program. The receiving system collects animal activities from the attached sensor and transmits these data in real-time through the receiving system to the software program (Brehme et al. 2008). The advantage of on-farm accelerometers and pedometer technologies is the continuous monitoring of animals in commercial systems, particularly when paired with other portable techniques (Moreau et al. 2009; O'Leary et al. 2020). The pitfalls of using these instruments for behavioral tracking include the expense of accessing sensor data and data processing. The risk of damage to sensor boxes due to continuous movements of the animal in the barn and growing labor demand for fixing and removing equipment from livestock are another disadvantage (Norling 1991). However, further modification including increasing battery life, memory capacity, and reducing the device size to be conveniently attached to the animal can efficiently transform on-farm accelerometers and pedometer technologies into functional behavioral measurements technologies.

7.6.4 On-Farm Global Positioning Systems (GPS) for Position Monitoring

On-farm GPS technology is designed to monitor the animal's location and remotely track the movement of the animal within a given area of the farm (Moen et al. 2001). The detailed methodology of GPS collars functioning is presented in Fig. 7.3. The GPS is designed to additionally inform the animal keeper of the distribution of animals on pastures (Griffin 2009). It can also track animal movement using a digital map for easy monitoring and submit information on animal status, such as animal behavior, delivery time, and position (Tang and Abplanalp 2014). Recently, low-cost GPS positioning sensors have been used in veterinary practice to assess the position status of animals (Godsk and Kjærgaard 2011). GPS sensors are attached to the animal collars to detect possible changes in animal behavior, such as feeding, walking, lying down, and standing, which may be associated with stressful situations or changes in animal health (Tang and Abplanalp 2014). Advances in GPS technology have produced lighter and more precise receivers, but the identification of numerous animals in various geographical regions is still difficult (Davis et al. 2011; Foley and Sillero-Zubiri 2020). The drawbacks of GPS technology include costs, battery life, and the frequency of animal location updates. Current technology enables the animal's location to be changed every second, but this rate exceeds the capacity resources available in most animal tracking barns (Tomkiewicz et al. 2010). These limitations restrict the ability to use GPS systems in behavioral tracking for longer periods and limit their use in small local zones. Furthermore, attaching the GPS to the animal collars may have an adverse effect, such as decreased power, tissue damage, and the death of the animal. Therefore, there is still much research needed to improve this technology concept and components used in these devices to be widely used on large scale.

7.6.5 On-Farm Automatic Milking Robot System for Monitoring Leg Health

The milking robot technology was used to track the leg health of the dairy animal in which the sensors were attached to the amplifier and the data was collected on a computer using advanced processing software (Pastell and Kujala 2007). This

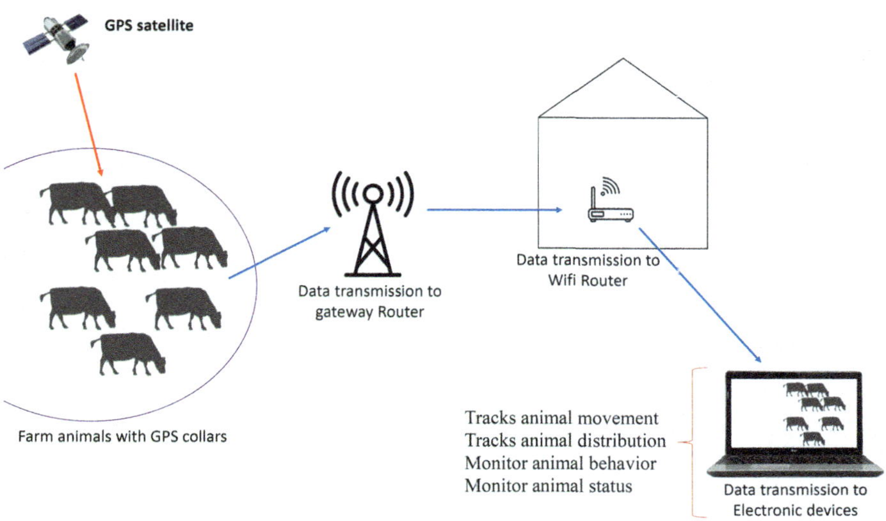

Fig. 7.3 Methodology of GPS collars functioning

technology can be used to automatically detect kick frequency during milking, which warns the operator of potential hoof diseases and other leg problems, but more work is required to improve accuracy (Borderas et al. 2008). Automated milking robots automatically collect data during the milking process and track poor health conditions in a regulated, confined barn up to three times a day. For example, lameness in dairy cows is evident in a reduced frequency of visits to robot system technology (Bach et al. 2007), but this appears to have a low specificity as many other cows are not lame (Borderas et al. 2008). Research focused on the weight distribution between the limbs and the probabilistic neural network showed that automated weight distribution measurements could classify lame cows markedly faster than those obtained by visual inspection applying gait score systems (Pastell et al. 2006). Moreover, these weight distribution steps are responsive to the degree of discomfort associated with lameness (Visen et al. 2002; Chapinal et al. 2011), but the relationship between input and output values is difficult to translate into mathematical representation (Visen et al. 2002). Therefore, the impact of different management factors and their interaction with the robot utilization should be carefully considered together to improve the system efficiency in livestock.

7.6.6 On-Farm Automated Feeder for Monitoring Feeding Behavior

Farm animals are increasingly housed in groups where disease is challenging to identify, leading to increased morbidity and care costs. Automatic detection of feed, water, and frequency of particular behaviors of animals, typically through radio frequency recognition, can be used for early identification of animals that are ill and can provide insight into potential improvements in the pain conditions of animals (Weary et al. 2009). Sick animals spend more time in the feed bunk than healthy ones, particularly the newly received animals (González et al. 2010). Data from such feeders can help recognize morbid animals from healthy ones based on differences in feeding behaviors (Sowell et al. 1999). Recording variations in feed behavior can also help track dairy cows suffering from periparturient diseases such as metritis, ketosis, or lameness (Proudfoot et al. 2010). Current limitation in monitoring the animal feeding behavior is the setup requirement, system maintenance, and the associated cost (Brown-Brandl and Eigenberg 2011). All of these possible drawbacks must be addressed when assessing automated feed intake and behavioral monitoring systems.

7.6.7 On-Farm Physical Inspection for Monitoring Health and Stress-Related Disorders

Any health and stress-related problems are reflected in the physiological parameters such as rectal temperature, pulse rate, and respiratory rate. Simple evaluation of these variables with physical or visual inspection for few minutes may explain the underlying health-related issues. However, quantification of these parameters involves animal contact, restraint, and human-interface (Jorquera-Chavez et al. 2019). Besides, monitoring these activities in large flocks is time-consuming, labor-intensive, and subjective, being inappropriate at the field level. Reddy et al. (2019) projected panting score as the facile measure for quantifying health problems, especially respiration-related issues. They attributed the technique's eminence to its non-invasive functioning of direct observation from two or three meters away from the animal (Nejad and Sung 2017). Therefore, digitalization of diagnostic technology to stress-related disorders through precision livestock farming technologies has the potential to address many aspects of animal health and related to animal welfare and public health.

7.6.8 On-Farm Monitoring Change in Core Body Temperature

Body temperature is the most useful measure of animal reactions to physical-environmental stimuli, occurrence of diseases, and physiological functions such as diet, lactation, and reproduction (Galan et al. 2018). Remote temperature monitoring may help to rapidly detect diseases, minimizing associated treatment costs, and improving animal efficiency (Liang et al. 2013). Core body temperature is determined by many factors such as health status, physiological status, environmental temperature, humidity, water consumption, and feed intake (Davis et al. 2003). Core body temperature can be measured using a variety of methods, such as rectal probe, tympanic membrane, implanted transmitter, infrared thermography, and rumen temperature bolus (Firk et al. 2002). These approaches reflect mechanized technologies, but we must ensure that they are the most reliable, practical, and economical (Firk et al. 2002). Rectal probe is an alternative to conventional rectal thermometers since thermometers are labor-intensive and time-consuming (Godyń et al. 2019). Measurement of body temperature using a rectal probe is useful by continuous measurement of changing temperature over time; however, existing measurement methods are only suitable for single or short-term continuous measurements (Reuter et al. 2007). The rectal probe approach causes the loss of certain data while removing it out of the rectum. Hence, the probe must be connected to the animal in a manner that is not feasible for large-scale use. On the other hand, tympanic temperatures are a reliable indicator of heat stress in cattle, with scales somewhat close to rectal temperatures (Gasim et al. 2013). To measure the tympanic temperature, thermistor is connected to the data logger and then inserted several cm down the ear canal until the tip is located near the tympanic membrane; nevertheless, the implementation of these techniques requires trained staff (Mader et al. 2002). The tympanic temperature procedure can only be used for limited numbers of animals and should not be used for long periods (Goodwin 1998). Animals with tympanic membrane recorders should also be closely monitored for infection at the insertion site (Davis et al. 2003).

Other methods for tracking core body temperature are infrared thermography (IRT) and rumen temperature bolus. Thermography imaging was used to detect nasal mucosal temperatures in a non-invasive manner (Willatt 1993) and corneal surface temperatures (Stewart et al. 2008). In bovine medicine, IRT is used mainly for diagnostic purposes, animal health evaluation, and feed quality monitoring (Montanholi et al. 2010). It has also been used for the prediction and early diagnosis of mastitis and lameness in dairy cattle (Hovinen et al. 2008; Alsaaod and Büscher 2012; Stokes et al. 2012). Additionally, IRT has been used for bovine respiratory disease diagnosis at an earlier stage (Schaefer et al. 2007). The key drawbacks of these technologies are that images must be obtained in direct sunlight, and animal hair coats must be free of soil, moisture, and foreign material (Stewart et al. 2005). Other drawbacks of infrared thermography include the cost of infrared cameras, size, angle of measurement, and animal activity (Johnson et al. 2011). Therefore, IRT requires some adjustment before it becomes practical for the livestock industry.

Developing rumen temperature bolus for tracking core body temperature can be the most appropriate way for producers to monitor body temperature. These boluses could also periodically collect data at different time intervals to monitor sudden or subtle changes in the ruminal environment (Edwards 2010). The benefit of rumen temperature bolus over other approaches is the quick administration without side effects to the animal and the potential to track the large flock at once (Ipema et al. 2008). Rumen bolus is delivered by a baling gun and can be safely recovered after slaughter and does not irritate the epithelium of the reticulo-rumen wall (Ghirardi et al. 2006). This approach may have the longest duration in the body. Nonetheless, it may present some challenges for recovery in commercial harvesting facilities and the dynamics of rumen makes it susceptible to changes in the temperature from water intake and the type of diet due to fermentation (Timsit et al. 2011; Makinde 2020).

7.7 The Future Direction and Challenges of On-Farm and Point-of-Care Diagnostic Technologies

Advances in on-farm and point-of-care diagnostic technology have been linked to acceleration of mechanization, which could promote diagnostic validation, strengthen veterinary surveillance systems, increase livestock efficiency and improve animal welfare (Jones et al. 2019). Therefore, the potential effect of using this diagnostic technology depends on the number of livestock farming systems and the number of animals in each system to which these technologies will apply. These new devices will encourage investment and will support our society (Abuelo and Alves-Nores 2016). The biggest challenge for global economies is to bring new products that meet customer needs in a cost-effective and productive manner. Other critical problems currently faced by on-farm diagnostic technologies are the sluggish usage of these technologies on a broad commercial scale and lack of harmonization between the use of these technologies and the large-scale implementation of data analysis (Busin et al. 2016). The key explanation for this is the abundant data produced and the available technology's inability to promptly turn the ample data into useful knowledge that could be used in the decision-making process. Additional work on the system using these technologies will also contribute to implementing the next generation of an economical and appropriate health monitoring system, which will capture additional behavioral changes related to the level of operation and connect the collected animal data to predict animal health events. The new technology has now reached a stage where its application to biological processes has become practical. Moreover, the livestock industry requires a large number of animals and procedures, making it possible to manufacture personalized, applied low-cost technology (Berckmans 2014a). Of course, no major advancement in precision livestock medicine come without disadvantages. Point-of-care and on-farm diagnostic technologies are still in the early stages of implementation particularly in veterinary medicine, and several major issues will need further research before these techniques become practical with a wide application in livestock management systems.

7.8 Summary and Conclusion

This chapter explains basic on-farm and point-of-care diagnostic technologies as a valuable way of monitoring animal health, welfare, and disease in the livestock industry. The promise of using this technology is valid; however, some drawbacks need to be tackled. Advances in on-farm diagnostic technology will increase the market's efficiency and make these technologies more competitive in the livestock sector. This chapter outlines the attached and unattached type of technologies for tracking physiological and behavioral responses that are important for evaluating animal welfare. The choice of the appropriate diagnostic method depends on the anticipated gain relative to the cost of the device used. Behavioral evaluation must also be selected based on their importance to animal welfare rather than their ability to be reported automatically. Further work on the system using these types of technologies will contribute to creating a next-generation, feasible, and cost-effective animal health monitoring system to detect other critical activities that will enable producers to introduce prevention and treatment protocols against diseases at earlier times.

Conflicts of Interest The authors declare no conflict of interest.

References

Abuelo Á, Alves-Nores V (2016) Point-of-care testing in cattle practice: reliability of cow-side diagnostic tests. In Pract 38:293–302
Alsaaod M, Büscher W (2012) Detection of hoof lesions using digital infrared thermography in dairy cows. J Dairy Sci 95:735–742
Ambardar S, Gupta R, Trakroo D, Lal R, Vakhlu J (2016) High throughput sequencing: an overview of sequencing chemistry. Indian J Microbiol 56:394–404
Anderson PA, Berzins IK, Fogarty F, Hamlin HJ, Guillette LJ Jr (2011) Sound, stress, and seahorses: the consequences of a noisy environment to animal health. Aquaculture 311:129–138
Bach A, Dinarés M, Devant M, Carré X (2007) Associations between lameness and production, feeding and milking attendance of Holstein cows milked with an automatic milking system. J Dairy Res 74:40–46
Berckmans D (2014a) Precision livestock farming technologies for welfare management in intensive livestock systems. Rev Sci Tech 33:189–196
Berckmans D (2014b) Precision livestock farming technologies for welfare management in intensive livestock systems. Rev Sci Tech Off Int Epiz 33:189–196
Bisson I-A, Ssebide BJ, Marra PP (2015) Early detection of emerging zoonotic diseases with animal morbidity and mortality monitoring. EcoHealth 12:98–103
Bolboacă SD (2019) Medical diagnostic tests: a review of test anatomy, phases, and statistical treatment of data. Comput Math Methods Med 2019
Borderas T, Fournier A, Rushen J, De Passille A (2008) Effect of lameness on dairy cows' visits to automatic milking systems. Can J Anim Sci 88:1–8
Brehme U, Stollberg U, Holz R, Schleusener T (2008) ALT pedometer – new sensor-aided measurement system for improvement in oestrus detection. Comput Electron Agric 62:73–80
Brown-Brandl T, Eigenberg R (2011) Development of a livestock feeding behavior monitoring system. Trans ASABE 54:1913–1920
Busin V, Wells B, Kersaudy-Kerhoas M, Shu W, Burgess ST (2016) Opportunities and challenges for the application of microfluidic technologies in point-of-care veterinary diagnostics. Mol Cell Probes 30:331–341
Carpentier L, Berckmans D, Youssef A, Berckmans D, Van Waterschoot T, Johnston D, Ferguson N, Earley B, Fontana I, Tullo E, Guarino M, Vranken E, Norton T (2018) Automatic cough detection for bovine respiratory disease in a calf house. Biosyst Eng 173:45–56
Ceballos A, Sanderson D, Rushen J, Weary D (2004) Improving stall design: use of 3-D kinematics to measure space use by dairy cows when lying down. J Dairy Sci 87:2042–2050
Chapinal N, DE Passille AM, Pastell M, Hänninen L, Munksgaard L, Rushen J (2011) Measurement of acceleration while walking as an automated method for gait assessment in dairy cattle. J Dairy Sci 94:2895–2901
Chappell JG, Byaruhanga T, Tsoleridis T, Ball JK, Mcclure CP (2019) Identification of infectious agents in high-throughput sequencing data sets is easily achievable using free, cloud-based bioinformatics platforms. J Clin Microbiol:57
Cheon D-S, Chae C (2000) Comparison of virus isolation, reverse transcription-polymerase chain reaction, immunohistochemistry, and in situ hybridization for the detection of porcine reproductive and respiratory syndrome virus from naturally aborted fetuses and stillborn piglets. J Vet Diagn Investig 12:582–587
Chung Y, Oh S, Lee J, Park D, Chang H-H, Kim S (2013) Automatic detection and recognition of pig wasting diseases using sound data in audio surveillance systems. Sensors 13:12929–12942
Coghe J, Uystepruyst CH, Bureau F, Detilleux J, Art T, Lekeux P (2000) Validation and prognostic value of plasma lactate measurement in bovine respiratory disease. Vet J 160:139–146
Commission E (2011) Communication from the Commission to the European Parliament and the Council. Action plan against the rising threats from antimicrobial resistance
Craft ME (2015) Infectious disease transmission and contact networks in wildlife and livestock. Philos Trans R Soc B 370:20140107

D'souza MKA (2020) Point of care animal side simple phosphorous detection test kit in cattle. Institute of Chemical Technology (ICT), Department of Pharmaceutical Sciences & Technology, Deemed University, Elite status, Centre of excellence (GOM), Matunga (E), Mumbai, India-400019

Dai Y, Liu CC (2019) Recent advances on electrochemical biosensing strategies toward universal point-of-care systems. Angew Chem 131:12483–12496

Das J, Cross G, Qu C, Makineni A, Tokekar P, Mulgaonkar Y, Kumar V (2015) Devices, systems, and methods for automated monitoring enabling precision agriculture. In: 2015 IEEE international conference on automation science and engineering (CASE). IEEE, pp 462–469

Daszak P, Cunningham AA, Hyatt AD (2000) Emerging infectious diseases of wildlife – threats to biodiversity and human health. Science 287:443–449

Davis J, Vanzant E, Purswell J, Green A, Bicudo J, Gates R, Holloway L, Smith W (2003) Methods of remote, continuous temperature detection in beef cattle. ASAE annual meeting, 2003, vol 1. American Society of Agricultural and Biological Engineers

Davis JD, Darr MJ, Xin H, Harmon JD, Russell JR (2011) Development of a GPS herd activity and well-being kit (GPS HAWK) to monitor cattle behavior and the effect of sample interval on travel distance. Appl Eng Agric 27:143–150

den Uijl I, Gómez Álvarez CB, Bartram D, Dror Y, Holland R, Cook A (2017) External validation of a collar-mounted triaxial accelerometer for second-by-second monitoring of eight behavioural states in dogs. PloS One 12(11):e0188481

Edwards T (2010) Control methods for bovine respiratory disease for feedlot cattle. Vet Clin Food Anim Pract 26:273–284

Elelu N, Ferrolho J, Couto J, Domingos A, Eisler MC (2016) Molecular diagnosis of the tick-borne pathogen Anaplasma marginale in cattle blood samples from Nigeria using qPCR. Exp Appl Acarol 70:501–510

Elolimy A, Alharthi A, Zeineldin M, Parys C, Loor JJ (2020) Residual feed intake divergence during the preweaning period is associated with unique hindgut microbiome and metabolome profiles in neonatal Holstein heifer calves. J Anim Sci Biotechnol 11:13

Enstipp MR, Ciccione S, Gineste B, Milbergue M, Ballorain K, Ropert-Coudert Y, Kato A, Plot V, Georges J-Y (2011) Energy expenditure of freely swimming adult green turtles (Chelonia mydas) and its link with body acceleration. J Exp Biol 214:4010–4020

Exadaktylos V, Silva M, Aerts J-M, Taylor CJ, Berckmans D (2008) Real-time recognition of sick pig cough sounds. Comput Electron Agric 63:207–214

Felton C, Colazo M, Bench C, Ambrose D (2013) Large variations exist in prepartum activity among dairy cows continuously housed in a tie-stall barn. Can J Anim Sci 93:435–444

Ferrari S, Silva M, Guarino M, Aerts JM, Berckmans D (2008) Cough sound analysis to identify respiratory infection in pigs. Comput Electron Agric 64:318–325

Ferrari S, Piccinini R, Silva M, Exadaktylos V, Berckmans D, Guarino M (2010) Cough sound description in relation to respiratory diseases in dairy calves. Prev Vet Med 96:276–280

Firk R, Stamer E, Junge W, Krieter J (2002) Automation of oestrus detection in dairy cows: a review. Livest Prod Sci 75:219–232

Fleishman LJ, Mcclintock WJ, D'eath RB, Brainard DH, Endler JA (1998) Colour perception and the use of video playback experiments in animal behaviour. Anim Behav 56:1035–1040

Foley CJ, Sillero-Zubiri C (2020) Open-source, low-cost modular GPS collars for monitoring and tracking wildlife. Methods Ecol Evol 11:553–558

Frost A, Parsons D, Stacey K, Robertson A, Welch S, Filmer D, Fothergill A (2003) Progress towards the development of an integrated management system for broiler chicken production. Comput Electron Agric 39:227–240

Galan E, Llonch P, Villagra A, Levit H, Pinto S, Del Prado A (2018) A systematic review of non-productivity-related animal-based indicators of heat stress resilience in dairy cattle. PLoS One 13:e0206520

Gasim GI, Musa IR, Abdien MT, Adam I (2013) Accuracy of tympanic temperature measurement using an infrared tympanic membrane thermometer. BMC Res Notes 6:194

Gattani A, Singh SV, Agrawal A, Khan MH, Singh P (2019) Recent progress in electrochemical biosensors as point of care diagnostics in livestock health. Anal Biochem 579:25–34
Ghirardi J, Caja G, Garín D, Casellas J, Hernández-Jover M (2006) Evaluation of the retention of electronic identification boluses in the forestomachs of cattle. J Anim Sci 84:2260–2268
Glaser L, Carstensen M, Shaw S, Robbe-Austerman S, Wunschmann A, Grear D, Stuber T, Thomsen B (2016) Descriptive epidemiology and whole genome sequencing analysis for an outbreak of bovine tuberculosis in beef cattle and white-tailed deer in northwestern Minnesota. PLoS One 11:e0145735
Godsk T, Kjærgaard MB (2011) High classification rates for continuous cow activity recognition using low-cost GPS positioning sensors and standard machine learning techniques. In: Industrial conference on data mining. Springer, pp 174–188
Godyń D, Herbut P, Angrecka S (2019) Measurements of peripheral and deep body temperature in cattle – a review. J Therm Biol 79:42–49
González L, Schwartzkopf-Genswein K, Caulkett N, Janzen E, Mcallister T, Fierheller E, Schaefer A, Haley D, Stookey J, Hendrick S (2010) Pain mitigation after band castration of beef calves and its effects on performance, behavior, Escherichia coli, and salivary cortisol. J Anim Sci 88:802–810
Goodwin SD (1998) Comparison of body temperatures of goats, horses, and sheep measured with a tympanic infrared thermometer, an implantable microchip transponder, and a rectal thermometer. J Am Assoc Lab Anim Sci 37:51–55
Griffin TW (2009) Whole-farm benefits of GPS-enabled navigation technologies. Reno, Nevada, June 21-June 24, 2009. American Society of Agricultural and Biological Engineers, p 1
Gubala V, Harris LF, Ricco AJ, Tan MX, Williams DE (2012) Point of care diagnostics: status and future. Anal Chem 84:487–515
Handcock RN, Swain DL, Bishop-Hurley GJ, Patison KP, Wark T, Valencia P, Corke P, O'Neill CJ (2009) Monitoring animal behaviour and environmental interactions using wireless sensor networks, GPS collars and satellite remote sensing. Sensors 9:3586–3603
Helwatkar A, Riordan D, Walsh J (2014) Sensor technology for animal health monitoring. In: 8th international conference on sensing technology, Liverpool, pp 266–271
Hennessy DA, Wolf CA (2018) Asymmetric information, externalities and incentives in animal disease prevention and control. J Agric Econ 69:226–242
Hovinen M, Siivonen J, Taponen S, Hänninen L, Pastell M, Aisla A-M, Pyörälä S (2008) Detection of clinical mastitis with the help of a thermal camera. J Dairy Sci 91:4592–4598
Ipema A, Goense D, Hogewerf P, Houwers H, Van Roest H (2008) Pilot study to monitor body temperature of dairy cows with a rumen bolus. Comput Electron Agric 64:49–52
Iraguha B, Hamudikuwanda H, Mushonga B, Kandiwa E, Mpatswenumugabo JP (2017) Comparison of cow-side diagnostic tests for subclinical mastitis of dairy cows in Musanze district, Rwanda. J S Afr Vet Assoc 88:e1–e6
Ishiwata T, Kilgour R, Uetake K, Eguchi Y, Tanaka T (2007) Choice of attractive conditions by beef cattle in a Y-maze just after release from restraint. J Anim Sci 85:1080–1085
Iwersen M, Falkenberg U, Voigtsberger R, Forderung D, Heuwieser W (2009) Evaluation of an electronic cowside test to detect subclinical ketosis in dairy cows. J Dairy Sci 92:2618–2624
Johnson SR, Rao S, Hussey SB, Morley PS, Traub-Dargatz JL (2011) Thermographic eye temperature as an index to body temperature in ponies. J Equine Vet 31:63–66
Jones G, Bork O, Ferguson SA, Bates A (2019) Comparison of an on-farm point-of-care diagnostic with conventional culture in analysing bovine mastitis samples. J Dairy Res 86:222–225
Jorquera-Chavez M, Fuentes S, Dunshea FR, Warner RD, Poblete T, Jongman EC (2019) Modelling and validation of computer vision techniques to assess heart rate, eye temperature, ear-base temperature and respiration rate in cattle. Animals 9(12):1089
Karapinar T, Hayirli OKA, Kom M (2013) Evaluation of 4 point-of-care units for the determination of blood L-lactate concentration in cattle. J Vet Intern Med 27:1596–1603
Katsoulos PD, Minas A, Karatzia MA, Pourliotis K, Christodoulopoulos G (2011) Evaluation of a portable glucose meter for use in cattle and sheep. Vet Clin Pathol 40:245–247

Krieger S, Oczak M, Lidauer L, Berger A, Kickinger F, Öhlschuster M, Auer W, Drillich M, Iwersen M (2019) An ear-attached accelerometer as an on-farm device to predict the onset of calving in dairy cows. Biosyst Eng 184:190–199

Kumar KR, Cowley MJ, Davis RL (2019) Next-generation sequencing and emerging technologies. Seminars in thrombosis and hemostasis. Thieme Medical Publishers, pp 661–673

Kumar P, Chakraborty S, Nagar D, Birader K, Suman P (2020) Application of biosensors to enhance reproductive efficiency and production of livestock and poultry by diverse antigen analysis. Immunodiagnostic technologies from laboratory to point-of-care testing. Springer

Lahdenoja O, Hurnanen T, Kaisti M, Koskinen J, Tuominen J, Vähä-Heikkilä M, Parikka L, Wiberg M, Koivisto T, Pänkäälä M (2019) Cardiac monitoring of dogs via smartphone mechanocardiography: a feasibility study. Biomed Eng Online 18(1):1–14

Liang D, Wood C, Mcquerry K, Ray D, Clark J, Bewley J (2013) Influence of breed, milk production, season, and ambient temperature on dairy cow reticulorumen temperature. J Dairy Sci 96:5072–5081

Lokhorst C, Ipema A (2010) Precision livestock farming for operational management support in livestock production chains. In: Trienekens J, Top J, van der Vorst J, Beulens A (eds) Towards effective food chains: models and applications. Wageningen Academic Publishers, pp 293–308

Løvendahl P, Chagunda M (2010) On the use of physical activity monitoring for estrus detection in dairy cows. J Dairy Sci 93:249–259

Macmillan K, Lopez Helguera I, Behrouzi A, Gobikrushanth M, Hoff B, Colazo MG (2017) Accuracy of a cow-side test for the diagnosis of hyperketonemia and hypoglycemia in lactating dairy cows. Res Vet Sci 115:327–331

Mader T, Holt S, Hahn G, Davis M, Spiers D (2002) Feeding strategies for managing heat load in feedlot cattle. J Anim Sci 80:2373–2382

Mahen PJ, Williams HJ, Smith RF, Grove-White D (2018) Effect of blood ionised calcium concentration at calving on fertility outcomes in dairy cattle. Vet Rec 183:263

Makinde A (2020) Investigating perceptions, motivations, and challenges in the adoption of precision livestock farming in the beef industry. Masther thesis presented to The University of Guelph

Manteuffel G, Puppe B, Schön PC (2004) Vocalization of farm animals as a measure of welfare. Appl Anim Behav Sci 88:163–182

Martiskainen P, Järvinen M, Skön J-P, Tiirikainen J, Kolehmainen M, Mononen J (2009) Cow behaviour pattern recognition using a three-dimensional accelerometer and support vector machines. Appl Anim Behav Sci 119:32–38

Megahed AA, Constable PD (2020) Technical note: evaluation of a colorimetric point-of-care test for measuring urine ammonium concentration in periparturient dairy cattle. J Dairy Sci 103:8655–8660

Megahed AA, Hiew MW, Townsend JR, Messick JB, Constable PD (2015) Evaluation of an electrochemical point-of-care meter for measuring glucose concentration in blood from periparturient dairy cattle. J Vet Intern Med 29:1718–1727

Megahed AA, Hiew MWH, Grunberg W, Constable PD (2016) Evaluation of 2 portable ion-selective electrode meters for determining whole blood, plasma, urine, milk, and abomasal fluid potassium concentrations in dairy cattle. J Dairy Sci 99:7330–7343

Megahed AA, Hiew MWH, Townsend JR, Constable PD (2017) Characterization of the analytic performance of an electrochemical point-of-care meter for measuring beta-hydroxybutyrate concentration in blood and plasma from periparturient dairy cattle. Vet Clin Pathol 46:314–325

Moen R, Pastor J, Cohen Y (2001) Effects of animal activity on GPS telemetry location attempts. Alces 37:207–216

Mohr S, Beard R, Nisbet AJ, Burgess ST, Reeve R, Denwood M, Porphyre T, Zadoks RN, Matthews L (2020) Uptake of diagnostic tests by livestock farmers: a stochastic game theory approach. Front Vet Sci 7:36

Montanholi Y, Swanson K, Palme R, Schenkel F, Mcbride B, Lu D, Miller S (2010) Assessing feed efficiency in beef steers through feeding behavior, infrared thermography and glucocorticoids. Animal 4:692–701

Moreau M, Siebert S, Buerkert A, Schlecht E (2009) Use of a tri-axial accelerometer for automated recording and classification of goats' grazing behaviour. Appl Anim Behav Sci 119:158–170

Naqvi AN (2007) Application of molecular genetic technologies in livestock production: potentials for developing countries. Adv Biol Res 34:72–84

Neethirajan S (2020). Transforming the adaptation physiology of farm animals through sensors. Animals 10(9):1512

Neethirajan S, Tuteja SK, Huang S-T, Kelton D (2017) Recent advancement in biosensors technology for animal and livestock health management. Biosens Bioelectron 98:398–407

Nejad JG, Sung K-I (2017) Behavioral and physiological changes during heat stress in Corriedale ewes exposed to water deprivation. J Animal Sci Technol 59(1):1–6

Nilsson M, Herlin A, Ardö H, Guzhva O, Åström K, Bergsten C (2015) Development of automatic surveillance of animal behaviour and welfare using image analysis and machine learned segmentation technique. Animal 9:1859–1865

Norling B (1991) Accelerometers: current and emerging technology. Kinematic systems in geodesy, surveying, and remote sensing. Springer

O'Leary N, Byrne D, O'Connor A, Shalloo L (2020) Invited review: cattle lameness detection with accelerometers. J Dairy Sci

Pastell ME, Kujala M (2007) A probabilistic neural network model for lameness detection. J Dairy Sci 90(5):2283–2292

Pastell M, Takko H, Gröhn H, Hautala M, Poikalainen V, Praks J, Veermäe I, Kujala M, Ahokas J (2006) Assessing cows' welfare: weighing the cow in a milking robot. Biosyst Eng 93:81–87

Pereira AMF, Alves A, Infante P, Titto EA, Baccari F, Afonso Almeida JA (2010) A device to improve the schleger and turner method for sweating rate measurements. Int J Biomet 54(1):37–43

Proudfoot K, Weary D, Von Keyserlingk M (2010) Behavior during transition differs for cows diagnosed with claw horn lesions in mid lactation. J Dairy Sci 93:3970–3978

Quandt JE, Barletta M, Cornell KK, Giguere S, Hofmeister EH (2018) Evaluation of a point-of-care blood glucose monitor in healthy goats. J Vet Emerg Crit Care (San Antonio) 28:45–53

Reddy PR, Rajeev Kumar B, Srinivasa Prasad Ch, Venkataseshiah Ch, Hyder I (2019) Erythrocyte fragility based assessment of true thermal resilience in tropical small ruminants. Biol Rhythm Res 1–12

Reuter R, Carroll J, Dailey J, Chase C Jr, Coleman S, Riley D, Spiers D, Weaber R, Galyean M (2007) Development of an automatic, indwelling rectal temperature probe for cattle research. J Anim Sci 85:12

Reuter JA, Spacek DV, Snyder MP (2015) High-throughput sequencing technologies. Mol Cell 58:586–597

Revathi Poonati PCM, Punati RD, Maity SN, Alapati KS, Polavarapu KKB, Polavarapu R (2020) Development of rapid, sensitive and in-expensive point of care diagnostic method for brucellosis in dairy cattle at resource-limited areas. Indian J Publ Health Res Dev:11

Ringgenberg N, Bergeron R, Devillers N (2010) Validation of accelerometers to automatically record sow postures and stepping behaviour. Appl Anim Behav Sci 128:37–44

Robinson C, Creedon N, Sayers R, Kennedy E, O'Riordan A (2020) Electrochemical detection of bovine immunoglobulins G to determine passive transfer of antibodies to calves. Anal Methods

Rose-Dye T, Burciaga-Robles L, Krehbiel C, Step D, Fulton R, Confer A, Richards C (2011) Rumen temperature change monitored with remote rumen temperature boluses after challenges with bovine viral diarrhea virus and Mannheimia haemolytica. J Anim Sci 89:1193–1200

Rothwell ES, Bercovitch FB, Andrews JR, Anderson MJ (2011) Estimating daily walking distance of captive African elephants using an accelerometer. Zoo Biol 30:579–591

Rushen J, Chapinal N, De Passille A (2012) Automated monitoring of behavioural-based animal welfare indicators. Anim Welfare UFAW J 21:339

Saegerman C, Porter S, Humblet M (2011) The use of modelling to evaluate and adapt strategies for animal disease control. Revue Scientifique et Technique-OIE 30:555
Sala JE, Quintana F, Wilson RP, Dignani J, Lewis MN, Campagna C (2011) Pitching a new angle on elephant seal dive patterns. Polar Biol 34:1197–1209
Sargeant JM, O'Connor AM (2020) Scoping reviews, systematic reviews, and meta-analysis: applications in veterinary medicine. Front Vet Sci 7:11
Schaefer AL, Cook NJ, Church JS, Basarab J, Perry B, Miller C, Tong AK (2007) The use of infrared thermography as an early indicator of bovine respiratory disease complex in calves. Res Vet Sci 83:376–384
Schwabe C (1982) The current epidemiological revolution in veterinary medicine. Part I. Prev Vet Med 1:5–15
Shao B, Xin H (2008) A real-time computer vision assessment and control of thermal comfort for group-housed pigs. Comput Electron Agric 62:15–21
Shepley E, Berthelot M, Vasseur E (2017) Validation of the ability of a 3D pedometer to accurately determine the number of steps taken by dairy cows when housed in tie-stalls. Agriculture 7:53
Shirley MW, Charleston B, King DP (2010) New opportunities to control livestock diseases in the post-genomics era. J Agric Sci 149:115–121
Singh B, Mal G, Gautam SK, Mukesh M (2019) Next-generation sequencing vis-à-vis veterinary health management. Advances in animal biotechnology. Springer
Sowell B, Branine M, Bowman J, Hubbert M, Sherwood H, Quimby W (1999) Feeding and watering behavior of healthy and morbid steers in a commercial feedlot. J Anim Sci 77:1105–1112
Stewart M, Webster J, Schaefer A, Cook N, Scott S (2005) Infrared thermography as a non-invasive tool to study animal welfare. Anim Welf 14:319–325
Stewart M, Stafford K, Dowling S, Schaefer A, Webster J (2008) Eye temperature and heart rate variability of calves disbudded with or without local anaesthetic. Physiol Behav 93:789–797
Stokes J, Leach K, Main D, Whay H (2012) An investigation into the use of infrared thermography (IRT) as a rapid diagnostic tool for foot lesions in dairy cattle. Vet J 193:674–678
Stone AE, Tsai I-C, Bewley JM (2017) Precision dairy monitoring of fresh cows. In: Proceedings from the Western Dairy Management conference, pp 120–133
Sun AC, Hall DA (2019) Point-of-care smartphone-based electrochemical biosensing. Electroanalysis 31:2–16
Tang L, Abplanalp P (2014) GPS guided farm mapping and waypoint tracking mobile robotic system. In: 9th IEEE conference on industrial electronics and applications. IEEE, pp 1676–1681
Theurer ME, Amrine DE, White BJ (2013) Remote noninvasive assessment of pain and health status in cattle. Vet Clin Food Anim Pract 29:59–74
Timsit E, Assie S, Quiniou R, Seegers H, Bareille N (2011) Early detection of bovine respiratory disease in young bulls using reticulo-rumen temperature boluses. Vet J 190:136–142
Tomkiewicz SM, Fuller MR, Kie JG, Bates KK (2010) Global positioning system and associated technologies in animal behaviour and ecological research. Philos Trans R Soc B 365:2163–2176
Tschandl P, Codella N, Akay BN, Argenziano G, Braun RP, Cabo H, Gutman D, Halpern A, Helba B, Hofmann-Wellenhof R (2019) Comparison of the accuracy of human readers versus machine-learning algorithms for pigmented skin lesion classification: an open, web-based, international, diagnostic study. Lancet Oncol 20:938–947
Van Arendonk JAM (2011) The role of reproductive technologies in breeding schemes for livestock populations in developing countries. Livest Sci 136:29–37
Van Schyndel SJ, Bogado Pascottini O, Leblanc SJ (2018) Comparison of cow-side diagnostic techniques for subclinical endometritis in dairy cows. Theriogenology 120:117–122
Van Veen TS (1997) Sense or nonsense? Traditional methods of animal parasitic disease control. Vet Parasitol 71:177–194
Vidic J, Manzano M, Chang CM, Jaffrezic-Renault N (2017) Advanced biosensors for detection of pathogens related to livestock and poultry. Vet Res 48:11
Visen N, Paliwal J, Jayas D, White N (2002) Ae – automation and emerging technologies: specialist neural networks for cereal grain classification. Biosyst Eng 82:151–159

Von Holst, D. 1998. The concept of stress and its relevance for animal behavior
Walker P, Subasinghe RP (2000) DNA-based molecular diagnostic techniques: research needs for standardization and validation of the detection of aquatic animal pathogens and diseases. Food & Agriculture Org
Wang J (2006) Electrochemical biosensors: towards point-of-care cancer diagnostics. Biosens Bioelectron 21:1887–1892
Wathes C, Kristensen H, Aerts J, Berckmans D (2005) Is precision livestock farming an engineer's daydream or nightmare, an animal's friend or foe, and a farmer's panacea or pitfall? In: Cox S (ed) Precision livestock farming'05. Proceedings of the 2nd European conference on precision livestock farming, pp 33–46
Weary D, Huzzey J, Von Keyserlingk M (2009) Board-invited review: using behavior to predict and identify ill health in animals. J Anim Sci 87:770–777
Willatt D (1993) Continuous infrared thermometry of the nasal mucosa. Rhinology 31:63–67
Yanase J, Triantaphyllou E (2019) A systematic survey of computer-aided diagnosis in medicine: past and present developments. Expert Syst Appl 138:112821
Zeineldin M, Yassein AE-R, Hassam E-A, Mohamed G (2016) Lung ultrasonography and computer-aided scoring system as a diagnostic aid for bovine respiratory disease in feedlot cattle. Global Veterinaria 17:588–594
Zeineldin M, Lowe J, De Godoy M, Maradiaga N, Ramirez C, Ghanem M, Abd El-Raof Y, Aldridge B (2017a) Disparity in the nasopharyngeal microbiota between healthy cattle on feed, at entry processing and with respiratory disease. Vet Microbiol 208:30–37
Zeineldin M, Ghanem M, Abd El-Raof Y, Elattar H (2017b) Clinical utilization of point-of-care blood L-lactate concentrations in naturally occurring respiratory disease in feedlot cattle. Pak Vet J 37
Zeineldin M, Aldridge B, Lowe J (2019a) Antimicrobial effects on swine gastrointestinal microbiota and their accompanying antibiotic resistome. Front Microbiol 10:1035
Zeineldin M, Lowe J, Aldridge B (2019b) Contribution of the mucosal microbiota to bovine respiratory health. Trends Microbiol 27:753–770
Zin TT, Kobayashi I, Tin P, Hama H (2016) A general video surveillance framework for animal behavior analysis. In: 2016 third international conference on computing measurement control and sensor network (CMCSN). IEEE, pp 130–133
Zumla A, Al-Tawfiq JA, Enne VI, Kidd M, Drosten C, Breuer J, Muller MA, Hui D, Maeurer M, Bates M, Mwaba P, Al-Hakeem R, Gray G, Gautret P, Al-Rabeeah AA, Memish ZA, Gant V (2014) Rapid point of care diagnostic tests for viral and bacterial respiratory tract infections – needs, advances, and future prospects. Lancet Infect Dis 14:1123–1135

Chapter 8
Biotechnological Applications in Poultry Farming

S. M. Lutful Kabir and S. K. Shaheenur Islam

Abstract Global meat consumption is growing exponentially because of rapid urbanization, population growth including lifestyle changes in the developing countries. Amongst the sources of meat consumption, maximum proportion comes from poultry birds as a white meat. Advancement of biotechnology revolution triggers a numerous opportunities in poultry industry to mitigate such emerging demand of meat and meat products. Biotechnology in poultry that never be regarded only for better farming, but its multifaceted impacts in the poultry production food systems in recent times. Here we review the activities related with the poultry sector, for example, breed development, health, nutrition, disease prevention and control, preservation of poultry origin meat and meat products along with the food safety are mostly achieved enormous improvements. The biotechnological tools based on modern techniques offer enormous prospective for the production of biologics and vaccines, medicinal products including disease diagnostic assays as a part of disease control and prevention. The advancement and application of these technologies are mostly used in the developed countries, however, these are being used moderately in developing countries in different segments of poultry science. Both conventional and modern biotechnologies in poultry have contributed enormously to enhance productivity and optimize income generation activities and reduce poverty, lessen the burdens of diseases and guarantee to environmental sustainable livestock production in developing countries.

Findings

In conformity with in-depth literature review on the above captioned title, we may conclude that biotechnology is attending to the upgradation of veterinary science especially in the light of poultry production and processing of poultry origin

S. M. L. Kabir (✉)
Department of Microbiology and Hygiene, Bangladesh Agricultural University, Mymensingh, Bangladesh
e-mail: lkabir79@bau.edu.bd

S. K. S. Islam
Department of Microbiology and Hygiene, Bangladesh Agricultural University, Mymensingh, Bangladesh

Department of Livestock Services, Krishi Khamar Sarak, Dhaka, Bangladesh

V. K. Yata et al. (eds.), *Sustainable Agriculture Reviews 54*, Sustainable Agriculture Reviews 54, https://doi.org/10.1007/978-3-030-76529-3_8

food that entwined with food safety issues. There is a need to publicize these issues since the approaches are almost new, and to provide handsome training for laboratorians from both private and government sectors targeted to getting best benefits on disease diagnostic areas. It is pertinent to instituting of inclusive and effective regulatory framework considering the country perspectives and the safe application of technologies regarding biosafety, biosecurity and to disperse the confusion among the mass people in this evolving discipline. The government with the partners of the food systems, and the biotechnology industry should completely utilize this science-based devise to corroborate food safety, and to stance at the back of public health decisions.

Keywords Poultry production and processing · Poultry origin food · Food safety · Food systems · Biotechnology · Disease diagnostic techniques · Regulatory framework · Biotechnology industry · Developing countries

Abbreviations

AI	Artificial insemination
AMR	Antimicrobial resistance
AMU	Antimicrobial uses
BAC	Bacterial artificial chromosome
BLAST	Basic local alignment sequence tool
BULP	Best linear unbiased prediction
CCAC	Canadian council of animal care
CCPS	Combination of crossbred and purebred selection
CLA	Conjugated linoleic acid
DNA	Deoxyribonucleic acid
EBU	Expected breeding value
EMO	Effective microorganism
EST	Expressed sequence tag
FASTA	Fast alignment
FCR	Feed conversion ratio
GAH	Good animal husbandry
GEBV	Genomic estimate breeding value
GPH	Good personal health
HA	Hemagglutinin
LBM	Live bird market
LD	Linked disequilibrium
MAS	Marker assisted selection
NA	Neuraminidase
NCBI	National Center for Biotechnology Information
NGS	Next generation sequencing

PCR	Polymerase chain reaction
PLS	Pure-line selection
QTL	Quantitate trait loci
QTL	Quantitate trait locus
SNP	Single nucleotide polymorphism
SRRC	Sheep red blood cell
TVC	Total viable count
US	United States
WGS	Whole genome studies

8.1 Introduction

As a diverse domain of scientific discipline, biotechnology has been associated with the human development when the life has advanced. Archaeologically, this was evolved when humans used microbes to prepare foods like bread and beverage, it may be dated before 4000 BC. The word "Biotechnology" is usually specified as the ability to use living microbes or their ingredients to promotion or to modernize a product, to develop fauna or flora or to change microorganisms for desired objectives (Raju et al. 2015; Vijayakumar and Sasikala 2012). Principally, this makes association among microorganism, human and animals, and with the effect of technology for the betterment of human life. The exploration in cell biology, agriculture, plant sciences, animal sciences, food science, medicine and environmental sciences are some paramount areas where biotechnology acts a key player (Sugumaran and Ponnusami 2017). A broad range of tools for the genetic upgradation of animal species, vaccines and biologics for disease control, and advancement of rapid test kits to diagnose the disease of poultry and livestock, including other animal species. Therefore, this discipline has already had a substantial influence on poultry production, control and prevention of diseases, and processing of poultry origin food (Pal et al. 2017) and also be linked with the safety standards of food and food products (Fig. 8.1).

As a part of global poultry meat production, 18% comes from Unites States (US) as a top producer, followed by China, Brazil and the Russian Federation. Nevertheless, considering global egg production, 42% is supplied by China as a top producer, followed by US and India. More than 60% of global egg production comes from Asia as considered to be the highest egg producing region in the World. From 1961 to 2017, the global poultry production increased about 14 times more (122 million tonnes) to mitigate incessant demand, however, egg production increased approximately 6 times (87 million tonnes) with 37% global contribution at present time. Interestingly, world egg production has been increased by more than 150% during past three decades. Enormous growth has been witnessed in Asia due to supply of cheapest animal source protein requirement for huge population in this region, where production increased nearly fourfolds. Approximately, 80% rural

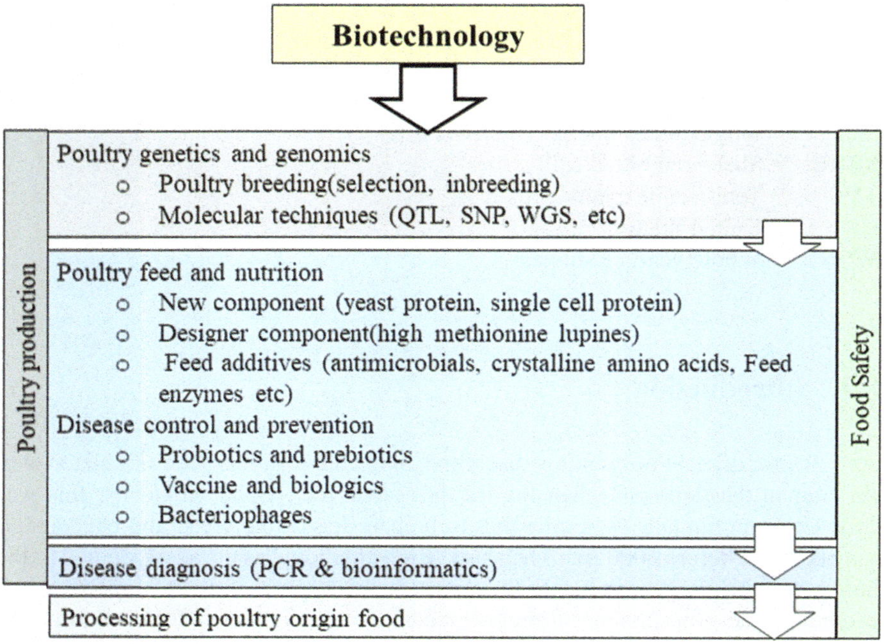

Fig. 8.1 Biotechnological application at poultry production, processing of poultry origin food and interlinking with food safety. (Adapted from Pal et al. 2017)

households rear poultry as their nutrition demand and livelihoods (FAO 2020) with biotechnologically developed high-yielding commercial layer or broiler poultry and low performance duel-purpose breeds in developing countries (Pym 2013). However, proportions in these two categories fluctuate widely as indigenous poultry shares around 90% of the poultry population in these countries (Pym et al. 2006). Moreover, this figure may vary due to rapid growth of the commercial poultry industry in developing countries at present time.

Nowadays, biotechnology is extensively applied in animal production along with a plentiful other potential uses in many countries in the world. This could have applied for enhance animal performance via better nutrition, upgraded production traits or better health condition through integrating antibodies or vaccines into feeds. However, this can efficiently defend the animals against infectious diseases and to enhance animal productivity economically (Daniel 2016). Over the last 75 years, a remarkable influence on poultry production through selection of production traits has been noticed. Moreover, genetic improvement of production parameters in poultry to be optimized in the next 20 years due to the emerging challenges of global food security (Mohammed et al. 2016). Nowadays, more emphasis on the demand of the producers and consumer like productivity, quality, food safety and animal welfare in addition to the necessities of the breeders on economically feasible options like low cost of feed and other production inputs (vaccine, biologics and

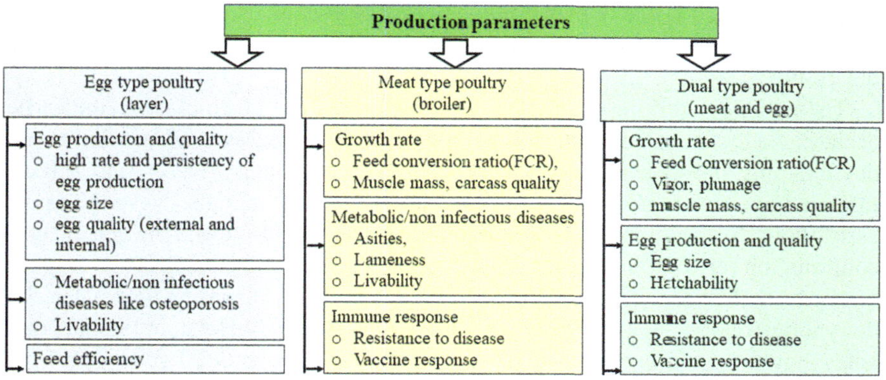

Fig. 8.2 Characteristics associated with the production parameters for selection of targeted poultry birds. (Adapted from Burt 2002, Engormix 2008, and Leeson and Summers 2010)

medicine) and higher fertility are the pivotal areas as prerequisite of poultry industry (Fig. 8.2). To obtain desire level of production, conventional genetic selection is unable to fulfill the targeted objectives unless the application of genomics to be functional. The poultry genomics has been influenced a lot as the progress of genetics technology of human and model organisms. At present, a number of techniques are well documented viz., quantitative trait locus (QTL), genetic markers and physical mapping, comparative mapping, bacterial artificial chromosome (BAC) resources, expressed sequence tag (EST) are tailoring to benefit to poultry industry (Burt 2002). The scope of this paper will glimpse on biotechnological approaches are being used for production and processing and to enable safety standard of poultry origin food. The paper is also focus on the potential limitation and challenges to be addressed to operationalize the biotechnological approaches in low resource settings. Moreover, this paper is also outlined the way forwards to mitigate the challenges as a long term goals.

8.2 Historical Perspectives

Biotechnology is a technology based biology through which the scientific and engineering principles applied in living animals for the production of biological materials targeted for the benefit of human and animal health. The animal biotechnology has been commenced approximately 8000 years before by means of animal domestication and artificial selection process. Nearly 48,000 mammalian species, less than 20 were successfully tamed in human captivity (Diamond 1999). Those animals were successfully tamed, and subsequently reared in captive condition, they shared and displayed a distinctive mixture of traits (Diamond 1999; Hale 1969). They were found to be comparatively quiet, and cooperative in their feed requirements, and could be raised and go maturity rapidly on natural available fodders, and could be easily reproduced in the human confinement. These animals had hierarchical societal norms and permitted humans to restrain and they were adapted to living within big group.

The paradigm shifting in animal production had been taken place in the earlier decades via the techniques, namely artificial selection process to accumulation production traits, immunization to prevent and control of animal diseases, and artificial insemination technique to increase production of food animals. Nevertheless, in early 1960s, a contemporary biotechnology was emerged followed the genetic code confirmation (NRC 2002). Actually, the updated breeds of livestock differ very distinctly from their ancestors is mirroring of how rapidly the breeding took place. The dairy industry of contemporary United States where modern Holstein Frisian dominates, however, a little similarity to its descendants of about a 50 years ago. From 1945 to 1995, milk production increased each cow nearly three times (Majeskie 1996) due to breeding program of selected proven cattle. They had undertaken significant adjustment in traits over a short duration of time from an intense selection process and to bring the diversity observed in present breeds as revealed a huge disparity in pool genetics in species (Wayne and Ostrander 1999). The selection method formed the farm animal species (cattle) that facilitated to the development of the modern farmed poultry breeds of chickens, ducks, turkeys and geese raised for meat, eggs, and other uses like feathers. Thus the broiler poultry was established that gained at the market weight by around 40 days (Pisenti et al. 1999) through feeding of less quantity feed (Lacy 2000). The use of probiotics supplementation has a significant impact on live weight gain, increase carcass weight including high output of cut up meat and immune response including heat tolerance capacity of poultry, and to ensure the safety of poultry origin food (Roy et al. 2015; Kabir 2009a, b; Kabir et al. 2004). This has been established as probiotics inhibit harmful microorganisms through living in the inner side of the intestine (Kabir et al. 2005, 2014) and the prebiotics used in feed eliminate the harmful organism from gut and found to be better growth (FAO 2013).

Different of types of genetically modified animals have been developed principally for food, and others mainly for nonfood materials like pharmaceuticals, fibers, vaccines, and other high value products. Introduction of biotechnology have an important influence on livestock sector as vitamins, nutraceuticals for nutrition, vaccines and antibiotics for animal health, therapeutic proteins for transgenic animals and breeding for genomics (Mishra and Singh 2013). A rapid progresses in genomics with the important impetus of human genome project observed during last 15 years (Lander and Weinberg 2000). Research on livestock genetics has gained from this advancement, along with the formation of comprehensive genetic marker trait-genes through genetic mapping (Georges and Andersson 1996). Notwithstanding, quality food production, fuel synthesis from various raw products, and selective breeding program are the paramount areas of biotechnology.

The key biotechnological applications namely tissue culture, DNA finger printing, fermentation, selective breeding, and recombinant DNA technology are being used nowadays (Ledoux and Antunes 2018; Krasznai et al. 2018; Nitschke and Silva 2018, Kamle et al. 2017; Lucarini et al. 2016). However, biotechnology can diagnose of various hereditary disorders including infectious diseases through complete genetic analysis and threat them accordingly (Šuster et al. 2017; Hamad et al. 2017; Lao et al. 2017; Kavousipour et al. 2017; Calvo-González 2016).

8.3 Biotechnology in Poultry Industry Through Genetics and Genomics

The great expansion of poultry industry was observed since last several decades due to scientific and technological progressions. The genetically improved superior stocks adept with higher productivity, even under extreme climate has shifted from rural farming to a complete form of poultry industry (Saxena and Kolluri 2018). The parameters were evaluated in layers birds with better feed conversion rate including other traits as high rate and persistent egg production, egg size and quality, and livability. In meat type poultry birds, growth rate, carcass quality, muscle mass, feed conversion rate and livability are essential parameters for consideration (Fig. 8.2). Since these traits are up gradated genetically in birds and achieved via selection process like crossbred and purebred selection (CCPS) that enhanced productivity per bird and production volume (Saxena and Kolluri 2018; Engormix 2008). In poultry, growth characters are connected with polymorphisms in various hormones and associated receptors like lambr1 ghrelin, growth hormone, growth hormone receptor, MC3R, MC4R, TGF-β and IGF-II, (Fang et al. 2007, Huang et al. 2007; Jiang et al. 2002; De Vries et al. 1998; Grobet et al. 1997; Feng et al. 1997). In animal production, genomic estimate of breeding value (GEBV) of an individual animal is anticipated through the whole genome as "genomic selection" was developed and considered to be as an innovative revolutionary technology using SNPs. (Saxena and Kolluri 2018). The impact of genetics and genomics at meat and egg type poultry production are as follows:

8.3.1 Impact on Broiler (Meat Type Poultry) Production

The past decade witnessed strong growth in the global per capita poultry consumption increased by 16% whereas the per capita beef and veal consumption decreased by nearly 5% between 2008 and 2017 (OECD 2018). At present nearly 100 million tonnes of poultry meat and 73 million tonnes of egg produced annually throughout the World (FAO 2016). However, backyard poultry contributes 2% of poultry meat demand, and 8% egg demand globally. The proportions of global poultry meat supply from broiler and layer poultry as 92% and 6% respectively (Mottet and Tempio 2017). The poultry especially broiler is playing a key role as the source of improved value protein and fats for the world people (Tavárez and Solis de los Santos 2016). Both in developing and developed countries the consumption of poultry meat is gradually increasing since the past decade due to its cheaper price as the broiler industry is capable to attain remarkable production and efficiency (Scanes 2007). A broiler bird weighed approximately 600 g at 42 day old where a feed conversion ratio (FCR) was recorded as 2.8 in 1957, however, today the FCR is deduced to 1.7 and weighted 2900 g at the same age (Zuidhof et al. 2014) or more favorite FCR is observed in recent time. Due to development of production techniques that included enhanced genetics and breeding, good knowledge of nutrition and feeding along with total better management practices are responsible for such advancement (Tavárez and Solis de los Santos 2016). Nowadays, maximum production enhances (80–90%) observed due to impact of improved genetics and breeding (Havenstein et al. 2003a, b).

8.3.2 *Impact on Egg (Layer Type Poultry) Production*

At present, global demand of eggs is 75 million tonnes where yearly growth of 1 million tonnes observed. To minimize the escalating demands, approximately 50 million layer hens to be included per year. A good management environments to support the genetic traits for 20 kg egg mass/each hen with an age of 20–76 weeks (Preisinger 2018). Constant upgrading of egg production quality in hen is the best selection parameter in layers. Annual increment of around two to three eggs/year in a production cycle with a duration of 13 months could be predictable and confirm through field results on genetic trend (Fig. 8.3). Simultaneously, feed efficiency had been enhanced relatively with the egg production mass in brown layers, moreover, daily feed requirement were significantly lessened over the years (Fig. 8.4).

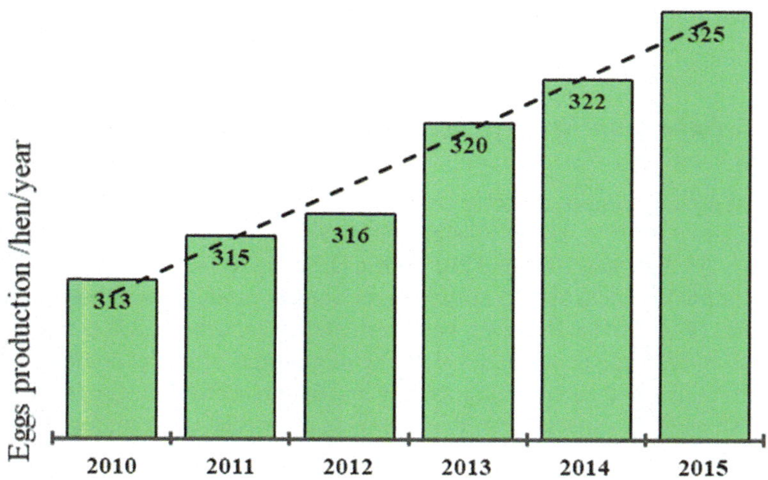

Fig. 8.3 Recent genetic trend with egg production output/year. (Adapted from Preisinger 2018)

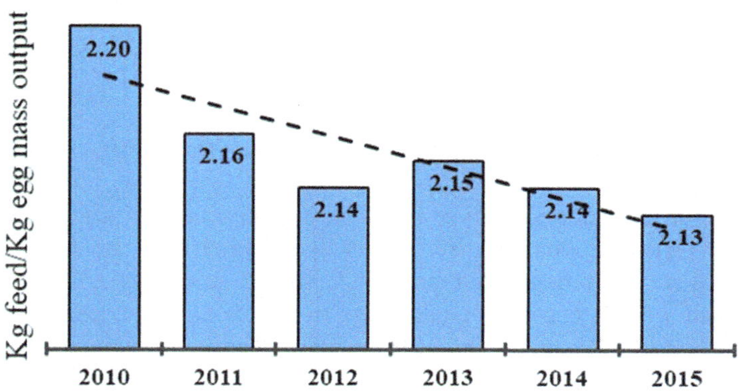

Fig. 8.4 Recent genetic trend with feed efficiency (kg feed/kg egg mass output). (Adapted from Preisinger 2018)

8.3.3 Biotechnology in Poultry Breeding

The genetic improvement in poultry is achieved through implementation of selection program from phenotypes of expected breeding Value (EBV) assessment. However, a lots of drawback are also challenged as capability to precisely and judiciously documenting the phenotypes on candidates and their near relatives; the expenses of documenting the data; and at beginning of major of the production traits in late life impedes genetic improvement each unit time. The heritability estimates of production parameters are important in determining the technique of selection. The additive genetic difference retains decreasing over the generations effecting in lowered evaluation of genetic upgradations and heritability for each generation (Saxena and Kolluri 2018).

8.3.3.1 Selection Process

The selection process in poultry breeding program has been adjusted as per the updated knowledge and the requirements. The birds those fulfilled desired traits were retained and the other birds eliminated from the flocks as a part selection process in early 1940s. After taken maize development program during 1980s, the knowledge of 2, 3, or 4 way breeding process was tailored in poultry birds. Interestingly, the purebred birds were also switched by hybrid commercial birds by terminal cross and the particular egg and or meat type poultry birds transformed to both type poultry birds. The adverse relationship among production and reproduction traits requires the improvement of particular male and female lines both in broiler and layer poultry birds. Above improvements enabled to established specialized lines in meat type (Chambers 1990) and egg type (Fairfull 1990) commercial poultry and also specific male and female lines characteristically with distinct basis of genetic source (O'Sullivan et al. 2010). The poultry breeding at present time includes combination of crossbred and purebred selection (CCPS) and pure-line selection (PLS) (Saxena and Kolluri 2018).

Development of Specialized Lines Through Pure-Line Breeding

An exceptional selection method centered on distinct set of characteristics of father and mother lines by which desired father and mother lines are developed. The mother lines are tactfully selected for reproductive parameters, viz., egg production, egg size and weight, egg-shell quality, age of sexual maturity, hatchability along with growth at young age. On the contrary, the growth rate, configuration of body, FCR, carcass texture, and fertility are considered predominately for selection of father lines. Thus developed commercial broiler poultry birds associated with particular lines had been achieved to lessen the production cost. The consequences of the gene recombination in crossing among the genetically diverse lines resulting a heterotic outcome in progeny with economic characters. A control stock having the same improvements in inbreeding in the targeted poultry population could be conserved to evaluate of impacts.

Egg Type Poultry (Layer)

The primary objective of layer selection is to gain highest number of marketable eggs/hen per raised by reduced feed cost/egg or kg egg mass. Further, the eggs should be with better inner and outer conditions. The layer stocks should be with minimum mortality and maximum adaptableness at diverse conditions. However, layer parent stock are selected to enhance more than 30 characteristics are vital for profitable egg production. Breeders (grandparent stock) are selected for the sexual maturity age, laying rate, viability, body weight, weight of egg, feed conversion ratio (FCR), color of shell, shell strength, albumen height, presence of inclusion body in egg (meat and or blood spots), and adaptability. The selection approaches to enhance egg yields comprises the consistency of lay, clutch duration, FCR, skeletal deformities like osteomalacia and osteoporosis (Thiruvenkadan et al. 2010).

Meat Type Poultry (Broiler)

The selection strategies in broiler poultry are focused on fast growing and carcass features. The best experienced strategy for broiler pure-line selection is related to be considered the commercial weight selection that gains the market weight and the age of selection complies relatively earlier as growth capacity improves. A salable age and period selection are also practiced for the selection broilers. Different biotechnological methods through breeding and selection approaches like trap nesting, mass selection, hybridization, artificial insemination, family feed conversion testing, selection index, individual feed conversion testing, Best Linear Unbiased Prediction (BLUP) breeding value estimation including DNA markers were adopted since early 1900 for the genetic development of poultry (Thiruvenkadan and Prabakaran 2017).

Evaluation of Breeding Program Under Particular Climatic Conditions

Since the phenotypic traits depends on genotype (G) and environmental factors (E), the environmental impacts also are necessary to emphasize during choosing the poultry stocks, and to confirm breeding policy. The vital objective of a breeding scheme is to establish a profitable crossbred animals are able in production at specific climatic conditions. For this reason, the output of breeding program requires to be focused under unlike climatic situation before its profitable use. The interaction between microenvironment and macro-environment and the genotypes is very crucial among inter-population and intrapopulation and. However, feeds, temperature, and climate changes between locations and seasons are the components of macro-environments. Moreover, unusual environmental variations are considered as a part of micro-environments within a population (Thiruvenkadan and Prabakaran 2017).

Artificial Insemination and Cryopreservation

Application of conventional biotechnology method like artificial insemination (AI) in poultry production has supported to a quick transfer of hereditary traits from a lesser quantity of superior males to a greater quantity of females where AI could ensure excellent fertility in poultry relating to natural service. Effective uses of this method requirements better quality semen to be inseminated to the exact location of sperm preservation tubules in the female reproductive organ to ensure chances of best fertility in chicken (Mohan et al. 2018). However, liquid nitrogen can be used as cryopreservation in domestic birds for future using in AI. Since the last decade of the twentieth century, appropriate methods of cryo-storage of chicken semen of different breeds have developed. The techniques using the cryoprotectants glycerol, sulfoxide, dimethyl acetamide and dimethyl, slow or fast freezing-thawing procedures, pellets/vials/straws packaging have been established (Blesbois 2007).

8.3.3.2 Molecular Techniques Used in Poultry Production

The introduction of the era of molecular genetics in 1970s providing novel prospects to improve breeding programs via the application of DNA markers related to the traits of interest. As advance of whole genome studies (WGS) in genomics science is used in many biological systems like poultry breeding (Lander and Weinberg 2000). Various tools namely quantitative trait loci (QTL), candidate gene technique and high-density SNP genotyping for whole-genome selection, identification and genome wide scans, are widely applied nowadays in poultry genomics (Saxena and Kolluri 2018; Burt 2002).

QTL Classification and Genome Wide Scans

A QTL is defined as the genetic control of quantitative characters is foreseeable to be scattered all over the genome especially various locations of the genome that regulate the quantitative characters of interest. The specified experimental crosses use to confirm the QTLs particularly designed for the scope. The documentation of QTL and formulation of DNA tests were the key phases at applied uses of QTL through marker-assisted selection (MAS) as the selection based on an arrangement of information resultant from genetic markers linked with QTL and the indication of phenotype. Above these QTL examinations had accomplished by 200–350 MS markers and mating between diverse breeds, such as heavy meat type poultry birds with specific inbred lines or lighter egg producing birds (Fulton 2014).

Candidate Gene Technique

The candidate genes are those genes with direct and great impact on the character of importance. Formation of past information of certain candidate gene could be expected to be accountable for an identified main hereditary impact. The dissimilarity in sequences in target gene are documented and finally different alleles are attached with deviation of trait(s). The genes are straightly associated with production characters namely growth hormone (cGH) gene, growth hormone receptor (cGHR) gene, insulin-like growth factor-1 (IGF-1), IGF-1R, myostatin, etc. had assayed candidate genes and genetic markers like InDel/Dels and SNPs identified (Nie et al. 2005; Fritz et al. 2004; Lau and Leung 2002; Amills et al. 2003; Feng et al. 1998). Moreover, three physiological candidate genes viz. growth hormone genes for cGH, neuropeptide Y (NPY) and gonadotropin-releasing hormone receptor (GnRHR) were studied to detect their association with few production parameters like egg production/bird, dual yolk egg, and age at first laying egg (Ngu et al. 2015; Xu et al. 2007).

High-Density SNP Genotyping for Whole-Genome Confirmation

Advancement of NGS techniques and high performance of genotyping platforms has directed to the construction of high-density SNP array as an upgraded tool applicable for analyses of genetics and genomics in domestic animals. The important applications of SNP arrays in agriculture for upgrading of significant economic traits through genomic selection (Fan et al. 2010). An innovative type of marker-assisted selection (MAS) focuses on total markers in whole genome (Calus 2010; Hayes and Goddard 2001). It exactly forecasts the breeding values of animals using the data linked to the location of plentiful SNPs throughout the genome with a hypothesis that these SNPs are linked disequilibrium (LD) relations between QTL and SNPs. An abundance of SNPs basically needed for the outline and structure of arrays and could be retrieved by various techniques and platforms, for example, genome sequencing and HapMap project, and reduced representation library (RRL) generate predicted SNPs sequencing, (Amaral et al. 2011; Matukumalli et al. 2009) via extracting SNP data from dbSNP of NCBI (http://www.ncbi.nlm.nih.gov/projects/SNP/).

8.4 Biotechnology in Poultry Feed and Nutrition

Innovation in biotechnology since the last decades has provided new prospects for improving the output and efficiency of animal production through ensuring improved feed and nutrition. At present, biotechnology have a comprehensive uses in animal feed and nutrition or formulation of feed-grade enzymes for availability of nutrients (Havenstein et al. 1992). Use of beneficial microorganisms (prebiotics and probiotics) made by recombinant DNA technology has an immense influence on poultry nutrition and production. However, food fermentation technologies ensure various nutrients, complete proteins/essential amino acids or to increase the digestibility of feeds that enhance animal production. As well as, recombinant bacteria produce improve nutrients through production of particular enzymes and hormones that can improve animal productivity including reduce environment pollution. Besides, uses of fiber degrading enzymes to increase animal production and decrease environmental impact (FAO 2010). A few of the aforementioned approaches are currently being used shown in Table 8.1. However, others are identified as likely to be potential, due to technical glitch, and food safety and public health issue are not applied at wider scale (Table 8.2).

Table 8.1 Biotechnological approaches are currently used in animal nutrition adapted from FAO (2013)

Approaches/ technologies	Objectives
(1) New ingredients	Produce microbial proteins as a novel feed source for livestock feed and nutrition, for example protein derived from single cell and yeast
(2) Designer ingredients	Improve nutrition status e.g. high-oil enriched maize, high-methionine enriched lupins
	Decrease the status of anti-nutritive ingredients in a common feed stuff, like low-phytate maize.
(3) Feed additives	
(a) Antimicrobials	Diminish the growth of pathogenic bacteria and to support to create a desired equilibrium of gastro intestinal flora like antimicrobials.
(b) Crystalline amino acids	Increase the feed supply of particular amino acids and to enhance the protein balance in formulations of animal ration.
(c) Feed enzymes	Improve the accessibility of nutrients like energy, phosphorus, amino acids, etc. in feed components by lessening the detrimental effects of anti-nutritive ingredients like xylanases actions on arabinoxylans in wheat and microbial phytases actions on phytate,

Table 8.2 Biotechnological approaches are future potential in animal nutrition adapted from FAO (2013) and Harlander (1990)

Approaches/ technologies	Objectives
(1) Adaptation of gut microbes	Modify microorganisms genetically the existing naturally into gastro intestinal tract, to improve their ability for desired functions or to include new functions
(2) Inclusion of novel gut microbes or bateriocins	Inclusion novel genre of microbes into the gastro intestinal tract. Broad spectrum bacteriocins could be produced through engineering of nonpathogenic *Salmonella* spp. or new gut microbes. These strains could be included in feed and water for biological control of poultry pathogenic of *Salmonella* spp. or microbes in the gut.
(3) Bioactive peptides	Enhanced growth and efficiency like growth hormone-releasing peptides, better gut function, immunomodulation
(4) Antimicrobial replacers	Antimicrobial enzymes (like lysozyme), to deliver specific antibodies via spray-dried plasma and egg substances
(5) Transgenesis	Alter nutrient metabolism and enhance growth effectiveness by removal of genes.

8.4.1 Probiotics and Prebiotic

The role of biotechnology in poultry nutrition is enormous nowadays, as it shows the pivotal role in the poultry feed production. Since a wide reports on emergence of antimicrobial resistance (AMR) among the pathogens, the antimicrobial uses (AMU) as a protective measurement found to be flawed. In this situation, impact of antibiotics uses as growth promotors or therapeutic agents has created the alternative options where both consumer and producers will be gained. Therefore, probiotics was identified to cover this space. Nowadays, many farmers use them as antibiotics alternatives (Griggs and Jacob 2005; Nava et al. 2005; Trafalska and Grzybowska 2004). The probiotics organisms are used as gut ecosystem enhancers that improve to create desirable gut environment through proliferation of beneficial organism (fed-microbial). The prebiotics organisms direct eliminate harmful pathogens competitively (FAO 2013) and to improve FCR (Midilli et al. 2008) and improvement of host's health by particularly accelerating the growth and/or function of probiotic bacteria prevail in the intestinal tract (Borroto 2009) like Mannan oligosaccharides (Fig. 8.5).

Poultry nutritionists are constantly toiling their attempts into better feed formulation considering economical point of view. A "good feed" only not fulfill the target but also ensure substantial utilization. Nutritional modifications along with healthy feed can impact on equilibrium of the microflora in the intestine that prompting indigestion. Therefore, an optimum balanced ration with energy and nutrients components is needed to keeping a healthy gut. A lot of concerns have recently been highlighted from veterinary consultants and poultry nutritionists relating to proper utilization of nutrients along with uses of probiotics for growth promotion in poultry feed (Kabir 2009a, b).

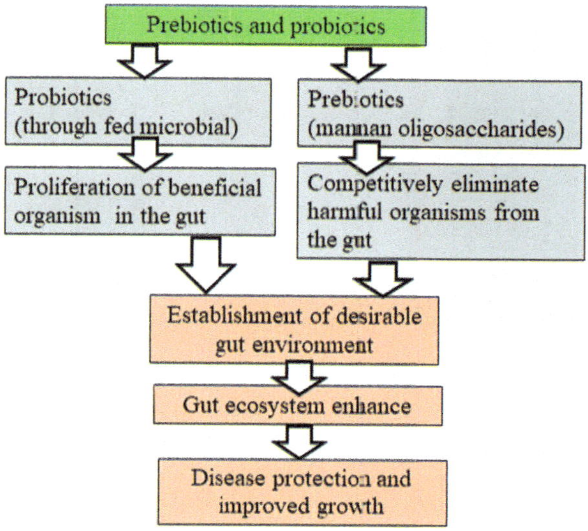

Fig. 8.5 Prebiotics and probiotics act as a gut ecosystem enhancer adapted from FAO (2013)

Probiotic is a broad term, are the products may contain culture bacteria, yeast cells, or together that influence microorganisms to adjusting the gastrointestinal tract related to health condition and enhance feed efficiency (Dierick 1989). In many ways these can affect in poultry are comprised of (1) keeping standard intestinal microflora by antagonism and competitive exclusion process (Kizerwetter-Swida and Binek 2009; Kabir et al. 2005; Schneitz 2005; La Ragione and Woodward 2003; Stern et al. 2001; Fritts et al. 2000; Netherwood et al. 1999; Jin et al. 1998; Line et al. 1998; Nisbet et al. 1998; Owings et al. 1990; Fuller 1989; Nurmi and Rantala 1973; Rantala and Nurmi 1973), (2) modifying metabolism through intensifying digestive enzyme activity and lessening ammonia production and bacterial enzyme action (Yoon et al. 2004; Han et al. 1999; Chiang and Hsieh 1995; Jonvel 1993; Cole et al. 1987), and (3) enhancing feed intake and digestion capacity (Awad et al. 2006; Horniakova 2005; Kumprecht and Zobac 1998; Yeo and Kim 1997; Tortuero and Fernandez 1995; Nahashon et al. 1993; Nahashon 1992; Dierick 1989), and finally (4) boosting the immune system of the host (Apata 2008; Brisbin et al. 2008; Nayebpor et al. 2007; Mathivanan and Kalaiarasi 2007; Haghighi et al. 2006; Haghighi et al. 2005; Dalloul et al. 2005; Kabir et al. 2004; Huang et al. 2004; Koenen et al. 2004). Therefore, the uses of probiotics in commercial broiler production even in low level of biosecurity in rural settings (Fig. 8.6).

8.4.1.1 Impact on Intestinal Microbiota and Intestinal Morphology

Probiotic supplementation in poultry feed have been observed as beneficial effect on growth performance at the different stages of age without relatedness of vaccination status (Awad et al. 2009; Sahin and Yardimci 2009; Roshanfekr and Mamooee

Fig. 8.6 (**a** and **b**) Uses of probiotics for safe broiler production that enables to no or minimum antibiotic uses under sector 3 (small scale commercial) poultry production system in rural Bangladesh with low/minimum level of biosecurity standard

2009; Willis and Reid 2008; Apata 2008; Willis et al. 2007; Nayebpor et al. 2007; Mountzouris et al. 2007; Khaksefidi and Ghoorchi 2006; Timmerman et al. 2006; Kamruzzaman et al. 2005; Gil De Los Santos et al. 2005; Hossain et al. 2005; Kabir et al. 2004; Islam et al. 2004; Kalavathy et al. 2003; Zulkifli et al. 2000; Jin et al. 1998). However, some changes in probiotics through inactivation by a high-pressure homogenizer that have a decisive impact on growth performance of broiler chicken as additives in concentered feed (Huang et al. 2004). Use of selective probiotics like FM-B11 in turkeys was found to increase the usual daily weight gain and salable body weight to be considered as a financial benefit to better turkey production (Torres-Rodriguez et al. 2007).

8.4.1.2 Impact of Probiotics on Immune Status

The immune response was found to be significantly higher in probiotic administered birds than other birds. The variation of the weights of different organs like bursa and spleen in probiotic use and usual feed in broiler production could be confirmed at the different stage of antibody production in reaction to sheep erythrocytes (SRBC) (Kabir et al. 2004). In the same way, it was confirmed as antibody titer was suggestively higher after post-immunization (PI) at 5 and 10 days with supplementation of probiotic 50 mg/kg group, related to control (Khaksefidi and Ghoorchi 2006). The use of probiotics improves serum and intestinal usual antibodies against a few foreign antigens in poultry (Haghighi et al. 2005). A several studies confirmed the possible impact of probiotic on adjustment of immune response at a desired level in poultry (Apata 2008; Nayebpor et al. 2007; Mathivanan and Kalaiarasi 2007; Haghighi et al. 2005; Dalloul et al. 2005; Koenen et al. 2004; Zulkifli et al. 2000).

8.4.1.3 Impact on Improve Water Quality and Improve Microbial Balance

Probiotics organism directly play role to improve water quality and microbial balance as their presence in the poultry gut (Lalloo et al. 2007, Hill et al. 2009, Picchietti et al. 2009) Supplementation of probiotics in broiler feed enhances quality of meat both at pre and post freezing storage time (Kabir 2009a, b; Kabir et al. 2005). Additionally, the physical appearance includes consistency, lusciousness and overall appropriateness of poultry meat substantially improves through probiotic (Lacto-Sacc) feeding as it declines total viable count (TVC) in poultry meat compared to untreated birds (Mahajan et al. 2000) or even meat tenderness could be enhanced by application *Saccharomyces cerevisiae* extract or whole yeast or (Zhang et al. 2005).

8.5 Biotechnology in Poultry Vaccine and Biologics

Immunization is an important method for maintaining animal health targeted for prevention and control of animal disease has been practiced for centuries and confirmed as a potential devise to lower animal suffering including the economic benefit for animal producers. A wide number of factors like capacity of national disease eradication programs and international trade policies along with cost-effectiveness of livestock and poultry production are responsible on availability of vaccines and biologics for disease control program. At present time, the techniques used in the development of vaccines have extended quickly due to improved understanding of the mechanisms at which protection is achieved, and the accessibility of genomic information on both hosts and pathogens (OIE 2010). Globally around 900 patent applications of poultry vaccine claims annually involving molecular techniques. The combined development of new tools in the area of molecular biology and immunology has a great influence on the formulation of novel vaccines including their quality. This is able to model of vaccines designed for the control and eradication of particular disease considering national, regional, and international aspects. As a cost-effective measure to control infectious diseases, vaccination has been proved widely for eradication of animal diseases. Nevertheless, a limited recombinant vaccines has been produce for commercial use, their utilization is insignificant in developing countries.

8.5.1 Reverse Genetics

As development of a reverse genetics system the area of virology through different techniques namely designed mutations, deletions and insertions into the viral genome of live viruses of different DNA and RNA viruses has been modernized. This is applied in a range of applications include attenuation of virus, the host specificity alteration and the generation of replication-deficient viruses. These protocols have been used to the development of new vaccines and are extensively used in the characterization of the function and structure of distinct viral genes and their sequences. The production of a cloned copy of complementary DNA (cDNA) from RNA by reverse transcription *in vitro* is widely used in reverse genetics; subsequently transfection in the permissive cells with the cloned DNA(s).The tool was first recognized in a positive-strand RNA virus termed as bacteriophage Q-Beta (Taniguchi 1978) and also has been used for formation of RNA viruses with segmented/unsegmented negative sense genomes, e.g. vaccine development against avian influenza virus (AIV) included an engineered haemagglutinin (HA) gene from H5N1 virus and a neuraminidase (NA) gene from a H2N3 virus, with in a H1N1 skeleton (Meeusen et al. 2007). Thus inactivated H5N3 virus vaccine was produced that confer full protection in birds against H5N1 challenge.

8.5.2 Recombinant Vector Technologies

Progresses in genomics, reverse genetics, and proteomics have enabled to the documentation of process of pathogenicity, host-pathogen action along with protective antigens from many organisms, and the advance of appropriate vector for transfer of these antigens to the host animal. The availability of bacterial and viral genome sequences has improved the quick structure of distinct deletions in the genomes of a diverse range of pathogens is not only the effect of attenuation, but also it makes place for the insertion of imported antigens coding genes for of heterologous microorganisms (OIE 2010). Normally, for virulence, deletion of genes attenuate bacterial vectors is mandatory for the important metabolic processes of organisms. *Salmonella* utilized as a role of vector for vaccine antigens delivering and finally applied for the production of live vaccine strains to be used poultry (Babu et al. 2004). There are numerous bacterial vectors are developed using commensal microorganisms like *Lactococcus Lactobacillus, Streptococcus,* and *Staphylococcus,* etc. However, pathogenic organisms are being used through attenuation namely *Shigella, Vibrio, Yersinia, Bacillus, Bordetella* and *Cornebacteria,* etc. as these are all evaluated for the capacity of production immunity at protection level.

The viral vectors were developed using those virus causing mild or no disease or even pathogenic viruses through deletion of virulence genes attenuation. Replication capacity of virus vectors can replicate progeny virus together with defective virus vectors replication and assessed as vaccine delivery vehicles. There are numerous commercial vaccines have been developed stand on DNA virus vectors technology and effectively approved for using in livestock and poultry practice like poxviruses and herpesviruses (Gerdts et al. 2006). However, a few viral vectors namely vaccinia virus, fowl pox virus, canary pox virus, and turkey herpes virus. Moreover, a number of viral vectors have been advanced or further evaluation is under way towards improvement. The information of certain virulence factor(s) of a pathogen and the accessibility of recombinant DNA technology has enabled the creation of particular gene-deleted organism for the application as a live vaccine. The method is used for marking and assaying of defined gene deletions benefits in lowering the virulence without affecting their immunogenic potentiality. Such organisms could be used as vaccines as they have the immunogenic characteristics, however they could not perform as causative agents of disease (OIE 2010). A several live attenuated vaccine strains of bacterial disease have been developed through this approach and found to be genetically potent, safe for use and make better immunity compare to killed vaccines, for example gene-deleted *Salmonella enterica* serovar *enteritidis* and serovar *typhimurium* and vaccines in poultry (Meeusen et al. 2007; Babu et al. 2004).

Subunit vaccines Since 1980s, using recombinant DNA technology semi-pure or purified proteins subunit vaccines produced and available commercially as (Ulmer et al. 1995; Rhodes et al. 1994; Cohen 1993; Ulmer et al. 1993). However, this vaccine do not uses live recombinant vector techniques that will deliver the recombinant proteins in vivo. Using the conventional biochemical or recombinant DNA technologies the subunit antigens are produced (OIE 2010). A range of prokaryotic and eukaryotic expression systems of insect cell, yeast, and plants are used (Chichester and Yusibov 2007) via a variety of combined or transient expression system. Biochemical methods are still helpful in some cases however the recombinant expression is not suitable, such as antigens required intricate assembly (like fimbriae), or at the stage of post-translational modification.

New vaccination strategy for immunization with DNA is founded on a clear-cut idea. The vaccines can be described as antigen-encoded bacterial plasmids are able to make specific immune reactions after inoculation into an appropriate host. Immunization is completed through the receipt of purified plasmid in the host cells and it remains outside of the chromosome.. From a technical perspective, DNA vaccines can be easy to engineered, produced and purified, structured and evaluated in animal models within months be. DNA vaccines are considered to be very stable for a longer time or even carried without a cool chain to a long distance. The safety of this vaccine has been evaluated in various field trials in different animal models as well as in humans (Kim et al. 2001; Bagarazzi et al. 1998).

8.6 Biotechnology in Disease Diagnosis

Biotechnologies have applied in livestock health practices to increase the accuracy of diagnosis of disease and treatment. Immunology based diagnostic conventional method are being used monoclonal antibodies like enzyme-linked immunosorbent assays (ELISA) and radioimmunoassays (RIA) along with microscopic examinations are not sufficient to distinguish between species and subspecies of disease producing pathogens (Maharana et al. 2016). However, these techniques may not suitable for distinguishing of vaccinated from infected animals (DIVA). Molecular diagnostic assays including single DNA sequences give high levels of sensitivity, specificity and even reliability in tests. Polymerase chain reaction (PCR) technique permits significantly higher level of specificity (Salih and Trifonov 2015).

The approaches in molecular biology discriminate targeted DNA sequences are now well familiar tools for detection of infectious agents (virus and bacteria) in poultry and other animals and widely used throughout the world even in developing countries. Mostly, PCR-based diagnostic assays are progressively used for timely diagnosis of causal agent of diseases, however, these are being used primarily in the labs of research organization and central national diagnostic laboratories due to incompatible facilities for diagnostic test (FAO 2010).

8.6.1 Polymerase Chain Reaction (PCR)

The PCR has an enormous impact on avian health. PCR can be used in diagnostic tools because of its higher specificity to detect microbial pathogens in clinical samples and epidemiological studies (Yamamoto 2002; Nagy 2001). This is an effective, simple and economical technique applied to amplify the preferred DNA sequence or sequences of into billions of identical copies. This assay is basically used for checking Short Tandem Repeats (STRs) and Single Nucleotide Polymorphisms (SNPs) in different livestock breeds, genetic mutation including parentage identification. The key component of the PCR are: (1) DNA template, (2) primers, (3) Taq polymerase enzyme, and (4) the PCR machine (Thermocycler) that devises the correct temperature for each step in every cycle. The PCR technique involves three major phases in each cycle that include; (1) denaturation, (2) annealing and (3) extension or elongation (Bartlett and Stirling 2003). To confirm as the PCR produced anticipated DNA fragment, agarose gel electrophoresis is often used for the confirmation of the sizes of PCR amplicons. The size of PCR amplicon is estimated by connection with a molecular weight marker (DNA ladder), that contains DNA fragments of known size, run on the gel quickly with the PCR amplicon (Jebakumar et al. 2012). In our most recent study, we had used *cdtC* gene-based multiplex PCR assay for detection of *Campylobacte*r *coli* contamination in broiler meat samples (Neogi et al. 2020) (Fig. 8.7).

8.6.2 Bioinformatics to Diseases Diagnosis

In bioinformatics, biological data are organized and understand using upgraded computing techniques (Akunaedozi 2014). Bioinformatics techniques and relevant algorithms have now been developed with an aim to use the sequence of biological information. Through the bioinformatics tools viz. BLAST (Basic Local Alignment Sequence Tool), Fast Alignment (FASTA) and Clustal W solutions for sequence search and analysis, together with other techniques enable to a cost-effective and time reduction strategy to obtain vital data on gene and protein levels, however,

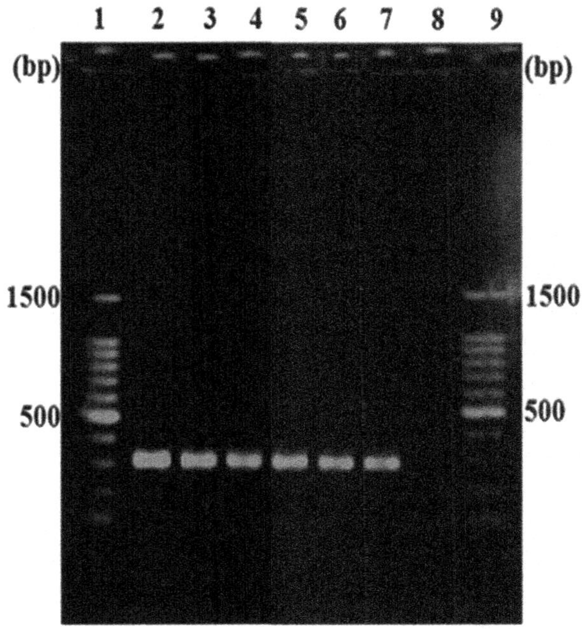

Fig. 8.7 *cdtC* gene-based multiplex PCR assay used for the detection of *Campylobacter coli* isolated from broiler meat samples of live bird market (LBM). Lanes: 1 and 9, 100 bp DNA ladder (Promega, USA); lane 2, positive control (*C. coli* ATCC33559); lanes 3–7, representative positive isolates; lane 8, negative control

these could not be achievable by other methods (Kaikabo and Kalshingi 2007). The phylogenetic analysis of the amplified nucleic acid sequences gives new information on the evolution of organisms and assists the studies of molecular epidemiology. Proof for interchange of DNA among different microorganisms with evolutionary ladder is reconsidering our knowledge on pathogens and reconfirming the classification. Nonetheless, the view of using this technology depends greatly on the accessibility of bioinformatics information. Present studies in Belgium, Western Europe confirmed the value of Next Generation Sequencing (NGS) technologies to veterinary science, with special emphasis for using in diagnosis of diseases and treatment capacities (Van Borm et al. 2015).

8.7 Biotechnology Application in Food Processing

Food processing sectors nowadays are using biotechnological tools for improvement of food products. The impacts of these techniques comprise increase food production, enrich the nutrition parameters, different food products produced through fermentation process, production of important enzymes, increase shelf life of the food, enhancing the organoleptic characteristics of food, and to improve safety compliances in food and food products (Lokko et al. 2018; Nguyen et al. 2017).

8.7.1 *Increase Food Production*

The use of transgenic technology is expanding poultry productions. The huge financial involvement of $38 billion in the United States at the commercial sector importance of biotechnology is evident. The aim for transfer of gene in commercial poultry is mostly to improve hereditary superior poultry birds for production of meat and egg, which has been found to be the same target of conventional selection programs taken during last 50 years (Petitte and Mozdziak 2014). The conventional selection program has been discussed elaborately in the subsection "selection process" under section biotechnological approaches in poultry breeding programs.

8.7.2 *Enhance Nutritional Parameters*

The idea of biotechnology to boost nutritional quality in food was first introduced in Japan during 1980s as targeted to certain health benefits spaced out from its own nutritional quality (Cruickshank 1934). There have been continuous interest in poultry biotechnology for changing the egg constitutes through manipulations of genetic and nutritional traits for human health advantages. This could be done by transforming cholesterol composition and its concentrations, fatty acids, lipids, amino acids, and minerals or by inclusion of therapeutic pharmaceutical elements (Alagawany et al. 2018). The designer egg approach was first introduced in 1934 by Cruickshank, who affirmed the adjustment of fatty acid contents in egg yolk by feed interventions (Arai 1996). Nowadays, designer eggs offer for vegetarian as immunity stimulant, safe or organic foodstuffs, better vitamins and minerals with adjusted ratio of fatty acids namely omega-6 and omega-3, lower level of total cholesterol, boost of antibodies including essential pigments like carotenoids (Alagawany et al. 2018). However, researchers had advanced to a designer egg with rich of antioxidants and omega-3 fatty acids (Sim and Sunwoo 2002). Further improvements in designer egg which were confirmed with enriched conjugated linoleic acid (CLA) (Raes et al. 2002), vitamin A and β-carotene content (Jiang et al. 1994). Chicken, beef and pork enriched with selenium (Se) can also be produced through dietary feed supplementation of organic Se in food animals and birds (Fisinin et al. 2009). Currently, the designer food or functional foods are receiving greater attention due to their role in health benefits and disease prevention (Fisinin et al. 2009).

8.7.2.1 Fermentation Process to Food Processing

Since hundred years ago, genetic upgradated microorganisms like bacteria, yeasts and molds have been used for the production of fermented foods. Conventional strain improvement techniques related to mutation and selection are erroneous and unmanageable. However, it is difficult to screening for all mutations that happen and screening their process is challenging. Therefore, gene engineering keeps a tool to overcome several of these challenges as it allows for the selection to transfer of a single, particular defined traits from basically any living organism in an exact, manageable way. Suppression of pathogens and spoilage microorganism through 'Bacteriocin' production in poultry meat sausage is a unique example of genetic improvement of food-grade microorganisms in fermented food (Harlander 1990).

To yield different value added fermented food staffs through commercial fermentation processes, the starter cultures have been created for using either a single or mixed strains of organisms presented to be beneficial (Holzapfel 2002). Inhibitory action of these microorganisms was observed for the creation of one or several products like bacteriocins, hydrogen peroxide, organic acids and diacetyl (Hutkins 2006). Moreover, *Lactobacillus* strains can be applied as essential probiotics for the processing of meat and dairy products, and having abundance of health benefits (Pennacchia et al. 2006). Therefore, the biotechnological technologies can be fitted to obtain better-quality strains of microorganisms (bacteria, yeast, and moulds) and could be used for the processing of meat and meat products through fermentation.

8.7.3 Increase Shelf Life of Food

The shelf life of beverages and food can be prolonged through bacterial fermentation process (Brochu 2018; Paramithiotis 2017). Most of the food fermentations include lactic acid that is produced from sugars by the lactic acid producing bacteria under the genera of *Lactococcus, Streptococcus, Lactobacillus* and *Pediococcus*. However, nowadays, *Lactobacilli* have been considered due to the production of bacteriocins (Collins et al. 2017). These substances could be used food industry as natural preservatives to increase shelf life. The uses bacteria producing lactic acid and of their metabolites are recognized as safe (Patel and Prajapat 2013). Nisin is a bacteriocin used officially in the food industry and recognized globally (Kaškonienė et al. 2017).

Numerous processed food products in refrigerated vacuum-packaged of meat, fish, dairy, and vegetable category contain usually gram-positive bacterial strains under the genera *Brochothrix Lactobacillus, Clostridium, Leuconostoc, Carnobacterium* (Marth 1998). They can able to spoilage of the product as they multiply at refrigerated temperature. By adding could be lessen. Through addition of *Lactobacillus sakei* culture in chilled raw ground meat and chilled cured pasteurized sliced vacuum-packed meats the *Listeria* count could be decreased, therefore, this facilitates to increase shelf life (Devi and Halami 2011; Hugas et al. 1998; Ennahar et al. 1998; Yang and Ray 1994; Kroeckel and Schmidt 1994).

8.7.4 Enhance Organoleptic Characteristics of Food

The food organoleptic properties can substantially have influence on acknowledgment of food and food products by the consumers. The genetics and genomics methods (discussed in Sect. 8.3.3) play a key role in the development of quality of product primarily for the nutritional and user friendly technological aspect and some for organoleptic properties (Smaldone et al. 2017). The cultured microbes used in food production are generally termed as starter cultures that also improves the organoleptic character of different foods stuffs. An earlier study has confirmed that above 100 commercial aroma chemicals are produced through biotechnological approach (Berger 2009). The foods produced through fermented process are found to be more value addition with higher nutrients with extend shelf life, and easy digestible as these are more appropriate for the digestive tract (Smaldone et al. 2017). The organoleptic properties of such meat are greater regarding taste, flavor, aroma and also color (Singh et al. 2012). Some aroma producing volatile compounds like acetic, acetaldehyde, diacetyl, ethanol, acetone, etc. were formed from carbohydrate during fermentation process of sausage (Awan et al. 2003).

8.8 Biotechnology and Food Safety

At present time biotechnology as a tool to surveillance on food safety, to avoid food-borne illnesses and to confirm traceability of the foods. In most present time a wide biotechnological developments have set common accessibility of tools of identification with rapid and economic than the conventional methods PCR and ELISA based methods are nowadays used for the detection of key food-borne microorganisms (Velusamya et al. 2010). Biotechnology has involved with the food safety in poultry production system in following areas:

8.8.1 Biotechnology Critics Get Priority to Food Safety

The outputs of biotechnology as new food animals, animal origin foods or processed food to fulfill certain desired goals. The animal biotechnology research emphases on producing transgenic poultry will use feed effectively, gain growth to slaughter weights at a minimal age, and can able to prevent a wide range of diseases. Therefore, farmers get more benefits from lower production costs, improved productivity with higher quality products. Consumers get benefit as farmers can provide leaner meat and poultry grown with a less dependency on vaccines, medicines. Additionally, consumers prefer to get the products at a lower price as the farmers paid less in a shorter period of time. The biotechnologically animal products and their associated events are exciting, however, likelihoods of public concerns and worries on food safety issues are enormous. The concern for all foods or food products regarding to free of hazards like chemical or biologic that may affect safety of food for the human consumption. Consequently, food safety has got priority as an easy focus area for biotechnology critics. Food safety has been adopt in most recent time in developing countries, conversely, the consumers as well as mass people are not completely confident on poultry and poultry origin food as the production and processing along the value chain associated with biotechnical tools that creates public misperception. So, this is to be needed to take remedial actions properly through generating public awareness and motivation towards building confident on biotechnologically poultry or poultry goods.

8.8.2 Food Safety Measures in Poultry Production

8.8.2.1 Protect Poultry Flock with Good Biosecurity

Implementation good biosecurity measures to prevent the entry of pathogens into the farm and prevent the micro-organisms to transmission within the farm or from the farm if it enters and infects the poultry (Yap 2015). The biosecurity measures include the facility of fencing around the farm, netting in farm sheds, vehicles are cleaned and disinfect before entry of the farm, footwear to be clean before entry in the farm (Mridha et al. 2020; Akhter et al. 2018; Ratananakorn and Wilson 2011). Use safe production inputs like quality raw material, feed and day old chicks (DoC) and water that should be free of biological hazards (*Salmonella*) for production of safe poultry and poultry products.

8.8.2.2 Good Animal Husbandry (GAH) Practices

Good animal husbandry practices should be maintained at the farm level that include, vaccination against poultry diseases considering the country context, practice separation of age group e.g. should ensure buffer zones of >30 m, or with all-in-all-out practice in broiler rearing, suitable stocking density to minimize stress to poultry (1.6–2.0 sq.ft. per bird considering size of broiler). Use of probiotics for competitive exclusion to prevent the establishment of pathogens (discussed in Sect. 8.4.1) and to isolate or quarantine any sick poultry for treatment and observation, veterinary inspection of all poultry before marketing to certify on good health status, However, proper record keeping and documentation of farm activities are very much needed (Yap 2015).

8.8.2.3 Good Personal Hygiene (GPH) Practices

Practice good personal hygiene (GPH) to be maintained through ensuring that workers do not transmit any infectious disease e.g. *Salmonella* to poultry and vice versa, cut and wounds may be infected with *Staphylococcus aureus*, washing of hands after toilet use and handling live poultry or poultry waste, and before entering poultry sheds; use of personal protective equipment (PPE) like boots, gloves and mask during working in poultry sheds or cleaning poultry sheds; and poultry attendants who are sick, having cuts or wounds or immunological sick should not work in the farm.

8.8.2.4 Poultry Waste Management

Environmental friendly poultry waste management is very much needed to subsequent exposure disease producing pathogen from poultry to poultry or human. These practices involve good poultry waste & environment management practices e.g. bury poultry waste with lime, compost or incinerate and practice good pest control (Yap 2015). Poultry waste can be used for biogas production by large digesters can be an important source of energy to generate electricity or green energy for vehicle. Outside energy generation, the digester is a good source of biomaterials. A feather-degrading bacterium and the enzyme keratinase present in this biomaterials is beneficial to decompose feather meal and to enhance feed digestibility and facilitates to damage prion proteins (source of bovine spongiform encephalopathy) (Shih 2012). However, in low resource settings, poultry manure and including other organic waste materials can be transformed through anaerobic condition using effective microorganism (EMO) through Bokashi method to a more stable form within shorter period of time (2 weeks) to use as a soil improvement as organic source of fertilizer in vegetable gardening or crop cultivation (Fig. 8.8). The process

Fig. 8.8 (**a**) Demonstration of preparing compost of poultry manure using EMO in the course of an environment friendly procedure; (**b**) the compost is ready for use within shortest period of time (2 weeks) for vegetable gardening

is called 'composting' and the end product that is produced via composting is called 'compost' (Earth Care 2019). These technologies could be considered to utilize poultry waste through environment friendly waste management at low resource settings since the poultry drooping creates a huge nuisance nowadays among people as these are been dispose at open field without treatment.

8.8.2.5 Biotechnology Directs Food Safety

Unpredicted and unintentional composition variations occur with all forms of biotechnological foods. The bacteria utilized for feed production or other purpose as part of biotechnical process might be a hazard to human and animal health due to presence of resistance genes (EFSA 2007). To keep as satisfactory standard of food quality along with safety is completely obligatory for ensuring protection for consumers and to ease trade. Surveillance on microbial adulteration in the finished goods including the different segments of production value chains, cleaning and sanitation practices are the utmost priority factors of the manufacturing process in food technology and also in biotechnology (Van Belkum et al. 2017). More sensitive and specific tools to be applied for determination of microbial food impurities and their toxins by state-of-art technologies of proteomics and genomics. Several commanding techniques of biotechnology have now set massive advances like genetic engineering, PCR, random amplified polymorphic DNA, rDNA technology, amplified fragment length polymorphism, matrix associated laser desorption ionization-time of flight mass spectroscopy (Naveena et al. 2017). These techniques can also support to meat verification. However, development of new technique for rapid confirmation of high-risk food borne emerging pathogens in livestock food commodities is an appealing requirement in food safety connection.

8.9 Constraints and Limitations

Using new tools and techniques for disease prevention and control activities like novel and effective diagnostic methods, drugs including vaccine and biologics in low resource settings are not considered to be bottlenecks rather than to ensure their availability to the rural communities is important. However, there are several issues to be converged as limitations of biotechnology, and these are to be overcome to maximize optimum benefits from this discipline are as follows:

8.9.1 Animal Welfare and Ethical Issues

Biotechnology is considered to be indispensable part of the upcoming veterinary medicine, and husbandry practices. However, these novel technologies have involvement with animal species deserves an ethical obligation for an effective uses and the recognition. There are possible unexpected risks may result with the incredible benefits. Thus, ethical approval and animal welfare issues can arise at different phases of multiplication and longevity of a particular genetically transformed animal. A theory of "Three Rs" (reduce, refine, and replace alternatives in the use of animals in experiments) had been undertaken by the Canadian Council of Animal Care (CCAC) with an agreeable ethics of animal (Avey and Griffin 2016). Firstly, they do not consent for challenges to a scientist's purpose in operating research with animals, even if that purpose is doubtful. A second point, gaps in the Animal Welfare Act have permitted researchers to keeping aside application of the Three R's in action. Finally, the Three R's have no application to novel and emerging issues of biomedical research that have potential to use of animals in experiments, comprised of research of stem cell, genetic modification, cloning, bioterrorism defense and xenotransplantation (Ibrahim 2006).

8.9.2 Lack of Available Data Like Traceability, Risk Assessment of Animal Biotechnological Foods

Available enriched foods to market necessitates the logical guarantee on safety concerns. Animal products derived through biotechnology are likely to be safe, however, that safety should be confirmed and documented as traceability to the safeguard of public health and support to public confidence (Crawford 1990). A system is to be developed concerning how biotechnology in poultry affect the meat and poultry industry will be scrutinized and assessed, and to share the findings/recommendations all scientific aspects have been taken into consideration. This is to be needed to building trust and confidence among mass people on biotechnologically developed poultry goods.

8.9.3 Lack of Consciousness Among the Mass People

The public concern for all foods or food products relating to free of agents like chemical or biological agents that may affect safety of the food for the human that got priority as an easy focus area for biotechnology critics. Food safety has been adopt in most the countries, conversely, the consumers as well as mass people are not completely confident and creates public misperception on poultry and poultry origin food as the production and processing along value chain has gone through the biotechnical approaches (Borroto 2009). So, this is to be needed to address remedial approaches properly towards generating public awareness and building trust on biotechnologically poultry goods.

8.9.4 Legal Consideration and Government Inspections

Due to lack of inappropriate legal issues like relevant regulations along with government inspection to confirmation of safety standard in biotechnologically foods that has declined consumer confidence (Borroto 2009). The government inspection streamlines to intensify food product testing is the single best technique to improve food safety (Crawford 1990). To overcome food safety bottlenecks, government, consumers and industry should recognize their individual accountabilities to ensure safe food. The inspection facilities must be suitable and science based, and propose for their assessment of biotechnology products.

8.9.5 Lack of Resources

Due to lack of adequate resources in terms of skilled manpower, advance laboratory facilities in developing countries, biotechnological approach do not utilize its full potentials, for example, advance disease diagnosis tools (RT-PCR) are being utilized at the central level (Ruane and Zimmermann 2001). However, government facilities are not adequate to inspect the food safety parameters poultry origin biotechnologically improved food. Therefore, resource allocation to be needed to optimize the benefits of biotechnology as well to ensure food safety.

8.10 Conclusion

As a new technology has commenced at a time when consumers are worried on the application of technology in the food production system. General people desire natural products are devoid of synthetic chemical substances. They also like nutritious, appropriate, enhance quality, good-packaged foods that require technology. Biotechnology will help give consumers what they want. Therefore, food safety choices must be stand on the state of art science based knowledge. Psychological and socioeconomic parameters are crucial but could not confirm the safety in biotechnology products. There are several key issues to review the current insights in this paper on assessing the safety of poultry origin biotechnology poultry, poultry products and genetically modified poultry that bring a desired gene along with ingredients applied to poultry meat and meat products with a biotechnologically made enzyme/flavoring/other food additive, or use of prebiotics/probiotics in production of poultry as an alternatives of antibiotics or growth promotors. This current review confirms that the substance used in production and processing of poultry meat and its related food items are found to be harmless and effective in its scientific and intended use.

In conclusion, we have delved from an in-depth review that biotechnology will serve as modernization in veterinary science especially poultry production and processing of poultry origin food that interlinked with food safety parameters where applicable. Biotechnology will benefit not only to fight against bacteria that contaminate food but also this will be helpful to the appropriateness of the meat products through inclusion of fat components and to add nutrients to other products. Now it is clear that biotechnology is encouraging the people as its involvement in food production systems and to ensure food safety and security. Further, backing to strengthen of national abilities in the development and suitable application of pro-poor animal biotechnologies through intervention of international and donor agencies is demanding. The government with other partners of the food system and the biotechnology industry should utilize completely the science based technology to corroborate food safety standards and to stance at the back of public health decisions.

Acknowledgments The authors thank to Dr. AHM Taslima Akhter, Dr. Tuhin-Al-Ferdous, Dr. A. K.M. Ziaul Haque and Dr. Nazmul Haque for providing scientific inputs in the formulation of this book chapter.

References

Akhter AHMT, Islam SS, Sufian MA, Hossain M, Rahman SMM, Kabir SML, Uddin MG, Hossain MM (2018) Implemenation of Code of practices (CoP) in selected poultry farms in Bangladesh. Asian Australas J Food Saf Secur 2:45–55

Akunaedozi C (2014) Genomics (bioinformatics) and computational approaches to analyse large sets of biological data. J Res Opin 1:5. Retrieved from: http://www.researchopinion.info/index.php/jro/article/view/37

Alagawany M, Farag M, Dhama K, Patra A (2018) Nutritional significance and health benefits of designer eggs. World Poult Sci J 74:317–330

Amaral AJ, Ferretti L, Megens HJ, Crooijmans RP, Nie H, Ramos-Onsins SE, Perez-Enciso M, Schook LB, Groenen MA (2011) Genome-wide footprints of pig domestication and selection revealed through massive parallel sequencing of pooled DNA. PLoS One 6:e14782

Amills M, Jimenez N, Villalba D, Tor M, Molina E, Cubilo D, Marcos C, Francesch A, Sanchez A, Estany J (2003) Identification of three single nucleotide polymorphisms in the chicken insulin-like growth factor 1 and 2 genes and their associations with growth and feeding traits. Poult Sci 82:1485–1493

Apata D (2008) Growth performance, nutrient digestibility and immune response of broiler chicks fed diets supplemented with a culture of Lactobacillus bulgaricus. J Sci Food Agr 88:1253–1258

Arai S (1996) Studies on functional foods in Japan—state of the art. Biosci Biotech Bioch 60:9–15

Avey MT, Griffin G (2016) Pedagogical merit review of animal use for education in Canada. PLoS One 11:e0158002

Awad W, Böhm J, Razzazi-Fazeli E, Ghareeb K, Zentek J (2006) Effect of addition of a probiotic microorganism to broiler diets contaminated with deoxynivalenol on performance and histological alterations of intestinal villi of broiler chickens. Poult Sci 85:974–979

Awad W, Ghareeb K, Abdel-Raheem S, Böhm J (2009) Effects of dietary inclusion of probiotic and synbiotic on growth performance, organ weights, and intestinal histomorphology of broiler chickens. Poult Sci 88:49–56

Awan UF, Shafiq K, Mirza S, Ali S (2003) Mineral constituents of culture medium for lipase production by Rhizopus oligosporous fermentation. Asian J Plant Sci 2:913–915

Babu U, Dalloul R, Okamura M, Lillehoj H, Xie H, Raybourne R, Gaines D, Heckert R (2004) Salmonella enteritidis clearance and immune responses in chickens following Salmonella vaccination and challenge. Vet Immunol Immunopathol 101:251–257

Bagarazzi ML, Boyer JD, Ugen KE, Javadian MA, Chattergoon M, Shah A, Bennett M, Ciccarelli R, Carrano R, Coney L (1998) Safety and immunogenicity of HIV-1 DNA constructs in chimpanzees. Vaccine 16:1836–1841

Bartlett JM, Stirling D (2003) A short history of the polymerase chain reaction. PCR protocols. Springer. Retrived from: https://www.springer.com/gp/book/9781592593842

Berger RG (2009) Biotechnology of flavours – the next generation. Biotechnol Lett 31:1651

Blesbois E (2007) Current status in avian semen cryopreservation. Worlds Poult Sci J 63:213–222

Borroto CG (2009) Biotechnology and its application to veterinary science. Compendium of technical items presented to the International Committee or to Regional Commissions of the OIE 2008–2009:231–250

Brisbin JT, Zhou H, Gong J, Sabour P, Akbari MR, Haghighi HR, Yu H, Clarke A, Sarson AJ, Sharif S (2008) Gene expression profiling of chicken lymphoid cells after treatment with Lactobacillus acidophilus cellular components. Dev Comp Immunol 32:563–574

Brochu SM (2018) Development and shelf life evaluation of a novel fermented seaweed sauerkraut utilizing commercially important maine seaweeds. MS thesis. https://digitalcommons.library.umaine.edu/etd/2891

Burt D (2002) Applications of biotechnology in the poultry industry. World Poult Sci J 58:5–13

Calus MP (2010) Genomic breeding value prediction: methods and procedures. Animal 4:157–164

Calvo-González E (2016) Low-complexity biotechnology and everyday aspects of "care:" neonatal testing and sickle cell diagnosis in Brazil. História, Ciências, Saúde-Manguinhos 23:79–94

Chambers J (1990) Genetics of growth and meat production in chicken. In: Chambers JR (ed) Poultry breeding and genetics. Agriculture Canada Research Station, Ottawa, ONT, Canada. Animal Research Centre

Chiang S, Hsieh W (1995) Effect of direct-fed microorganisms on broiler growth performance and litter ammonia level. Asian Aust J Anim 8:159–162

Chichester JA, Yusibov V (2007) Plants as alternative systems for production of vaccines. Hum Vacc Immunother 3:146–148

Cohen J (1993) Naked DNA points way to vaccines. Science 259:1691–1693

Cole C, Fuller R, Newport M (1987) The effect of diluted yoghurt on the gut microbiology and growth of piglets. Food Microbiol 4:83–85

Collins FW, O'Connor PM, O'Sullivan O, Gómez-Sala B, Rea MC, Hill C, Ross RP (2017) Bacteriocin gene-trait matching across the complete *Lactobacillus* Pan-genome. Sci Rep 7:1–14

Crawford LM (1990) Biotechnology and food safety. NABC. Retrived from: http://nabc.cals.cornell.edu/Publications/Reports/nabc_02/nabc_02_15_Crawford.pdf

Cruickshank EM (1934) Studies in fat metabolism in the fowl: the composition of the egg fat and depot fat of the fowl as affected by the ingestion of large amounts of different fats. Biochem J 28:965

Dalloul RA, Lillehoj HS, Tamim NM, Shellem TA, Doerr JA (2005) Induction of local protective immunity to Eimeria acervulina by a Lactobacillus-based probiotic. Comp Immunol Microbiol Infect Dis 28:351–361

Daniel DS (2016) Biotechnology for animal (poultry) production in a changing world: a must for developing nations. In: Proceedings of the 6th global veterinary summit, November 14–16, 2016, Atlanta, USA

De Vries A, Sosnicki A, Garnier J, Plastow G (1998) The role of major genes and DNA technology in selection for meat quality in pigs. Meat Sci 49:S245–S255

Devi SM, Halami PM (2011) Detection and characterization of pediocin PA-1/AcH like bacteriocin producing lactic acid bacteria. Curr Microbiol 63:181–185

Diamond J (1999) Guns, germs and steel: the fates of human societies. WW Norton & Company, New York/London

Dierick N (1989) Biotechnology aids to improve feed and feed digestion: enzymes and fermentation. Arch Anim Nutr 39:241–261

Earth Care (2019) EMO composting-organic waste composting process. Retrived from: https://www.ecepl.com/all/organic-waste-composting-process/

EFSA (2007) Introduction of a qualified presumption of safety (QPS) approach for assessment of selected microorganisms referred to EFSA-Opinion of the Scientific Committee. EFSA J 5:587

Engormix (2008) The impact of biotechnology on poultry genetics and breeding. Cornell Cooperative Extension Newsletter. Retrived from: https://en.engormix.com/MA-poultry-industry/news/the-impact-biotechnology-poultry-t11887/p0.htm

Ennahar S, Assobhei O, Hasselmann C (1998) Inhibition of Listeria monocytogenes in a smear-surface soft cheese by Lactobacillus plantarum WHE 92, a pediocin AcH producer. J Food Prot 61:186–191

Fairfull R (1990) Genetics of egg production in chickens. Poult Breed Genet:705–759

Fan B, Du ZQ, Gorbach DM, Rothschild MF (2010) Development and application of high-density SNP arrays in genomic studies of domestic animals. Asian Aust J Anim 23:833–847

Fang M, Nie Q, Luo C, Zhang D, Zhang X (2007) An 8 bp indel in exon 1 of Ghrelin gene associated with chicken growth. Domest Anim Endocrinol 32:216–225

FAO (2010) Agricultural biotechnologies in developing countries, Guadalajara, Mexico. Retrived from: http://www.fao.org/fileadmin/user_upload/abdc/documents/fara.pdf

FAO (2013) FAO poultry development review. Retrived from: http://www.fao.org/3/i3531e/i3531e.pdf

FAO (2016) Global livestock environmental assessment model (GLEAM). FAO, Rome. Retrived from: http://www.fao.org/gleam/en/

FAO (2020) Gateway to poultry production and products. FAO. Retrived from: http://www.fao.org/poultry-production-products/en/

Feng X, Kuhnlein U, Aggrey S, Gavora J, Zadworny D (1997) Trait association of genetic markers in the growth hormone and the growth hormone receptor gene in a White Leghorn strain. Poult Sci 76:770–1775

Feng X, Kuhnlein U, Fairfull R, Aggrey S, Yao J, Zadworny D (1998) A genetic marker in the growth hormone receptor gene associated with body weight in chickens. J Hered 89:355–359

Fisinin VI, Papazyan TT, Surai PF (2009) Producing selenium-enriched eggs and meat to improve the selenium status of the general population. Crit Rev Biotechnol 29:18–28

Fritts C, Kersey J, Motl M, Kroger E, Yan F, Si J, Jiang Q, Campos M, Waldroup A, Waldroup P (2000) *Bacillus subtilis* C-3102 (Calsporin) improves live performance and microbiological status of broiler chickens. J Appl Poult Res 9:149–155

Fritz DT, Liu D, Xu J, Jiang S, Rogers MB (2004) Conservation of BMP2 post-transcriptional regulatory mechanisms. J Biol Chem 279:48950–48958

Fuller R (1989) Probiotics in man and animals. J Appl Bacteriol 66:365–378

Fulton J (2014) Genomic selection for poultry breeding. Anim Front 2:30–36

Georges M, Andersson L (1996) Livestock genomics comes of age. Genome Res 6:907–921

Gerdts V, Mutwiri GK, Tikoo SK, Babiuk LA (2006) Mucosal delivery of vaccines in domestic animals. Vet Res 37:487–510

Gil De Los Santos J, Storch O, Gil-Turnes C (2005) *Bacillus cereus* var. toyoii and *Saccharomyces boulardii* increased feed efficiency in broilers infected with *Salmonella enteritidis*. Br Poult Sci 46:494–497

Griggs J, Jacob JP (2005) Alternatives to antibiotics for organic poultry production. J Appl Poult Res 14:750–756

Grobet L, Martin LJR, Poncelet D, Pirottin D, Brouwers B, Riquet J, Schoeberlein A, Dunner S, Ménissier F, Massabanda J (1997) A deletion in the bovine myostatin gene causes the double–muscled phenotype in cattle. Nat Genet 17:71–74

Haghighi HR, Gong J, Gyles CL, Hayes MA, Sanei B, Parvizi P, Gisavi H, Chambers JR, Sharif S (2005) Modulation of antibody-mediated immune response by probiotics in chickens. Clin Diagn Lab Immunol 12:1387–1392

Haghighi HR, Gong J, Gyles CL, Hayes MA, Zhou H, Sanei B, Chambers JR, Sharif S (2006) Probiotics stimulate production of natural antibodies in chickens. Clin Vaccine Immunol 13:975–980

Hale EB (1969) Domestication and the evolution of behavior. The bevavior of domestic animals, pp 22–42

Hamad GM, Taha TH, Hafez E, El Sohaimy SA, Ali SH (2017) Yeast-Probiotic coctile strains supplemented babies' Cerelac induce highly potential aflatoxins detoxification both in vitro and in vivo in mother and babies albino rats. J Sci Food Agric 98:707–718

Han S, Lee K, Lee B, Sung C (1999) Effect of feeding *Aspergillus oryzae* culture on fecal microflora, egg qualities, and nutrient metabolizabilities in laying hens. Asian Aust J Anim 12:417–421

Harlander SK (1990) Food processing biotechnology. NABC. Reftrived from: https://creativecommons.org/licenses/by-nc-nd/4.0/

Havenstein G, Crittenden L, Petitte J, Qureshi H, Foster D (1992) Application of biotechnology in the poultry industry. Anim Biotechnol 3:15–36

Havenstein G, Ferket P, Qureshi M (2003a) Growth, livability, and feed conversion of 1957 versus 2001 broilers when fed representative 1957 and 2001 broiler diets. Poult Sci 82:1500–1508

Havenstein G, Ferket P, Qureshi M (2003b) Carcass composition and yield of 1957 versus 2001 broilers when fed representative 1957 and 2001 broiler diets. Poult Sci 82:1509–1518

Hayes B, Goddard M (2001) Prediction of total genetic value using genome-wide dense marker maps. Genetics 157:1819–1829

Hill J, Baiano J, Barnes A (2009) Isolation of a novel strain of *Bacillus pumilus* from penaeid shrimp that is inhibitory against marine pathogens. J Fish Dis 32:1007–1016

Holzapfel W (2002) Appropriate starter culture technologies for small-scale fermentation in developing countries. Int J Food Microbiol 75:197–212

Horniakova E (2005) The influence of *Enterococcus faecium* M-74 bacteria on bone mineralization in chickens. In: Proceedings of the proceedings of the 15th European symposium on poultry nutrition, Balatonfüred, Hungary, 25–29 September, 2005, pp 206–208

Hossain M, Ali M, Chowdhury S, Haque M, Kabir SML (2005) Effect of yoghurt and protexin boost on gut microflora and broiler performance. The Agriculturists 3:24–29

Huang M, Choi Y, Houde R, Lee JW, Lee B, Zhao X (2004) Effects of Lactobacilli and an acidophilic fungus on the production performance and immune responses in broiler chickens. Poult Sci 83:788–795

Huang Y, Du X, Deng X, Qiu X, Wang C, Chen W, Li N, Wu C (2007) Single nucleotide polymorphisms in chicken lmbr1 gene were associated with chicken growth and carcass traits. Sci China Ser C Life Sci 50:62–69

Hugas M, Pages F, Garriga M, Monfort J (1998) Application of the bacteriocinogenic *Lactobacillus sakei* CTC494 to prevent growth of *Listeria* in fresh and cooked meat products packed with different atmospheres. Food Microbiol 15:639–650

Hutkins RW (2006) Fermented vegetables. Microbiology and technology of fermented foods 223–259

Ibrahim DM (2006) Reduce, refine, replace: the failure of the three R's and the future of animal experimentation. U. Chi. Legal F. 195 Retrived from: https://heinonline.org/HOL/LandingPage?handle=hein.journals/uchclf2006&div=9&id=&page=

Islam M, Rahman M, Kabir SML, Kamruzzaman S, Islam M (2004) Effects of probiotics supplementation on growth performance and certain haemato-biochemical parameters in broiler chickens. Bangl J Vet Med 2:9–43

Jebakumar AZ, Nondo HS, George SK (2012) Current research in molecular biology. J Sci 2:38–55

Jiang YH, McGeachin RB, Bailey CA (1994) α-Tocopherol, β-carotene, and retinol enrichment of chicken eggs. Poult Sci 73:1137–1143

Jiang SW, Jacobsson L, Kerje S, Andersson L, Xiong YZ (2002) Studies of relationship between the melanocortin-3 receptor gene and body weight in chicken for high and low weight lines' intercross. Yi chuan xue bao=. Acta Genet Sin 29:322–325

Jin L, Ho Y, Abdullah N, Jalaludin S (1998) Growth performance, intestinal microbial populations, and serum cholesterol of broilers fed diets containing *Lactobacillus* cultures. Poult Sci 77:1259–1265

Jonvel S (1993) Use of yeast in monogastrics. Feed Mix 1:1–4

Kabir SML (2009a) The dynamics of probiotics in enhancing poultry meat production and quality. MS thesis, Department of Microbiology and Hygiene, Faculty of Veterinary Science, Bangladesh Agricultural University, Mymensingh-2202

Kabir SML (2009b) The role of probiotics in the poultry industry. Int J Mol Sci 10:3531–3546

Kabir SML, Rahman MM, Rahman M, Rahman M, Ahmed S (2004) The dynamics of probiotics on growth performance and immune response in broilers. Int J Poult Sci 3:361–364

Kabir SML, Rahman M, Rahman M, Hosain M, Akand M, Das S (2005) Viability of probiotics in balancing intestinal flora and effecting histological changes of crop and caecal tissues of broilers. Biotechnology 4:325–330

Kabir SML, Sumon MH, Amin M, Yamasaki S (2014) Isolation, identification and antimicrobial resistance patterns of *Campylobacter* species from broiler meat sold at KR Market of Bangladesh Agricultural University Campus, Mymensingh. J Agric Food Technol 4:15–21

Kaikabo A, Kalshingi H (2007) Concepts of bioinformatics and its application in veterinary research and vaccines development. Niger Vet J 28:39–46

Kalavathy R, Abdullah N, Jalaludin S, Ho Y (2003) Effects of *Lactobacillus* cultures on growth performance, abdominal fat deposition, serum lipids and weight of organs of broiler chickens. Br Poult Sci 44:139–144

Kamle M, Kumar P, Patra JK, Bajpai VK (2017) Current perspectives on genetically modified crops and detection methods. 3 Biotech 7:219

Kamruzzaman S, Kabir SML, Rahman M, Islam M, Reza M (2005) Effect of probiotics and antibiotic supplementation on body weight and haemato-biochemical parameters in broilers. Bangl J Vet Med 3:100–104

Kaškonienė V, Stankevičius M, Bimbiraitė-Survilienė K, Naujokaitytė G, Šernienė L, Mulkytė K, Malakauskas M, Maruška A (2017) Current state of purification, isolation and analysis of bacteriocins produced by lactic acid bacteria. Appl Microbiol Biotechnol 101:1323–1335

Kavousipour S, Khademi F, Zamani M, Vakili B, Mokarram P (2017) Novel biotechnology approaches in colorectal cancer diagnosis and therapy. Biotechnol Lett 39:785–803

Khaksefidi A, Ghoorchi T (2006) Effect of probiotic on performance and immunocompetence in broiler chicks. J Poult Sci 43:296–300

Kim JJ, Yang JS, Nottingham LK, Tang W, Dang K, Manson KH, Wyand MS, Wilson DM, Weiner DB (2001) Induction of immune responses and safety profiles in rhesus macaques immunized with a DNA vaccine expressing human prostate specific antigen. Oncogene 20:4497–4506

Kizerwetter-Swida M, Binek M (2009) Protective effect of potentially probiotic Lactobacillus strain on infection with pathogenic bacteria in chickens. Pol J Vet Sci 12:15–20

Koenen M, Kramer J, Van Der Hulst R, Heres L, Jeurissen S, Boersma W (2004) Immunomodulation by probiotic lactobacilli in layer-and meat-type chickens. Br Poult Sci 45:355–366

Krasznai DJ, Champagne Hartley R, Roy HM, Champagne P, Cunningham MF (2018) Compositional analysis of lignocellulosic biomass: conventional methodologies and future outlook. Crit Rev Biotechnol 38:199–217

Kroeckel L, Schmidt U (1994) Hemmung von Listeria monocytogenes in vakuum-verpacktem Bruehwurstaufschnitt durch bacteriocinoge Schutzkulturen. Mitteilungsblatt-Bundesanstalt Fur Fleischforschung Kulmbach 1:428–428

Kumprecht I, Zobac P (1998) The effect of probiotic preparations containing *Saccharomyces cerevisae* and *Enterococcus faecium* in diets with different levels of beta-vitamins on chicken broiler performance. Czech J Anim Sci ISSN:0044–4847

La Ragione RM, Woodward MJ (2003) Competitive exclusion by *Bacillus subtilis* spores of *Salmonella enterica* serotype Enteritidis and *Clostridium perfringens* in young chickens. Vet Microbiol 94:245–256

Lacy MP (2000) Broiler production-past, present, and future. Poult Digest 59:24–26

Lalloo R, Ramchuran S, Ramduth D, Görgens J, Gardiner N (2007) Isolation and selection of *Bacillus* spp. as potential biological agents for enhancement of water quality in culture of ornamental fish. J Appl Microbiol 103:1471–1479

Lander ES, Weinberg RA (2000) Journey to the center of biology. Science 287:1777–1782

Lao TD, Nguyen DH, Nguyen TM, Le TAH (2017) Molecular screening for Epstein-Barr virus (EBV): detection of genomic EBNA-1, EBNA-2, LMP-1, LMP-2 among Vietnamese patients with nasopharyngeal brush samples. Asian Pac J Cancer P 18:1675–1679

Lau J, Leung F (2002) Characterization of the chicken growth hormone receptor gene polymorphisms. In: Proceedings of the proceedings of the 7th world congress on genetics applied to livestock production, pp 121–123

Ledoux JB, Antunes A (2018) Beyond the beaten path: improving natural products bioprospecting using an eco-evolutionary framework–the case of the octocorals. Crit Rev Biotechnol 38:184–198

Leeson S, Summers JD (2010) Broiler breeder production. Nottingham University Press

Line JE, Bailey JS, Cox NA, Stern NJ, Tompkins T (1998) Effect of yeast-supplemented feed on *Salmonella* and *Campylobacter* populations in broilers. Poult Sci 77:405–410

Lokko Y, Heijde M, Schebesta K, Scholtès P, Van Montagu M, Giacca M (2018) Biotechnology and the bioeconomy towards inclusive and sustainable industrial development. New Biotechnol 40:5–10

Lucarini S, Fagioli L, Campana R, Cole H, Duranti A, Baffone W, Vllasaliu D, Casettari L (2016) Unsaturated fatty acids lactose esters: cytotoxicity, permeability enhancement and antimicrobial activity. Eur J Pharm Biopharm 107:88–96

Mahajan P, Sahoo J, Panda P (2000) Effect of probiotic (Lacto-Sacc) feeding, packaging methods and seasons on the microbial and organoleptic qualities of chicken meat balls during refrigerated storage. J Food Sci Technol (Mysore) 37:67–71

Maharana BR, Tewari AK, Saravanan BC, Sudhakar NR (2016) Important hemoprotozoan diseases of livestock: challenges in current diagnostics and therapeutics: an update. Vet World 9:487–495

Majeskie JL (1996) Status of United States dairy cattle. In the National Cooperative Dairy Herd Improvement Program. In: Un ited Sates Department of Agricultre (USDA) Fact Sheet K-7, Washington, DC

Marth EH (1998) Extended shelf life refrigerated foods: microbiological quality and safety. Food Technol Chicago 52:57–62

Mathivanan R, Kalaiarasi K (2007) Panchagavya and Andrographis paniculata as alternatives to antibiotic growth promoters on haematological, serum biochemical parameters and immune status of broilers. J Poult Sci 44:198–204

Matukumalli LK, Lawley CT, Schnabel RD, Taylor JF, Allan MF, Heaton MP, O'Connell J, Moore SS, Smith TP, Sonstegard TS (2009) Development and characterization of a high density SNP genotyping assay for cattle. PLoS One 4:e5350

Meeusen EN, Walker J, Peters A, Pastoret PP, Jungersen G (2007) Current status of veterinary vaccines. Clin Microbiol Rev 20:489–510

Midilli M, Alp M, Kocabach N, Muglah O, Turan N, Yilmaz H, Cakir S (2008) Effects of dietary probiotic and prebiotic supplementation on growth performance and serum IgG concentration of broilers. S Afr J Anim Sci 38:21–27

Mishra S, Singh RB (2013) Physiological and biochemical significance of genetically modified foods: an overview. Open Nutraceuticals J 6:18–26

Mohammed B, Malang S, Mailafia S, Agbede R (2016) Application of biotechnology towards diagnosis and treatment in veterinary medicine in Africa: potentials and future developments. J Biotechnol Biomater 6:2

Mohan J, Sharma S, Kolluri G, Dhama K (2018) History of artificial insemination in poultry, its components and significance. Worlds Poult Sci 74:475–488

Mottet A, Tempio G (2017) Global poultry production: current state and future outlook and challenges. Worlds Poult Sci J 73:245–256

Mountzouris K, Tsirtsikos P, Kalamara E, Nitsch S, Schatzmayr G, Fegeros K (2007) Evaluation of the efficacy of a probiotic containing *Lactobacillus*, *Bifidobacterium*, *Enterococcus*, and *Pediococcus* strains in promoting broiler performance and modulating cecal microflora composition and metabolic activities. Poult Sci 86:309–317

Mridha D, Uddin MN, Alam B, Akhter AHMT, Islam SS, Islam MS, Khan MSR, Kabir SML (2020) Identification and characterization of *Salmonella* spp. from samples of broiler farms in selected districts of Bangladesh. Vet World 13:275–283

Nagy É (2001) Genetic engineering of bacteria and viruses. Worlds Poult Sci J 57:391–400

Nahashon S (1992) Effect of direct-fed microbials on nutrient retention production parameters of laying pullets. Poult Sci 71:111

Nahashon S, Nakaue H, Mirosh L (1993) Effect of direct fed microbials on nutrient retention and production parameters of single comb white leghorn pullets. Poult Sci 72:87

Nava G, Bielke L, Callaway T, Castaneda M (2005) Probiotic alternatives to reduce gastrointestinal infections: the poultry experience. Anim Heal Res Rev 6:105–118

Naveena BM, Jagadeesh DS, Babu AJ, Rao TM, Kamuni V, Vaithiyanathan S, Kulkarni VV, Rapole S (2017) OFFGEL electrophoresis and tandem mass spectrometry approach compared with DNA-based PCR method for authentication of meat species from raw and cooked ground meat mixtures containing cattle meat, water buffalo meat and sheep meat. Food Chem 233:311–320

Nayebpor M, Farhomand P, Hashemi A (2007) Effects of different levels of direct fed microbial (Primalac) on growth performance and humoral immune response in broiler chickens. J Anim Vet Adv 6:1308–1313

Neogi SB, Islam MM, Islam SKS, Akhter AHMT, Sikder MMH, Yamasaki S, Kabir SML (2020) Risk of multi-drug resistant *Campylobacter* spp. and residual antimicrobials at poultry farms and live bird markets in Bangladesh. BMC Infect Dis 20:278

Netherwood T, Gilbert H, Parker D, O'donnell A (1999) Probiotics shown to change bacterial community structure in the avian gastrointestinal tract. Appl Environ Microbiol 65:5134–5138

Ngu NT, Xuan NH, Vu CT, An NT, Dung TN, Nhan NTH (2015) Effects of genetic polymorphisms on egg production in indigenous Noi chicken. J Exp Biol 3:487–493

Nguyen TT, Barber AR, Corbin K, Zhang W (2017) Lobster processing by-products as valuable bioresource of marine functional ingredients, nutraceuticals, and pharmaceuticals. Bioresour Bioprocess 4:27

Nie Q, Lei M, Ouyang J, Zeng H, Yang G, Zhang X (2005) Identification and characterization of single nucleotide polymorphisms in 12 chicken growth-correlated genes by denaturing high performance liquid chromatography. Genet Sel Evol 37:339

Nisbet DJ, Tellez GI, Lowry VK, Anderson RC, Garcia G, Nava G, Kogut MH, Corrier DE, Stanker LH (1998) Effect of a commercial competitive exclusion culture (Preempt™) on mortality and horizontal transmission of *Salmonella gallinarum* in broiler chickens. Avian Dis:651–656

Nitschke M, Silva SSE (2018) Recent food applications of microbial surfactants. Crit Rev Food Sci 58:631–638

NRC (2002) Animal biotechnology: science-based concerns. National Research Council (NRC). National Academies Press

Nurmi E, Rantala M (1973) New aspects of *Salmonella* infection in broiler production. Nature 241:210–211

O'Sullivan N, Preisinger R, Koerhuis A (2010) Combining pure-line and cross-bred information in poultry breeding. In: Proceedings of the proceedings of the world congress on genetics

applied to livestock production, volume genetic improvement programmes: design of selection schemes exploiting additive and/or non-additive effects–lecture sessions, p 0984
OECD (2018) OECD-FAO agricultural outlook 2018–2027. OECD Publishing
OIE (2010) The application of biotechnology to the development of veterinary vaccines. World Organization for Animal Health (OIE)
Owings W, Reynolds D, Hasiak R, Ferket P (1990) Influence of dietary supplementation with Streptococcus faecium M-74 on broiler body weight, feed conversion, carcass characteristics, and intestinal microbial colonization. Poult Sci 69:1257–1264
Pal M, Patel A, Bariya A, Godishala V, Kandi V (2017) A review of biotechnological applications in food processing of animal origin. Am J Food Sci Tech 5:143–148
Paramithiotis S (2017) Lactic acid fermentation of fruits and vegetables. CRC press
Patel A, Prajapat J (2013) Food and health applications of exopolysaccharides produced by lactic acid bacteria. Adv Dairy Res 1:107
Pennacchia C, Vaughan E, Villani F (2006) Potential probiotic *Lactobacillus* strains from fermented sausages: further investigations on their probiotic properties. Meat Sci 73:90–101
Petitte JN, Mozdziak PE (2014) Production of transgenic poultry. Transgenic animal technology. Elsevier, pp 335–357. https://doi.org/10.1016/B978-0-12-410490-7.00012-8
Picchietti S, Fausto AM, Randelli E, Carnevali O, Taddei AR, Buonocore F, Scapigliati G, Abelli L (2009) Early treatment with *Lactobacillus delbrueckii* strain induces an increase in intestinal T-cells and granulocytes and modulates immune-related genes of larval Dicentrarchus labrax (L.). Fish Shellfish Immunol 26:368–376
Pisenti J, Delany M, Taylor R, Abbott U, Abplanalp H, Arthur J, Bakst M, Baxter-Jones C, Bitgood J, Bradley F (1999) Avian genetic resources at risk: an assessment and proposal for conservation of genetic stocks in the USA and Canada. Avian Poult Biol Rev 12:1–102
Preisinger R (2018) Innovative layer genetics to handle global challenges in egg production. Br Poult Sci 59:1–6
Pym RAE (2013) Poultry genetics and breeding in developing countries. Poult Dev Rev FAO:80–83
Pym RAE, Guerne Bleich E, Hoffmann I (2006) Contribution of indigenous chicken breeds to poultry meat and egg production and consumption in the developing countries of Africa and Asia. In: The Proceedings of the XII European poultry conference
Raes K, Huyghebaert G, De Smet S, Nollet L, Arnouts S, Demeyer D (2002) The deposition of conjugated linoleic acids in eggs of laying hens fed diets varying in fat level and fatty acid profile. J Nutr 132:182–189
Raju S, Suma M, Nalina M, Chandrashekara K (2015) Basic concept of biotechnology. Laxmi Book Publication, India, pp 240–322
Rantala M, Nurmi E (1973) Prevention of the growth of Salmonella infantis in chicks by the flora of the alimentary tract of chickens. Br Poult Sci 14:627–630
Ratananakorn L, Wilson D (2011) Zoning and compartmentalisation as risk mitigation measures: an example from poultry production. Revue Scientifique et Technique-OIE 30:297
Rhodes G, Abai A, Margalith M, Kuwahara-Rundell A, Morrow J, Parker S, Dwarki V (1994) Characterization of humoral immunity after DNA injection. Dev Biol Stand 82:229–236
Roshanfekr H, Mamooee M (2009) Effect of dietary antibiotic, probiotic and prebiotic as growth promoters, on growth performance, carcass characteristics and hematological indices of broiler chickens. Pak J Biol Sci 12:52–57
Roy BC, Chowdhury SD, Kabir SML (2015) Effects of feeding *Bacillus subtilis* to heat stressed broiler chickens with or without an antibiotic growth promoter. Asian J Med Biol Res 1:80–88
Ruane J, Zimmermann M (2001) Agricultural biotechnology for developing countries: results of an electronic forum. FAO. Retrived from: http://www.fao.org/3/Y2729E/Y2729E00.htm
Sahin EH, Yardimci M (2009) Effects of kefir as a probiotic on growth performance and. J Anim Vet Adv 8:562–567
Salih BTV, Trifonov EN (2015) Visible periodicity of strong nucleosome DNA sequences. J Biomol Struct Dyn 33:1–9

Saxena VK, Kolluri G (2018) Selection methods in poultry breeding: from genetics to genomics. Appl Genet Genomics Poult Sci 19. https://doi.org/10.5772/intechopen.77966
Scanes CG (2007) The global importance of poultry. Poult Sci 86:1057–1058
Schneitz C (2005) Competitive exclusion in poultry 30 years of research. Food Control 16:657–667
Shih J (2012) From biogas energy, biotechnology to new agriculture. Worlds Poult Sci J 68:409–416
Sim JS, Sunwoo HH (2002) Designer eggs: nutritional and functional significance. Eggs Health Promot:19–35. https://doi.org/10.1002/9780470376973.ch3
Singh V, Pathak V, Verma AK (2012) Fermented meat products: organoleptic qualities and biogenic amines–a review. Am J Food Technol 7:278–288
Smaldone G, Marrone R, Zottola T, Vollano L, Grossi G, Cortesi ML (2017) Formulation and shelf-life of fish burgers served to preschool children. Ital J Food Saf 6:6373
Stern N, Cox N, Bailey J, Berrang M, Musgrove M (2001) Comparison of mucosal competitive exclusion and competitive exclusion treatment to reduce *Salmonella* and *Campylobacter* spp. colonization in broiler chickens. Poult Sci 80:156–160
Sugumaran K, Ponnusami V (2017) Review on production, downstream processing and characterization of microbial pullulan. Carbohyd Polym 173:573–591
Šuster K, Podgornik A, Cör A (2017) Quick bacteriophage-mediated bioluminescence assay for detecting *Staphylococcus* spp. in sonicate fluid of orthopaedic artificial joints. New Microbiol 40:190–196
Taniguchi T (1978) Site-directed mutagenesis on bacteriophage Qbeta RNA (author's transl). Tanpakushitsu Kakusan Koso 23:159–169
Tavárez MA, Solis de los Santos F (2016) Impact of genetics and breeding on broiler production performance: a look into the past, present, and future of the industry. Anim Front 6:37–41
Thiruvenkadan A, Prabakaran R (2017) Recent approaches in poultry breeding. Appro Poult Dairy Vet Sci 2(2). https://doi.org/10.31031/APDV.2017.02.000533
Thiruvenkadan A, Panneerselvam S, Prabakaran R (2010) Layer breeding strategies: an overview. Worlds Poult Sci J 66:477–502
Timmerman H, Veldman A, Van den Elsen E, Rombouts F, Beynen A (2006) Mortality and growth performance of broilers given drinking water supplemented with chicken-specific probiotics. Poult Sci 85:1383–1388
Torres-Rodriguez A, Donoghue A, Donoghue D, Barton J, Tellez G, Hargis B (2007) Performance and condemnation rate analysis of commercial Turkey flocks treated with a Lactobacillus spp.-based probiotic. Poult Sci 86:444–446
Tortuero F, Fernandez E (1995) Effects of inclusion of microbial cultures in barley-based diets fed to laying hens. Anim Feed Sci Technol 53:255–265
Trafalska E, Grzybowska K (2004) Probiotics-an alternative for antibiotics? Wiadomosci lekarskie (Warsaw, Poland: 1960) 57:491
Ulmer JB, Donnelly JJ, Parker SE, Rhodes GH, Felgner PL, Dwarki V, Gromkowski SH, Deck RR, DeWitt CM, Friedman A (1993) Heterologous protection against influenza by injection of DNA encoding a viral protein. Science 259:1745–1749
Ulmer JB, Donnelly JJ, Deck RR, DeWITT CM, Liu MA (1995) Immunization against viral proteins with naked DNA. Ann N Y Acad Sci 772:117–125
Van Belkum A, Welker M, Pincus D, Charrier JP, Girard V (2017) Matrix-assisted laser desorption ionization time-of-flight mass spectrometry in clinical microbiology: what are the current issues? Ann Lab Med 37:475–483
Van Borm S, Belák S, Freimanis G, Fusaro A, Granberg F, Höper D, King DP, Monne I, Orton R, Rosseel T (2015) Next-generation sequencing in veterinary medicine: how can the massive amount of information arising from high-throughput technologies improve diagnosis, control, and management of infectious diseases? In: Veterinary infection biology: molecular diagnostics and high-throughput strategies, vol 1247. Springer/Humana Press, New York. https://doi.org/10.1007/978-1-4939-2004-4_30
Velusamya V, Arshak K, Korostynska O, Oliwa K, Adley C (2010) An overview of foodborne pathogen detection: in the perspective of biosensors. Biotechnol Adv 28:232–254

Vijayakumar S, Sasikala M (2012) Application of biotechnology: a current review. Int Res J Pharm 2:59–66

Wayne RK, Ostrander EA (1999) Origin, genetic diversity, and genome structure of the domestic dog. BioEssays 21:247–257

Willis WL, Reid L (2008) Investigating the effects of dietary probiotic feeding regimens on broiler chicken production and Campylobacter jejuni presence. Poult Sci 87:606–611

Willis W, Isikhuemhen O, Ibrahim S (2007) Performance assessment of broiler chickens given mushroom extract alone or in combination with probiotics. Poult Sci 86:1856

Xu W, Li HF, Yan MJ, Tang QP, Chen KW, Wang JY, Gao YS, Tu YJ, Yu YB, Zhu WQ (2007) Associations of gonadotropin-releasing hormone receptor (GnRHR) and neuropeptide Y (NPY) genes' polymorphisms with egg-laying traits in Wenchang chicken. Agric Sci China 6:499–504

Yamamoto Y (2002) PCR in diagnosis of infection: detection of bacteria in cerebrospinal fluids. Clin Diagn Lab Immunol 9:508–514

Yang R, Ray B (1994) Factors influencing production of bacteriocins by lactic acid bacteria. Food Microbiol 11:281–291

Yap T (2015) Guidelines on improving food safety in poultry value chain in Bangladesh. Department of Livestock Services (DLS) & Food and Agriculture Organization -Food Safety Program (FAO-FSP), Bangladesh

Yeo J, Kim KI (1997) Effect of feeding diets containing an antibiotic, a probiotic, or yucca extract on growth and intestinal urease activity in broiler chicks. Poult Sci 76:381–385

Yoon C, Na C, Park J, Han S, Nam Y, Kwon J (2004) Effect of feeding multiple probiotics on performance and fecal noxious gas emission in broiler chicks. Korean J Poult Sci 31:229–235

Zhang AW, Lee BD, Lee SK, Lee KW, An GH, Song KB, Lee CH (2005) Effects of yeast (Saccharomyces cerevisiae) cell components on growth performance, meat quality, and ileal mucosa development of broiler chicks. Poult Sci 84:1015–1021

Zuidhof MJ, Schneider BL, Carney VL, Korver DR, Robinson FE (2014) Growth, efficiency, and yield of commercial broilers from 1957, 1978, and 2005. Poult Sci 93:2970–2982

Zulkifli I, Abdullah N, Azrin NM, Ho Y (2000) Growth performance and immune response of two commercial broiler strains fed diets containing *Lactobacillus* cultures and oxytetracycline under heat stress conditions. Br Poult Sci 41:593–597

Index

A
Abdelatty, A.M., 154–162
Abdelmegeid, M., 210–225
Akbar, S.J., 96, 97
Al-Bulushi, S., 87, 90, 96
Alharthi, A.S., 154–162
Almutairi, S.E., 59
Al-Sobayil, K.A., 59
Ambatipudi, K., 168–198
Amini, R.K., 175
Ampere (mA), 79
Anderson, K.J., 24
Annual herd fertility (AF), 66
Anouassi, A., 84, 95
Antral follicles, 40–41, 93
Artificial insemination (AI), v, 1–42, 57–99, 112, 115, 117, 123, 124, 238, 242, 243
Artificial vagina (AV), 77
Ayad, E.H., 172

B
Bakr, M.H., 154–162
Baldassini, W.A., 161
Best Linear Unbiased Prediction (BLUP), 59
Bhaskar, V., 132–144
Billah, M., 95
Biotechnological applications, v, vi, 117, 168–170, 197, 198, 236, 238
Biotechnologies, v–vii, xi, xii, 3, 5, 6, 31, 32, 58, 60, 99, 110, 178, 180, 189–192, 198, 235–262
Biotechnology industry, 262
Blood, 134, 139, 141, 155, 158–162, 211, 213, 215, 217, 218, 242
Bonilha, S.F.M., 159
Bovines, 5, 12, 36, 110, 116–118, 134, 143, 179, 185, 196, 217, 224, 258
Breeding, vi, 3–5, 7, 21, 23, 25, 26, 32, 33, 35–37, 58–69, 71–77, 79, 84, 86, 97, 99, 109–119, 121–125, 238–245, 254
Broleze, D.F., 159
Bund, R.K., 175
Burt, D., 237

C
Calves, 5, 7, 21, 26, 27, 29, 32, 33, 66, 97, 159, 160, 173
Calving interval (CI), 37, 63–66, 75, 161, 162
Calving to first mating interval (CFM), 66
Camel bulls, 61, 65, 67, 69, 76, 77, 79, 81–84, 90, 93
Carvalho, E.B., 161
Cattle, v, vi, xi, xii, 1–42, 109–112, 115–117, 119, 123, 124, 133, 137, 143, 144, 153–162, 215, 217–219, 223, 224, 238
Chettri, D., 109–125
Clemmons, B.A., 155, 158, 159
Computer Assisted Sperm Analysis (CASA), 87
Controlled Internal Drug Releasing (CIDR), 69
Cow, 57, 137, 159, 174, 175, 189, 194, 217, 218, 220, 238
Cruickshank, E.M., 254

V. K. Yata et al. (eds.), *Sustainable Agriculture Reviews 54*, Sustainable Agriculture Reviews 54, https://doi.org/10.1007/978-3-030-76529-3

D
Dairy products, vi, 167–198, 255
Davis, M.P., 161
Deen, A., 95
Degrees Celsius (°C), 80
Den Uijl, I, 215
Derar, R., 66
Developing countries, 109, 110, 124, 214, 236, 249, 252, 257, 261
Dholpuria, S., 69
Diagnosis, v, vi, 33, 35, 58, 63, 65, 73, 191–192, 210, 211, 213, 215, 219, 224, 251–253, 261
Diseases, vi, 5, 60, 65, 76, 77, 98, 111, 112, 115–118, 120, 123, 124, 134, 137, 141, 143, 144, 178, 190, 191, 194, 210, 212–215, 217, 219, 222–225, 235, 236, 238, 239, 249–254, 256–259, 261
Dromedary camel (DC), 99
Dussault-Chouinard, I., 173

E
Elghandour, M.M.M.Y., 154–162, 210–225
Elolimy, A., 210–225
Elolimy, A.A., 154–162
Embryonic stem cells (ESC), 116, 117, 133–135, 137
Embryo transfer (ET), 58, 115–118
Equine chorionic gonadotropin (eCG), 70
Estrous cycle, v, 6–13, 15, 16, 18, 19, 35, 40, 42, 115
Ethylenediaminetetraacetic acid (EDTA), 90

F
Falomo, M.E., 143
Female camel (FC), 61, 62, 66, 69, 71, 78, 81, 95–96, 98
Fertility, v, 7, 16, 19, 21–31, 36, 38–41, 58, 59, 65, 66, 71, 88, 90, 97–99, 112, 115, 118, 123, 124, 237, 242, 243
Follicular cycle, 60, 61
Food and Agriculture Organization (FAO), 63
Food industries, 168–170, 182, 189, 190, 193–198, 255
Food packaging, 168, 173, 176, 183, 187
Food safety, 109, 123, 168–170, 174, 176–177, 182, 183, 193, 194, 196–198, 210, 236, 245, 256–262
Food systems, 168, 170, 262

G
Gherissi, D.E., 57
GnRH-PGF2 α-GnRH (GPG), 73
Gonadotropin releasing hormone (GnRH), 61

H
Harlander, S.K., 246
Hashem, N.M., 154–162
Hindgut, 159
Hour (*h*), 85
Human chorionic gonadotropin (hCG), 72

I
Institut de médecine vétérinaire (IMV), 90
Institut national de la recherche agronomique (INRA), 90
International Institute for Environment and Development (IIED), 63
International Livestock Research Institute (ILR), 62
Islam, S.K.S., 235–262

K
Kabir, S.M.L., 235–262
Kanawjia, S.K., 175
Kapoor, A., 168–198
Kitts, D.D., 182
Kumar, S., 132–144

L
Lamraoui, R., 57
Lee, S.H.I., 174
Liver, 134, 155, 158, 161
Livestock production, 3, 109–125, 210–225
Loor, J.J., 154–162
Lorenzetti, E., 24
Luteinizing hormone (LH), 61

M
Maiting1,Maiting 2...(M1,M2), 73
Malakar, D., 132–144
Malo, C., 92
Mambelli, L.I., 143
Manjunatha, B.M., 73, 74

Marker assisted selection (MAS), v, 119–125, 244
Marques, M.O., 31, 36
McKinnon, A.O., 95
Medan, M.S., 95
Mellado, M., 210–225
Mesenchymal stem cells (MSC), vi, 133–144
Metabolomics, 155, 158–160, 190, 214
Milliliter (ml), 83
Minutes (min), 80
Mistry, V.V., 182
Morin, P., 181
Morotti, F., 14
Morton, K.M., 96
Muscles, 155, 158, 161, 162, 239
Myer, P.R., 158

N
Ndife, J., 178
Nerve growth factor (NGF), 61
Noci, F., 188
Non-conventional technologies, 196
Non-thermal processing, 174, 195
Nutrition, v, vi, 6, 21, 24, 62–64, 118, 170, 178, 189, 192, 196, 215, 236, 238, 240, 245–249, 253

O
Oestrus synchronization, 75, 98, 99
On-farm, vi, 5, 210–214, 216–225
Ovulation inductor factor (OIF), 61

P
Pal, M., 236
Pandey, R.S., 177
Pandit, A.B., 175
Pasteurization, 169, 171, 172, 174, 179, 181, 182, 186, 188, 189, 194, 195, 197
Petrus, R.R., 187
Poduval, V.S., 182
Point-of-care, vi, 210–225
Poultry origin food, 235–238, 257, 260, 262
Poultry production and processing, 262
Pregnancies, xii, 3, 4, 9, 19–21, 23, 24, 26, 28–33, 35–40, 63–68, 72–74, 76, 77, 86, 87, 93, 95–98, 117, 177
Progeny History Testing (PHT), 62
Progesterone Relasing Intravaginal Device (PRID), 69
Prostaglandin F2α (PGF2α), 72
Proteomics, xii, 153–162, 181, 250, 259

R
Ravi Kanth Reddy, P., 210–225
Residual feed intake (RFI), 154, 155, 158–162
Resynchronization, 4, 32–37, 42, 74
Rizvi, S.S., 175
Rojas, M.L., 196
Ross, C.F., 177
Rumen, 155, 158, 162, 213, 223, 224

S
Salem, A.Z.M., 154–162, 210–225
Samtlebe, M., 172
Schlossareck, C., 177
Selenium nanoparticles (SeNPs), 92
Seneda, M.M., 41
Sexed semen, 26–31, 42
Sharma, A., 177
Sharma, B., 109–125
Sharma, R.B., 175
Sharma, V.K., 160
Silva, F.R.N., 174
Skidmore, J.A., 95
Small and smooth ovaries (SSO), 73
Somatic cell nuclear transfer (SCNT), v, 110–112
Sperm, 16, 26–30, 58, 60, 75–77, 79–93, 97, 98, 111, 112, 118, 123, 243
Stem cells, vi, 116, 117, 125, 131–144, 260
Stem cell technology, v, 116–117, 125
Synchronization, v, 3, 4, 13–20, 22–24, 26, 27, 29, 30, 32, 33, 37, 58, 71, 73, 74, 99, 115, 117

T
Technologies, vi, xi, xii, 27, 30, 58, 99, 110, 112, 117, 119, 121, 122, 124, 125, 131–144, 155, 168, 169, 171–176, 180, 181, 184, 187–190, 192–198, 210–225, 235, 237–239, 245, 250–251, 253–255, 259–262

Therapeutic applications, 134
Tibary, A., 84
Timed artificial insemination (TAI), v, 3, 4, 16–42, 70, 98

V
Verkempinck, S., 196
Verma, A., 168–198
Verma, A.K., 109–125
Volt (*V*), 79

W
Welker, D.L., 180
Wong, P.Y.Y., 182

Y
Yadav, M., 168–198

Z
Zeineldin, M., 154–162, 210–225
Zinc oxide nanoparticles (ZnONPs), 92

The manufacturer's authorised representative in the EU is Springer Nature Customer Service Centre GmbH, Europaplatz 3, 69115 Heidelberg, Germany. If you have any concerns regarding our products, please contact ProductSafety@springernature.com

Printed and bound by CPI Group (UK) Ltd, Croydon, CR0 4YY
30/09/2026
02225491-0006